# Concise Guide to Computation Theory

Akira Maruoka

# Concise Guide
# to Computation
# Theory

 Springer

Akira Maruoka
Faculty of Science and Engineering
Ishinomaki Senshu University
Shinmito Minamisakai 1
Ishinomaki, Japan
amaruoka@isenshu-u.ac.jp

ISBN 978-1-4471-5816-5 ISBN 978-0-85729-535-4 (eBook)
DOI 10.1007/978-0-85729-535-4
Springer London Dordrecht Heidelberg New York

British Library Cataloguing in Publication Data
A catalogue record for this book is available from the British Library

*Cover design*: VTeX UAB, Lithuania

Printed on acid-free paper

Springer is part of Springer Science+Business Media (www.springer.com)

# Endorsements

# Foreword

This book is unusual: ambitious in scope, yet compact. Drawing on his broad research interests and extensive teaching experience, Professor Maruoka distills some of the major topics in the theory of computation into a relatively slim textbook accessible to undergraduates. His aim is to combine intuitive descriptions and illustrations with rigorous arguments and detailed proofs for key topics. The book is self-contained in that it briefly introduces any mathematical prerequisites (in Part I), and it would be ideal for a one- or two-semester undergraduate course in the theory of computation. The principal material is in three sections: Automata and Languages, Computability, and Complexity of Computation. These represent the foundations of the theory of computation as understood today, underpinning the rich developments of the subject which have become important in so many areas of science, technology, and business.

Early progress toward defining models and establishing interesting correspondences between classes of finite automata and formal languages were made in the 1940s and 1950s through the work of such thinkers as McCulloch, Pitts, Kleene, Moore, Mealy, Shannon, Chomsky, Rabin, and Scott. Some of the striking highlights, in concepts and theorems, from this area are presented clearly and concisely in Chaps. 3, 4 and 5 (Part II). The foundations of computability, covered in Chaps. 6 and 7 (Part III), were established earlier, from the 1930s on, by scientists such as Goedel, Church, Markov, Post, and Turing. A rich literature on topics in computability had been developed by the 1960s, with much emphasis on "reducibility" (introduced in Chap. 7) between computational problems, showing that some "unsolvable" problems were more unsolvable than others.

Prominent since the 1970s, studies of computational "complexity" such as those of Cook and Levin focused attention on the distinction between problems that were solvable in a "feasible" amount of time and those which had to be considered infeasible, for example if the computation time had to grow exponentially in the size of the input instance. A key concept, discussed in Chap. 8, is the class NP of problems solvable in polynomial time by a "nondeterministic" Turing machine. The concepts needed to understand this class are introduced gradually throughout the earlier chapters. Chapter 9 introduces the further computational model of "Boolean circuits," an abstraction of the logic circuits at the heart of computers, thoroughly explaining close links between Boolean function complexity and Turing machine complexity. Chapter 10 gives the definition of NP-completeness. Karp, Garey, and Johnson were among the many early researchers to establish the abundance of NP-complete problems arising in practical applications. Many thousands of apparently difficult problems in operations research, mathematics, network design, etc., can be shown by reductions to be all equivalent in their complexity. The notion of reduction is key here. It is a celebrated conjecture that $P \neq NP$, i.e., that NP-complete problems cannot be solved in polynomial time.

A more superficial treatment could certainly outline some of the main ideas in the theory of computation, but this textbook provides both a thorough foundation for the subject, and a much deeper insight into it through a selection of results obtained from the theory. With Professor Maruoka's guidance, the student can develop a real understanding of this fascinating subject and a solid basis for carrying on to pursue research in the theory of computation, maybe even to prove that $P \neq NP$.

University of Warwick, UK                                                     Mike Paterson

# Preface to the English Edition

I am both gratified and grateful to present this English edition of my book titled *Concise Guide to Computation Theory*. The original Japanese edition was written with the intention of enabling readers to understand, both deeply and intuitively, the profound results achieved in this field. In translating the book, I revised it extensively and thoroughly, adding a number of problems to the end of each chapter. I sincerely hope that this revision makes it both easy and enjoyable to read about these fascinating topics. I could not have completed the translation without the help of my friends and colleagues. Rüdiger Reischuk and John Savage advised me to publish a translated edition of the book that had been written in Japanese. Itsuo Takanami helped me enormously in preparing the translated edition. Satoshi Okawa read the translated manuscript and gave me helpful comments. Brad Fast nicely revised a part of the translated manuscript, improving its readability. I am indebted to Kazuko Takimoto for typing a large volume of the manuscript. I am also grateful to Eiji Takimoto for providing technical support for typing of the manuscript. Finally, I would like to thank Wayne Wheeler and Simon Rees of Springer-Verlag, London, and Alfred Hofmann and Ronan Nugent of Springer-Verlag, Heidelberg, for their excellent suggestions and thoughtful guidance in preparing this book.

Just before the publication of this textbook, this area of Japan was struck by a disaster of historic scale: the devastating earthquake and tsunami. It will surely often come to my mind that our recovery from this devastation had just started when this book was published. As we heal and rebuild, all here appreciate the best wishes extended to us from all over the world.

Sendai, Japan                                                                                       Akira Maruoka

# Preface

What is the theory of computation all about? The theory of computation embodies the principle by which computers have become the basis of modern digital technology, which makes a computer perform as desired, and which, consequently, has led to the prosperity of our advanced information society (just as physical sciences constitute the principle by which physical phenomena are elucidated, thereby bringing all the prosperity of our material civilization). Throughout this book, the term "computation" points to all tasks that are performed by a computer.

A computer not only performs basic arithmetic operations; it tells us what the weather is going to be, and it can inform us of the most popular product by extracting useful information from enormous amounts of customer purchasing data. The IBM computer *Deep Blue* has even defeated a grand master chess player. In this book, we describe how a computer performs these tasks in principle, and we also clarify the limits on what computers can do. We think of what a computer executes as a problem, which is defined as a correspondence relation between input and output data. When we try to forecast weather or to compute the most popular product based on available data, there might be cases in which the information is not sufficient to accomplish these tasks. In such cases, we need to somehow infer a conclusion based on insufficient data, but this type of discussion will be omitted from this book.

We will take up the game of chess to explain what computation time is. Given an arrangement of pieces on one side of a chessboard, to decide the best move, one must trace all possible sequences of moves until the game is completed. To do so, all possible moves must first be enumerated; then all possible countermoves of an opponent for each move of the player must be envisioned. Those processes must be repeated until the end of the game. Tracing all the possible moves in this way, the first player can choose the best move based on all the moves traced. In principle, a program that plays chess executes those tasks. But it is impossible in practice because it takes too much time, even for a computer, to read ahead all the moves to the end of the game. If even a supercomputer takes, say, 1,000,000 years to compute the next move, it is of no use. So, to beat a grand master of chess, it becomes crucial for a chess program to render a sufficiently appropriate, but perhaps not perfect, judgment about the next move based on some proper criterion of judgment.

This book consists of four parts. In Part IV, taking the computation time into account, we divide computable problems into those that are feasibly computable in a reasonable time and those that are considered to be intractable, requiring an inordinate amount of time. We describe the structure of problems of the latter group based on their mutual relations. There even exist problems of correspondence between inputs and output that cannot be computed no matter how much time we spend. In Part III, we explain these problems that are incomputable in principle.

To develop the arguments presented in Parts III and IV, we are required to give a model of computing machines by removing many functionalities from computers, retaining only the indispensable fundamental functions. A computer that confronts a chess opponent proceeds through its computations almost identically as a human being might look ahead one move, and study arrangements of pieces one after another using a paper, a pencil, and an eraser. However, to develop the argument strictly, we must express the set of tasks which can be executed by a computer in one step as a clearer and more concise representation than that which can be made using a paper, a pencil, and an eraser. The expression given in this way turns out to be a computation model. In Part III, we introduce a Turing machine as a computational model of a computer; in Part II, we mention other computational models with restrictions imposed on the Turing machine. Part I provides an introduction to and a preparation for the theory of computation.

A computer executes whatever commands we give it, provided that they can be written as a program. So it seems that there exist no underlying laws that govern what takes place in a computer. But in fact such laws do exist. As described above, we can classify computable problems as those that can and cannot be computed feasibly in a reasonable time. The characteristic features distinguishing these two groups lurk in the problems around the boundary separating these groups. Moreover, there are problems that cannot be computed no matter how much time is allocated to compute them. Inherent in such problems are the aspects that make these problems unsolvable. By "exposing computation to extreme constraints," we can see that the computation reveals its substantial intrinsic nature. This is just like the situation where phenomena in the extremely high-speed world cannot be explained using Newtonian dynamics, thereby beckoning the principle of relativity. In this book, we study substantial mechanisms and underlying principles of computation that can be revealed by examining computational difficulties posed by unsolvable problems and by exploiting those impediments that feasible computation must overcome.

In Parts II, III, and IV of this book, we deal respectively with the topics of automata and language theory, computability theory, and complexity theory. We select the important primary results from these fields together with their proofs, emphasizing the intuition behind them. These results are rich and beautiful. However, those who find these topics fascinating might be limited to a handful of sophisticated researchers of related fields. To improve the situation, in this book we present materials from the theory of computation without omitting any important points of the arguments so that students and engineers just beginning to study the topics for the first time can capture essential aspects intuitively. Researchers in this field often enjoy discussing research topics with each other because they can capture the results along with the genuine intuition behind them. Ideas are often exchanged with intuitive expressions such as "A calls B to make work," "A tries to check exhaustively," "A recognizes it as a proof," and so on. The researchers can communicate through these expressions because their meaning is shared among them.

The textbooks of the field often seem to be produced for students who are supposed to become researchers in the future. The author himself was taught in his school days that he had to read books deeply, i.e., to "read between the lines." But now, the achievements in the field should be open to not only specialized researchers, but to a larger number of people outside the field. With that in mind, this book is written so that even readers who are learning about this field for the first time can understand material intuitively and thoroughly, saying (we hope) "Ah yes! I've got it now!" by virtue of viewing many figures, examining examples, and developing arguments, just as if they were attending a lecture in the classroom. The author, however, is not quite sure whether or not he has written this book as it was supposed to be; it is the readers who ultimately judge whether or not the original goal is attained. I would like to encourage the readers to practice the numerous examples and exercises throughout the book and to examine the figures carefully because the illustrative ideas might not be obtained otherwise. The readers will come to have a deep understanding of modern computers, including their limitations, if they read through this book and capture, intuitively, the principles that govern the computation by studying the examples and solving the exercises.

I hope that the readers take enjoyment in this book and thus equip themselves for life in our information society.

Sendai, Japan                                                                                   Akira Maruoka

# Contents

# Part I

# The Theory of Computation

# Everything Begins with Computation

1

Computer science deals with the issue of what can and cannot be computed and, if possible, how it can be computed. We refer to what a computer does, whatever it is, as "computation." What to compute is formalized as a *problem*, whereas how to compute it is formalized as a *mechanical procedure* or an *algorithm*. What is defined as a field within which an algorithm works is a *computational model*. Once a computational model is defined, a set of basic moves that are performed is fixed as one step. Under these settings, the theory of computation is intended to uncover the laws that govern computation, as physical sciences discover the laws that control physical phenomena.

## 1.1    Computational Barrier

The progress of computer performance is remarkable. It can be said that computer performance has been the key to success in nearly every field, in that a supercomputer, which has appeared as the extreme upper model of the computer, simulates world-scale weather phenomena, defeats the reigning master of chess, finds the most popular articles from vast stores of client information, and so on. However, computation time has come to increase according to the decrease of mesh size so as to improve the predictive accuracy of weather phenomena simulation. Furthermore, although a program exists that has enabled a computer to win against the master of chess, no such program that reaches the level of chess exists for shogi or go. The reason is that, in general, as the depth of reading ahead in a game increases, the number of cases that must follow increases explosively. In this respect, computers cannot yet cope well with shogi and go, as they can with chess. We must confront the ultimate constraints and overcome them when we seek to solve practical concrete problems that might have substantial computational difficulty, either requiring sufficient accuracy or dealing with reasonably large input data. It might happen that a computer must run for the life time of the universe to solve a problem. It cannot be said that such a problem is computed in a feasible amount of time.

A. Maruoka, *Concise Guide to Computation Theory*,
DOI 10.1007/978-0-85729-535-4_1, © Springer-Verlag London Limited 2011

Worse still, given a certain problem, there are cases in which no program to solve the problem exists. After all, we will always face a variety of computational limitations no matter how much computers are improved.

The theory of computation, which was in its infancy before and after the dawn of the present computer age, has clarified limits of computation performed in principle and limits of computations executed in a feasible amount of time, and has explored the underlying computational laws, thereby making computers perform as desired. The theory of computation is intended to explore the laws that govern computation performed by computers, just as physics has uncovered the laws that control phenomena that are observed in the physical world. Computer science, which has the theory of computation as a nucleus, is expected to develop substantially, eventually serving as the basis of an advanced information society, as physics has made a role of laying the basis of our material civilization.

To begin with, we take the word computation as meaning all the tasks that a computer performs rather than just calculation involving the four basic arithmetic operations. The computational model is primarily what describes how computation proceeds. Stated more concretely, it specifies what is allowed as the basic individual steps performed. Once a model of computation is taken to be fixed, we are necessarily confronted by impediments posed by the constraints of the computational model, as well as those that come from computational resource bounds. The latter come to divide computation into feasible and infeasible tasks. Let us say that computational barriers, or simply barriers, divide computations. As such a computational barrier, let us discuss the barrier imposed by the computational time bounds. Given a problem we want to solve, generally too many ways exist to solve it if we are allowed to spend a considerable amount of time solving it. On the other hand, only a few plausible ways exist to solve the problem if we are allowed to spend only somewhat more or less than the minimum computational time. One hopes that genuine computational mechanisms or substantial impediments lurk around the computational barrier, inside or outside, caused by imposition of the severe computational constraints. Once the barrier is identified, not only can we avoid futile efforts to pursue essentially impossible computation outside the barrier; we can also seek the most efficient computation close to, but inside, that barrier.

The notion of the computational barrier is crucial for the study of the theory of computation in the same sense that the theory of General Relativity, along with Newtonian dynamics, is indispensable for the study of physics. Three types of computational barriers will be discussed in this book:

- A barrier defined in terms of computational intractability in principle.
- A barrier defined based on the bounds of computational resources, such as time and storage, for tractable computation.
- A barrier that arises from constraints of a computational model such as those on how to write into and read from memory.

In Parts II, III, and IV of this book, we take up themes related to the barrier entailed by the constraint of the computational models, the barrier imposed by the principal limitation of computation, and the barrier caused by computational time bounds, respectively.

Before we proceed to a detailed discussion of computational impediments, we briefly mention a few related topics that have been omitted from this book. The first is an approach to give up finding an accurate answer in preference to an approximate answer, thereby coping realistically with a computation that is outside the barrier. In such an approximate computation, techniques to sample input data and to control, probabilistically, the way that a computation proceeds are utilized. Another topic is that of cryptosystems, in which secret codes are designed based on a computationally hard problem so as to guarantee that a formidably enormous amount of time is required to break the code.

Finally, a theme omitted in this book is parallel computation. There are cases in which the time needed to compute a problem could be reduced significantly if multiple operations were employed simultaneously rather than as single operations. In fact, multiple processors working in parallel have been introduced for practical use in many modern computers.

Next, we shall mention three types of computational barriers to be treated in this book.

## Computational Limit in Principle

Suppose we are given a collection of instructions that describe, unambiguously, what to execute so that all we need to do is to follow the instructions mechanically. We shall call such a set of instructions a *mechanical procedure* or an *algorithm*. Although this mechanical procedure is an intuitive concept, Turing (*A. Turing*, 1912–1954) tried to specify the notion by introducing a computational model called a *Turing machine*. Here, the term "machine" does not mean a real machine that is built from wheels and gears; rather, it denotes a theoretical computational model that is defined in terms of the basic individual steps performed.

As described in Chap. 6, a Turing machine is built from a control part and a tape of infinite length, which is divided into squares. The tape works as memory, whereas the control part, taking one of a finite number of states, controls how to read or rewrite a symbol on the square that the control part looks at currently. These things are done through successive steps according to a program written in the control part. A step of the program consists of updating (1), (2), and (3), listed below, of the subsequent step depending on (1) and (2) of the present step.
(1) A state of the control part.
(2) A symbol on the square that is examined in the present step.
(3) A square that is looked at in the subsequent step (the square is assumed to be next to the right or left of the square that is looked at in the present step).
To update these items, we have a table that describes how to determine (1), (2), and (3) at the next step depending on (1) and (2) at the present step. The table turns out to be the program in the control part. Once we are given all the settings at the starting time, that is, a state in the control part, a location of the square looked at, and the content of the tape, updating items (1), (2), and (3) is continued repeatedly, as described above, until possibly encountering a distinguished state that is supposed to

stop the update. The sequence of these moves can be thought of as the computation of a Turing machine.

Turing asserted that what can be computed by a mechanical procedure is equivalent to what can be computed by a Turing machine. This is called the *Church–Turing thesis*, which asserts that the intuitive concept of being computable by a mechanical procedure is equivalent to the precise notion defined mathematically in terms of a Turing machine. All varieties of tasks can be executed using mechanical procedures including not only the four basic arithmetic operations, but also game playing by repeatedly choosing the next move provided that the rules of the game are given. It is therefore surprising that the notion of a mechanical procedure that can deal with a large variety of problems is replaced by the notion of a Turing machine that can be specified in terms of a collection of simple updated rules over a few items.

If the Church–Turing thesis is assumed to hold, the computational barrier, in principle, comes to separate what can, and cannot, be described as a program of a Turing machine. As described in Chap. 7, a problem called the *halting problem* exists among those that cannot be solved by a Turing machine. This is a problem that, given as input to a Turing machine's program, decides whether the Turing machine halts eventually or continues to compute forever, where a Turing machine halts when it enters into one of the designated states for which the next move is not defined.

## Practical Limit of Computation

We might confront a problem that can be computed in principle, but which requires time comparable to the earth's duration of existence, even if we use the most powerful computer; hence, the problem does not seem to be computable practically. Therefore, we consider what separates those problems that can be computed in a feasible amount of time from those problems that require an infeasible amount of time. As computational resources, we can take memory as well, which might constitute a barrier to computation in practice, but which is omitted from discussion in this book.

Next, we will explain how to deal with practical computational time.

A Turing machine is used as a computational model, and its computation time is measured in terms of the number of steps required to compute a problem. In general, the computation time increases as the size of the input data increases. Hence, the computation time is grasped as a function $T(n)$ of the length $n$ of an input, and the function $T(n)$, which gives the number of steps when given an input of length $n$, is called the *time complexity*. The time complexity can be considered to be the function that indicates how fast the computation time increases as the size of an input $n$ increases.

Having defined the time complexity as above, a problem which is computable in practice is considered to be one whose time complexity is described as a function polynomial in input size $n$, whereas a problem that is not computable in practice is one whose time complexity is greater than *any* function polynomial in $n$, where a function $T(n)$ is a polynomial in $n$ if $T(n) \leq an^d$ holds for constants $a$ and $d$, chosen appropriately.

Dividing problems according to whether they are bounded by a polynomial or not is more or less equivalent to partitioning them by focusing on the degree of the increase when the length $n$ of an input becomes large.

It had long been argued that many practical problems that require enormous amounts of time to solve have certain intrinsic features in common that make the problems intractable. The controversy continued until *S. Cook* (1939–) and *L. Levin* (1948–) succeeded in introducing the notion of *NP-completeness*, which underpins the computational intractability of many concrete problems in practice. A typical example of an NP-complete problem is the *Hamiltonian path problem*, which is a problem to decide, for a given map of cities, whether there exists a route in the map that starts in a city, passes through each of the cities in the map just once, and returns to the starting city. The fact that the Hamiltonian path problem is NP-complete roughly means that this problem is substantially equivalent to several thousands of such concrete problems that seem to require an enormous computation time.

Here, we note that, although many researchers conjecture that an NP-complete problem cannot be computed in polynomial time, it has not yet been proved whether this conjecture is valid or not. The problem to prove or disprove the conjecture is said to be the *P vs. NP problem*, which remains as the greatest unsolved problem of computer science. The P vs. NP problem is one of the seven millennium prize problems for which the Clay Mathematics Institute offers a $1 million prize to solve.

## Barrier from Computational Models

In this book, we deal with a variety of computational models: a pushdown automaton created by replacing a tape of a Turing machine with one of restricted read-write fashion; a finite automaton obtained by replacing a tape of a Turing machine by a read-only tape of finite length; a context-free grammar, which is defined as a set of rewriting rules on which a certain restriction is imposed; and a Boolean circuit. A Turing machine is a computational model that can describe any mechanical procedure, whereas the finite automaton and the pushdown automaton are computational models obtained by imposing various constraints on a Turing machine. Among such constraints are the strong ones: the constraint allowing only a finite amount of memory, and the constraint on the way of reading and writing to a memory device, called a stack. In the cases of the finite automaton and the pushdown automaton with these constraints imposed, respectively, we must be concerned with barriers that come from those constraints on computational models.

## 1.2    Propriety of Computational Models

Corresponding to the field of physics, in which physical phenomena occur, a computational model is specified artificially as a field in which computations are executed. Once a computational model and a problem to be solved are given, computing the problem in the model becomes a mathematical question that leaves no vagueness.

However, from the informal aspect of plausibility, it is necessary to verify whether the adopted computational model is suitable in the first place.

### Robustness of Computational Model

One criterion that can indicate whether a computational model adopted is appropriate is robustness of the model. A computational model is robust when the model's performance remains unchanged even after the details of the model parameters are changed somehow. For example, in the case of a Turing machine, the computational ability does not change even if a few additional tapes are appended to the original one-tape model. Moreover, as opposed to the deterministic model for which the way of specifying items (1), (2), and (3) is unique, as described in the previous section, there is a nondeterministic model in which the specification is not necessarily unique, and thereby an arbitrarily chosen one among those specified is executed. It turns out that whichever we take, the deterministic model or the nondeterministic one, the computational ability of the Turing machines does not change.

As evidence for the robustness of the Turing machine computational model, there is not only the fact that the computational power does not change in accordance with the model parameters, but also the fact that what can be computed does not change even if we adopt, instead of a Turing machine, a superficially dissimilar computational model that is defined in terms of functions, which is the notion of a recursive function introduced by *S. Kleene* (1909–1994). It was proved that the notion of a recursive function, which is superficially different, is in fact equivalent to the concept of a Turing machine, which turns out to show that both models are robust.

In studying how a human being comprehends languages, *N. Chomsky* (1928–) introduced what is called generative grammar. It is worth noting that a Turing machine and a generative grammar were introduced by examining deeply how a human being solves a problem and understands a language, respectively, thereby abstracting the process of solving a problem and the way of understanding a language.

Turing claimed, through detailed discussions, that the Turing machine is valid as a computational model by arguing that a human being normally solves a problem by writing certain symbols on paper with a pencil, by discussing why we can assume a one-dimensional tape divided into squares instead of assuming a two-dimensional paper, and by explaining why we can suppose that the number of symbols on a square is finite, and so on.

On the other hand, Chomsky thought that a human being possesses, as innate knowledge of languages, a *deep structure*, which is common to every language, as well as a *surface structure* that is obtained by transforming the deep structure according to a mother tongue. Thereby a human comes to understand the mother tongue.

He introduced a generative grammar consisting of rewriting rules to express these structures. In this book, among various generative grammars, we exclusively focus on a context-free grammar in Chap. 4. For example, when we try to generate all the correct English sentences by defining a grammar, we prepare many rules such as

$$\langle sentence \rangle \rightarrow \langle noun\ phrase \rangle \langle verb\ phrase \rangle,$$

$$\langle verb\ phrase \rangle \rightarrow \langle verb \rangle \langle noun\ phrase \rangle,$$

$$\langle noun\ phrase \rangle \rightarrow \langle article \rangle \langle noun \rangle,$$

$$\langle article \rangle \rightarrow the,$$

$$\langle noun \rangle \rightarrow boy,$$

$$\langle verb \rangle \rightarrow broke.$$

That is, a correct sentence is generated by starting from ⟨sentence⟩ and repeating to rewrite the left side of a rewriting rule by the right side. Actually, applying such rewriting rules produces not only correct English sentences, but incorrect ones as well, thereby necessitating exceptional rules. Furthermore, if a grammar is required to generate sentences that are correct not only syntactically but semantically as well, it is obvious that such grammars have inherent deficiencies in generating any sentences that are correct in the strict sense.

In the case of context-free grammars, a rewriting rule can always be applied provided that a symbol, corresponding to the term on the left-hand side of the rule, appears in a generated string. In other words, the rewriting does not depend on the surroundings, i.e., context. In order to make rewriting depend on context, and thereby enhance the ability of generating sentences, we can introduce phase-structured grammars, which turn out to be equivalent to the Turing machines in their power. This fact can be interpreted as further evidence that these computational models are robust.

## Development of Computational Models

Turing and Chomsky [33] introduced the computational models of the Turing machine and generative grammar, respectively, from the consideration of human information processing abilities. This is not merely a coincidence. Von Neumann (1903–1957), who should be acknowledged as the originator of computer science and who is a titan in this field, is also known for having devised the stored-program concept used by present day computers. Touching upon computer design, von Neumann highlights the contrast between computers and humans and began to think about a general theory concerning the design and control of computers and humans. He called the computer an artificial automaton, and humans and other living organisms natural automata. He also tried to build a *theory of automata* that deals with the structure and control of artificial and natural automata. It has been foreseen using the theory that it is indispensable to relate the complex large-scale structure of automata and their behavior. He obtained leading achievements in research concerning the construction of a reliable system overall from unreliable functional elements, research of a self-reproducing machine, and so on. Unfortunately, he passed away at a young age in 1957 without completing the theory of the system, structure, language, information, and control, which he had planned and devoted himself to.

The theory of computation to be treated in this book is abstracted and formalized, but its object relates deeply to the design of the actual computer and human thought processes as well. Recently, the concepts of "learning" and "discovery," which appear to be alien to mechanical procedures at first glance, have become research objects, with the aim of elucidating the information processing ability of humans from the point of view of computation. That is, based on the methodology described in this book, L. Valiant (1949–) started research on learning, which is considered to change a program to a smarter one [35, 36]. Furthermore, considering that even a flash of humanity can be displayed by a computer in principle, S. Arikawa (1941–) [37] sparked research to investigate computational aspects of discoveries that are expected to be developed in the future. Research into artificial intelligence is designed to explore human intellectual activity so as to make a computer perform as a human behaves intellectually. Before computers appeared, Turing investigated computing machinery and intelligence, and proposed the *Turing test* for intelligence: an interrogator who is interfaced with a machine and a human in different rooms are allowed to ask questions through teletype channels; in such a situation if the interrogator is unable to distinguish the machine from the human, then the machine is said to have intelligence.

Through studies of artificial intelligence lasting a half-century, the ideas of how to express human common sense and what should be introduced as a computer model to design it, etc., have been investigated. Even if we restrict ourselves to research from the computational viewpoint, models based on a new idea that is beyond the frame of the computational model discussed in this book have been investigated recently [32]. Much research has been done on quantum computers, which utilize effects related to quantum mechanics, and on neurocomputers, which have a neuron-like structure as well. The computational barriers might be overcome by means of the effect of the superposition of quantum-mechanical states in quantum computing. When we think of the phenomena of learning in terms of the computational framework, what is produced as the output of learning turns out to be an algorithm that is perhaps cleverer. In conclusion, in the various cases of quantum computing and neural computing, the computational framework might be changed somehow, but we believe that we can retain, unchanged, the methodology of the theory of computation consisting of three constituents: a *model* in which computation takes place, a *problem* that formalizes a task, and a *program* to solve the problem on the model.

In fact, [6] relates that the theory of computation will establish profound discipline, as mathematics and physics have done so far, while [7], entitled All Science Is Computer Science, advocates that computer science plays a central role among all disciplines.

The theory of computation has changed its research objects in sequence, focusing on the theory of computability, the theory of automata and languages, and the theory of computing complexity. In the near future, it is hoped that it will grow as the theory which von Neumann anticipated, which will find a unified discipline for the study of artificial and natural automata.

## 1.3  Using this Book Efficiently

This book presents important results that form the core of three fields of the theory of automata and languages, the theory of computability, and the theory of computing complexity. It requires quite a few pages for material in any one of the three fields to be compiled into a book.

Consequently, not a few results of these fields had to be excluded from this book. Nevertheless, because the three fields are assembled in this one presentation, the readers can read through along the main thread of these three fields at a stretch and can grasp the core of the computations that a computer carries out.

This book is compiled so that it can be read without preliminary knowledge. When used as a textbook for university lectures, two one-semester courses are ideal; one for Parts I and II, and the other for Parts III and IV. One might think of a course as addressing Chaps. 3, 4, 6, 7 and the outline of Part IV so that students can grasp the essentials of the theory of computation when it is difficult to spare two courses. To make a one-semester course effective, an instructor might think carefully how to teach the class: to assign as homework some parts of the material, and to encourage students to prepare before the class, or alternatively, to skip the formal proofs of theorems, focusing instead on the proof ideas.

As for the proof of a theorem, because the proof idea is explained using an example before the description of the theorem itself, the readers can figure out intuitively the essential part of the proof.

With the exception of Chap. 1, exercises are provided at the end of each chapter. Some of the contents which should be described in the text are given as exercises so that the argument flow is not spoiled. A sharp symbol ♯ is marked next to such exercises. Either no mark or marks of *, **, and *** are assigned to other exercises in order of their difficulty.

# Preliminaries to the Theory of Computation   2

In this chapter, we explain mathematical notions, terminologies, and certain methods used in convincing logical arguments that we shall have need of throughout the book.

## 2.1   Set

A *set* is an assembly of things where "things" are objects such as natural numbers, real numbers, strings of symbols, and so on. A "thing" that constitutes a set is called an *element* of the set. In particular, when an element that constitutes a set is again a set, then the former set is called a class. In this case, instead of saying a set of sets, we say a class of sets. A set is described by enclosing the elements in braces, like $\{3, 7, 12\}$. A set composed of a finite number of elements is called a *finite set*, whereas a set of an infinite number of elements is called an *infinite set*. The set of natural numbers is expressed as $\{1, 2, 3, \ldots\}$. The number of elements of a finite set $A$ is called the *size* of set $A$ and is denoted by $|A|$. A set whose size is zero is called an *empty set* and denoted by $\emptyset$. That is, $\emptyset$ denotes a set that has no elements. If $a$ is an element of set $A$, then we say that $a$ is contained in $A$, which is denoted by $a \in A$. On the other hand, $a \notin A$ means that $a$ is not an element of set $A$. The set of all the natural numbers, that of all the integers, and that of all the real numbers are denoted by $\mathbb{N}$, $\mathbb{Z}$, and $\mathbb{R}$, respectively.

Let $A$ and $B$ be sets. If any element of $A$ is also an element of $B$, then $A$ is a *subset* of $B$, which is denoted by $A \subseteq B$. Even if $A$ equals $B$, $A \subseteq B$ holds by the definition. In particular, if $A \subseteq B$ and $A \neq B$, $A$ is a *proper subset* of $B$, which is denoted by $A \subsetneq B$.

For sets $A$ and $B$, the set that consists of all the elements $A$ and those of $B$ is the *union* of $A$ and $B$, which is denoted by $A \cup B$. The set that consists of the elements that are contained in both $A$ and $B$ is the *intersection* of $A$ and $B$, which is denoted by $A \cap B$.

Furthermore, the set that consists of the elements contained in $A$ but not in $B$ is the *difference* obtained by subtracting $B$ from $A$, which is denoted by $A - B$. What is described above is illustrated in Fig. 2.1, in terms of figures which are called

A. Maruoka, *Concise Guide to Computation Theory*,
DOI 10.1007/978-0-85729-535-4_2, © Springer-Verlag London Limited 2011

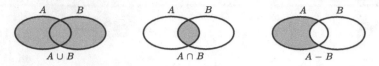

**Fig. 2.1**  Venn diagrams for union, intersection, and difference

**Fig. 2.2**  Cartesian product
$\{1, 2\} \times \{u, v, w\}$

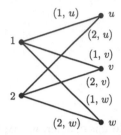

*Venn diagrams.* Let the universe, denoted by $U$, be the set of all elements under consideration. For a subset $A$ of a set $U$, the difference set obtained by subtracting $A$ from $U$ is the *complement* of $A$ in $U$, and when $U$ is obvious, it is denoted by $\overline{A}$ and is simply called the complement of $A$.

The set that consists of all the subsets of a set $A$ is called the *power set* of $A$ and is denoted by $\mathcal{P}(A)$. For example, for $A = \{1, 2, 3\}$,

$$\mathcal{P}(A) = \big\{\emptyset, \{1\}, \{2\}, \{3\}, \{1, 2\}, \{1, 3\}, \{2, 3\}, \{1, 2, 3\}\big\}.$$

Let $P(x)$ denote a certain condition on $x$. The set that consists of elements $x$'s that satisfy the condition $P(x)$ is denoted by

$$\big\{x \mid P(x)\big\}.$$

For example, $\{n \mid n$ is a natural number that is divided by 2$\}$ denotes the set that consists of positive even numbers. The power set of $A$ can be denoted by

$$\mathcal{P}(A) = \{B \mid B \subseteq A\},$$

in terms of this notation. What set does $\{(a, b) \mid a \in A, b \in B\}$ denote? For example, if $A = \{1, 2\}$ and $B = \{u, v, w\}$, the set denotes

$$\big\{(1, u), (1, v), (1, w), (2, u), (2, v), (2, w)\big\}.$$

If $P(a, b)$ in the notation above is taken to be the condition on $(a, b)$ that both $a \in A$ and $b \in B$ hold, $\{(a, b) \mid a \in A, b \in B\}$ means the set of all the elements that satisfy the condition $P(a, b)$. As illustrated in Fig. 2.2, it consists of six pairs, each consisting of one from $A$ and the other from $B$. In general, for sets $A$ and $B$, $\{(a, b) \mid a \in A, b \in B\}$ is called the *Cartesian product* of $A$ and $B$, and is denoted by $A \times B$. Since the Cartesian product is frequently used in this book, the readers are expected to have a firm image of it mind. The Cartesian product can be generalized for an arbitrary number of sets. For example, $A \times B \times C = \{(a, b, c) \mid a \in A, b \in B, c \in C\}$. In particular, the Cartesian product of $k$ As $A \times A \times \cdots \times A$ is denoted by $A^k$. In general, an element of $A^k$, denoted by $(a_1, a_2, \ldots, a_k)$, is called a $k$-tuple. In particular, an element of the Cartesian product of two sets is called a 2-tuple or a *pair*.

There is a difference between a $k$-tuple and a set of size $k$, depending on whether or not we take into account the order of the elements associated with them. For example, the elements of $\mathbb{N}^3$ $(3, 7, 12)$, $(12, 7, 3)$, and $(7, 12, 3)$ are different 3-tuples, whereas $\{3, 7, 12\}$, $\{12, 7, 3\}$, $\{7, 12, 3\}$ as well as $\{3, 7, 7, 12\}$ are the same set since any of them consists of 3, 7, and 12. On the other hand, you may want to take into account how many times an element appears in a set. In such a case, a set is called a *multi-set*. For example, the multi-sets $\{3, 7, 12\}$ and $\{3, 7, 7, 12\}$ are different from each other, while $\{3, 7, 7, 12\}$ and, say, $\{3, 7, 12, 7\}$ are the same set.

## 2.2    Strings and Languages

A *string* is a sequence of symbols, which will be discussed frequently throughout this book. Suppose we are given a finite set of symbols. The strings we deal with are assumed to be composed of symbols taken from the set. Such a set is called an *alphabet* and will be denoted by $\Sigma$ or $\Gamma$, etc. The number of symbols that appear in string $w$ is called the *length* of $w$ and is denoted by $|w|$. A string over $\Sigma$ is one of a finite length that consists of symbols chosen from $\Sigma$. In particular, a string of length 0 is called an *empty string* and is denoted by $\varepsilon$. Though it is not easy to figure out what the empty string really is, it is indispensable, just as we cannot do without the empty set. Since the empty string is of length 0, and hence can be thought of as invisible, we denote it by the special symbol $\varepsilon$. The reason why we introduce the empty sequence will become clear later in the context of how it will be used. For example, concatenating the empty sequence $\varepsilon$ and a sequence $w$ does not change the original sequence, so both $\varepsilon w$ and $w\varepsilon$ turn out to be $w$.

*Concatenation* is just the connection of two strings to obtain a string. That is, connecting strings $a_1 \cdots a_m$ of length $m$ and $b_1 \cdots b_n$ of length $n$ makes the string $a_1 \cdots a_m b_1 \cdots b_n$ of length $m + n$. When it is necessary to indicate the operation of concatenation explicitly, the symbol "·" is used. So when we need to express the operation of concatenation explicitly, the string $a_1 \cdots a_m b_1 \cdots b_n$ obtained by connecting the strings $a_1 \cdots a_m$ and $b_1 \cdots b_n$ will be denoted by $a_1 \cdots a_m \cdot b_1 \cdots b_n$. The string $a_i a_{i+1} \cdots a_j$ which is a consecutive part of $a_1 \cdots a_n$ is called a *substring* of the original string. Note that as special cases both the string $a_1 \cdots a_n$ and the empty string are substrings of the string $a_1 \cdots a_n$.

A *language* is a set that consists of strings over an alphabet. The reason that we use the terminologies a language or an alphabet comes from the following. Taking English as an example, the alphabet turns out to be $\{a, b, c, \ldots, y, z, \sqcup\}$ consisting of 26 symbols and the space symbol $\sqcup$, whereas the language turns out to be the set of correct sentences such as

*the*$_\sqcup$*boy*$_\sqcup$*broke*$_\sqcup$*the*$_\sqcup$*window.*

Since we pay no attention to aspects of English sentences such as correctness or meaning, we will use the terminology a string, rather than a sentence.

A set consisting of all the strings over an alphabet $\Sigma$ is denoted by $\Sigma^*$. For example, if $\Sigma = \{0, 1\}$,

$$\Sigma^* = \{\varepsilon, 0, 1, 00, 01, 10, 11, 000, \ldots\}.$$

A language over $\Sigma$ is a subset of $\Sigma^*$.

## 2.3    Functions and Problems

A *function* is what specifies the correspondence between elements in two sets. A function that defines the correspondence of elements of sets $A$ to those of $B$ is denoted by $f : A \to B$, and the element of $B$ corresponding to an element $a$ of $A$ is denoted by $f(a)$, where $a$ is said to be mapped to $f(a)$. A function is also called a *mapping*. The set $A$ of a function $f : A \to B$ is called the *domain*, and $B$ the *range*. If $f(a) \neq f(a')$ for any two different $a$ and $a'$ in $A$, the function $f : A \to B$ is called one-to-one. The condition can be interpreted as saying that different elements are differently named if element $a$ of $A$ is interpreted to be named $f(a)$. If for any element $b$ of $B$, there exists an element $a$ of $A$ such as $f(a) = b$, then $f : A \to B$ is called a function *onto* $B$. The condition can be interpreted as saying that every name in $B$ is used as a name of some element of $A$.

As examples of functions, consider the functions $f_{add} : \mathbb{N} \times \mathbb{N} \to \mathbb{N}$ and $f_{mult} : \mathbb{N} \times \mathbb{N} \to \mathbb{N}$ that represent addition and multiplication, respectively. These functions are defined by the equations $f_{add}(x, y) = x + y$ and $f_{mult}(x, y) = x \times y$, and are specified in Tables 2.1 and 2.2, respectively. In order to describe the correspondence completely in the form of tables as in Tables 2.1 and 2.2, the space for the tables must be infinite, which is impossible. Furthermore, notice that although when $a$ in $f(a)$ is taken to be a pair $(x, y)$ as in the case of addition we should write $f_{add}((x, y))$, but we simply denote it by $f_{add}(x, y)$.

Throughout this book, the term *problem* is used to mean function. Then, what is the difference between a function and a problem? These two notions differ only slightly in the sense that function is a concept in mathematics, while problem is used in the context where the correspondence associated with the problem might be excepted to be computed. But it will turn out that there is a problem, such as the halting problem described in the preceding chapter, that can be defined but cannot be computed. The function $f$ representing the halting problem is such that $f(\langle M \rangle) = 1$ if a Turing machine $M$ eventually halts and $f(\langle M \rangle) = 0$ otherwise, where $\langle M \rangle$ denotes an appropriately coded sequence that represents a Turing machine $M$. So,

**Table 2.1**  $f_{add}(x, y)$

| $x \backslash y$ | 1 | 2 | 3 | 4 | ... |
|---|---|---|---|---|---|
| 1 | 2 | 3 | 4 | 5 | ... |
| 2 | 3 | 4 | 5 | 6 | ... |
| 3 | 4 | 5 | 6 | 7 | ... |
| 4 | 5 | 6 | 7 | 8 | ... |
| $\vdots$ | $\vdots$ | $\vdots$ | $\vdots$ | $\vdots$ | $\ddots$ |

**Table 2.2**  $f_{mult}(x, y)$

| $x \backslash y$ | 1 | 2 | 3 | 4 | ... |
|---|---|---|---|---|---|
| 1 | 1 | 2 | 3 | 4 | ... |
| 2 | 2 | 4 | 6 | 8 | ... |
| 3 | 3 | 6 | 9 | 12 | ... |
| 4 | 4 | 8 | 12 | 16 | ... |
| $\vdots$ | $\vdots$ | $\vdots$ | $\vdots$ | $\vdots$ | $\ddots$ |

the halting problem can be represented as the function that decides whether a Turing machine halts eventually or continues to move around the states. The symbol $H$ that denotes the halting problem represents the associated function as well, so that $f(\langle M \rangle) = H(\langle M \rangle)$ holds.

The membership problem for a language $L \subseteq \Sigma^*$ is a problem that, given a string $w$ in $\Sigma^*$, asks whether $w \in L$ or $w \notin L$. So, the function that corresponds to the membership problem for a language $L$ is defined as follows:

$$f(w) = \begin{cases} 1 & \text{for } w \in L, \\ 0 & \text{for } w \notin L. \end{cases}$$

A language $L \subseteq \Sigma^*$ and the function $f : \Sigma^* \to \{0, 1\}$ defined from $L$ as above are substantially the same. In general, the set of elements of $A$ that are mapped to $b$ by a function $f : A \to B$ is denoted by $f^{-1}(b)$. That is, $f^{-1}(b) = \{a \mid f(a) = b\}$. Then, the equation $L = f^{-1}(1)$ holds between a language $L$ and the function defined based on the language $L$. As mentioned above, since a language and the function associated with the language are substantially the same, solving the membership problem for a language $L$ and computing the associated function $f : \Sigma^* \to \{0, 1\}$ are essentially the same task.

By the way, the answer for each question of either the halting problem or the membership problem is either affirmative or negative. Such a problem is called a decision problem or a *YES/NO problem*. In this book, affirmative and negative answers are expressed as 1 and 0, accepting and non-accepting, and YES and NO as well.

On the other hand, as in the case of addition and multiplication, there are problems that are not YES/NO problems. For example, for multiplication, $(x, y)$ is an input and $x \times y$ is an output, and what defines the correspondence is the function $f_{mult} : \mathbb{N} \times \mathbb{N} \to \mathbb{N}$. In general, a value $w$ or each instance $w$ for a problem $P$ that is substituted for $x$ of a function $f(x)$ is called an *input* of the function or the problem, and the value returned, denoted by $f(w)$ or $P(w)$, is called its *output*.

## 2.4  Relations and Graphs

We can think of relations, such as the larger-than relation and the divided-by relation, in terms of a collection of pairs of elements that satisfy the corresponding relation. The collection of pairs, in turn, can be represented as a graph whose nodes represent the elements and whose edges represent the pairs.

To begin with, we take the larger-than relation, denoted by $>$, for the set $\{0, 1, \ldots, 5\}$. The larger-than relation can be intuitively understood as the order on the number line. Such a relation can also be shown using arrows directing from a smaller number to a larger number, as shown in Fig. 2.3. If an arrow from $m$ to $n$ is denoted by $(m, n)$, the set of the arrows is the subset of the Cartesian product $\{0, 1, \ldots, 5\} \times \{0, 1, \ldots, 5\}$, that is, the set consisting of 15 pairs $\{(0, 1), (1, 2), (2, 3), (3, 4), (4, 5), (0, 2), (1, 3), \ldots, (0, 4), (1, 5), (0, 5)\}$.

Similarly, Fig. 2.4 shows the relation that indicates the difference of two numbers is even. That is, denoting the relation by $R$, $mRn$ is defined to be such that $|m - n|$ is even. Rearranging the nodes in Fig. 2.4, we have Fig. 2.5, in which $\{0, 1, \ldots, 5\}$ is partitioned into the two groups of even and odd numbers. Here, $R$ is used to connect the numbers that are in the relation such as $0R4$, $4R2$, $1R3$, and it is also used to denote the set consisting of pairs $(m, n)$ that are in the relation $mRn$. That is, in our case, $R$ denotes the set consisting of 18 pairs $\{(0, 0), (2, 2), (4, 4), (0, 2), (2, 0), (2, 4), (4, 2), \ldots, (5, 1), (1, 5)\}$ as well.

In general, a *relation* $R$ on a set $S$ is defined to be a subset of $S \times S$. Then, $(a, a') \in R$ is also denoted by $aRa'$. In this way, a subset of $S \times S$ is considered to be a relation on $S$. In particular, let us assume that a relation $R$ on $S$ satisfies the following three conditions.

*Reflexive law*: $xRx$ for all $x \in S$
*Symmetric law*: $xRy$ implies $yRx$ for all $x$ and $y \in S$
*Transitive law*: $xRy$ and $yRz$ implies $xRz$ for all $x$, $y$ and $z \in S$

If a relation $R$ satisfies the three conditions above, the relation $R$ is called an *equivalence relation*. An equivalence relation $R$ on a set $S$ partitions $S$ into blocks in such a way that any $x$ and $y$ with $xRy$ are placed in the same block, and any different blocks do not overlap. Precisely speaking, a partition is defined as follows. Let us denote the class of subsets of a set $S$ by $A = \{A_i \mid i \in I\}$, where $A_i \subseteq S$ and $I$ denotes the set of subscripts that identify the subsets of $S$. $A$ is called a *partition* if the following two conditions hold:

(1) For any $i$ and $j$ with $i \neq j$, $A_i \cap A_j = \emptyset$, and
(2) $S = \bigcup_{i \in I} A_i$

**Fig. 2.3** Size relation

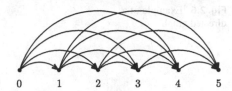

0    1    2    3    4    5

**Fig. 2.4** Relation that
difference is even

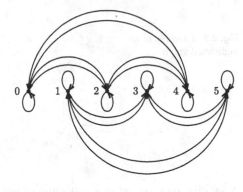

0    1    2    3    4    5

**Fig. 2.5** Another graph
drawing of the relation that
the difference is even

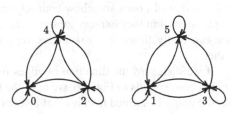

where a subset $A_i$ of $S$ is called a *block*. The details are left to Exercise 2.3**, in which it is proved that the equivalence relation $R$ and the partition $A$ are equivalent notions to each other. Namely, an equivalence relation $R$ induces the partition by placing elements in the relation $R$ into the same block. Conversely, a partition $A$ induces the equivalence relation $R$ by relating elements in the same block.

For the example above, the larger-than relation $>$ satisfies the transitive law, but does not satisfy the reflexive and symmetric laws. On the other hand, the relation $R$ indicating that the difference of two numbers is even satisfies the above three relations. Hence, $R$ is an equivalence relation which induces the partition $\{\{0, 2, 4\}, \{1, 3, 5\}\}$ of the set $\{0, 1, \ldots, 5\}$, as shown in Fig. 2.5.

A *directed graph* is a pair of a set of nodes and a set of edges connecting two nodes. A *node* is also called a *vertex*, and an *edge* is called an *arrow* or *arc*. Figure 2.6 shows an example of a directed graph in which the set of nodes is $\{1, 2, 3, 4\}$ and the set of edges is

$$\big\{(1, 2), (1, 3), (2, 3), (3, 2), (2, 4), (3, 4)\big\}.$$

**Fig. 2.6** Example of a
directed graph

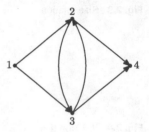

**Fig. 2.7** Example of an
undirected graph

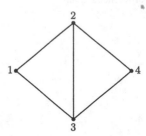

A relation represented as a collection of pairs and a directed graph represented as
a collection of nodes somehow connected with edges are slightly different in their
expressions, but they express substantially the same thing. That is, looking at graphs
we can easily figure out how nodes, representing elements, are related with each
other.

If we disregard the direction of edges in a directed graph, as shown in Fig. 2.7
which corresponds to Fig. 2.6, we have the notion of an undirected graph. An undirected graph is defined to consist of a set of nodes and a set of undirected edges. In
another words, in the case of an undirected graph, both $(i, j)$ and $(j, i)$ represent the
same edge that connects nodes $i$ and $j$. So to show that the direction is disregarded,
edge $(i, j)$ of an undirected graph is sometimes denoted by $\{i, j\}$ as well.

The **degree** of a node in an undirected graph is the number of edges that are
connected to the node. For the example shown in Fig. 2.7, the degree of node 1
is 2 and that of node 2 is 3. On the other hand, in the case of directed graphs, we
define notions of outdegree and indegree. The **outdegree** of a node in a directed
graph is the number of edges that leave that node, while the **indegree** of a node is
the number of edges that enter the node. For the example shown in Fig. 2.6, the
indegree of node 1 is 0 and its outdegree is 2, and the indegree and the outdegree of
node 2 are both 2.

Formally, a **graph** is defined to be a pair $(V, E)$ of a set $V$ of nodes and a set $E$
of edges. The directed graph shown in Fig. 2.6 is expressed as

$$(\{1, 2, 3, 4\}, \{(1, 2), (1, 3), (2, 3), (3, 2), (2, 4), (3, 4)\}),$$

and the undirected graph shown in Fig. 2.7 is expressed as

$$(\{1, 2, 3, 4\}, \{\{1, 2\}, \{1, 3\}, \{2, 3\}, \{2, 4\}, \{3, 4\}\}).$$

**Fig. 2.8** Example of a tree
where the root is 1 and the
leaves are 4, 5, 8, 9, and 7

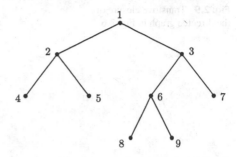

No matter whether it is a directed graph or an undirected graph, a **path** in a graph
is a sequence of consecutive edges which is represented by $(v_0, v_1), (v_1, v_2), \ldots,$
$(v_{m-1}, v_m)$. A **simple path** is a path in which any node appears at most once. The
**length of a path** is the number of edges in the path. If a path starts at a node and
ends at the same node, the path is called a **closed path** or **cycle**.

If for any nodes $v$ and $v'$ in a graph there exists a path that connects from $v$ to $v'$,
the graph is called a connected graph. Note that in the case of a connected directed
graph, it follows from the definition that for any node $v$ and $v'$ there exist a path
from $v$ to $v'$ as well as a path from $v'$ to $v$. If a connected undirected graph with
one distinguished node, called a **root**, is such that there exists no closed path, then
the graph is called a *rooted* tree, or simply a **tree**. Figure 2.8 shows an example of a
tree. In the case of this example, node 1 is the root, while nodes 4, 5, 8, 9, and 7 are
called leaves. As you can see from this example, a node in a tree is a **leaf** if a path
from the root to that node cannot be extended any further.

For a graph $G = (V, E)$ and a subset $V' \subseteq V$, the subgraph $G'$ of $G$ *induced*
from $V'$ is one obtained by leaving only edges that connect nodes, both from $V'$.
For example, let a graph $G$ be the directed graph given in Fig. 2.6; then the subgraph
$G'$ induced from $\{1, 2, 3\}$ is $(\{1, 2, 3\}, \{(1, 2), (1, 3), (2, 3), (3, 2)\})$. Formally, the
subgraph of $G = (V, E)$ induced from $V' \subseteq V$ is defined to be $(V', E \cap (V' \times V'))$.

In general, when we are given a relation that does not satisfy the transitive law,
we can transform it into a transitive relation by adding edges appropriately. The
relation obtained this way from relation $R$ by adding as few edges as possible is the
*transitive closure* of $R$.

By expressing a relation as the directed graph, we shall give relations and their
corresponding transitive closures. The transitive closure of the relation given as
Fig. 2.6 is illustrated in Fig. 2.9. Similarly, the transitive closure of the relation
given as Fig. 2.10 is illustrated in Fig. 2.3. Clearly the transitive closures obtained
above satisfy the transitive law.

Let a relation $R$ be expressed as the directed graph $G$ associated with the relation $R$. The *transitive closure* of $R$, denoted by $cl(R)$, is defined to be the graph
obtained from $G$ by adding every pair $(v, v')$ of nodes as an edge as long as there
exists a path in $G$ from $v$ to $v'$. It is easy to see that the transitive closure $cl(R)$ defined this way is the minimum in terms of the inclusion relation among the relations
that include $R$ and satisfy the transitive law.

**Fig. 2.9** Transitive closure of
the directed graph in Fig. 2.6

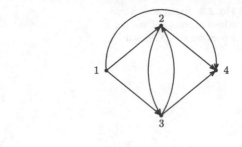

**Fig. 2.10** Transitive closure
of the relation given by this
figure is given by Fig. 2.3

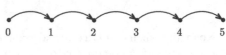

**Table 2.3** Incidence matrix
of the graph in Fig. 2.6

$$\begin{bmatrix} 0 & 1 & 1 & 0 \\ 0 & 0 & 1 & 1 \\ 0 & 1 & 0 & 1 \\ 0 & 0 & 0 & 0 \end{bmatrix}$$

**Table 2.4** Incidence matrix
of the graph in Fig. 2.7

$$\begin{bmatrix} 0 & 1 & 1 & 0 \\ 1 & 0 & 1 & 1 \\ 1 & 1 & 0 & 1 \\ 0 & 1 & 1 & 0 \end{bmatrix}$$

Before ending this section, we describe how to represent a graph. One way to
represent a graph is to give a set $V$ of nodes and a set $E$ of edges, each set being
expressed by listing its elements with appropriate punctuation marks between the
elements. We also need to consider how to represent elements in these sets. For example, when $V = \{1, \ldots, n\}$, node $i$ can be represented as the corresponding binary
number. Alternatively, a graph can also be represented as the $n \times n$ *adjacency matrix* defined as follows: the $(i, j)$ element of the incidence matrix is 1 if $(i, j)$ is an
edge and 0 otherwise. Tables 2.3 and 2.4 give the incidence matrices for the graphs
shown in Figs. 2.6 and 2.7, respectively. It is clear that an incidence matrix for an
undirected graph is a *symmetric* one, that is, the $(i, j)$ element is equal to the $(j, i)$
element for any $i$ and $j$.

## 2.5   Boolean Operations and Boolean Formulas

We shall explain *Boolean variables* that take the *Boolean values* of 0 or 1 and the
*Boolean operations* applied to them. The mathematical system based on these concepts is called a *predicative logic* or *Boolean algebra*. If 1 is related to the truth and
0 to the false, the propositional logic becomes the basis of discussing the correctness

of inferences as well as the basis of exploring the relationship between inputs and outputs of the gates that constitute the hardware of computers.

A *Boolean formula* is one obtained by applying operations $\vee$, $\wedge$, and $\neg$ to *Boolean variables* that take the values 0 or 1. The operations $\vee$, $\wedge$, and $\neg$ are called *disjunction*, *conjunction*, and *negation*, respectively, and are defined as follows:

$$0 \vee 0 = 0, \qquad 0 \wedge 0 = 0, \qquad \neg 0 = 1,$$
$$0 \vee 1 = 1, \qquad 0 \wedge 1 = 0, \qquad \neg 1 = 0,$$
$$1 \vee 0 = 1, \qquad 1 \wedge 0 = 0,$$
$$1 \vee 1 = 1, \qquad 1 \wedge 1 = 1.$$

We can interpret these operations by regarding Boolean values 1 and 0 as having a high or a low voltage, or indicating that a statement concerned holds or does not hold. That is, $x_1 \vee x_2$ means that "$x_1$ is true" *or* "$x_2$ is true," whereas $x_1 \wedge x_2$ means that "$x_1$ is true" *and* "$x_2$ is true." Furthermore, $\neg x$ means that the truth and falsity of $x$ are inverted. $\neg x$ is also denoted by $\overline{x}$. As an example of a Boolean formula, let us consider

$$\overline{(x_1 \wedge x_2) \vee (\overline{x}_1 \wedge \overline{x}_2)}.$$

According to the interpretations of $\vee$, $\wedge$, and $\neg$ mentioned above, it turns out that the Boolean formula can be interpreted as "$x_1$ is 0 and $x_2$ is 1" or "$x_1$ is 1 and $x_2$ is 0." This is because the formula is interpreted as the negation of "both $x_1$ and $x_2$ are 1's, or both $x_1$ and $x_2$ are 0's." In the following, this fact is derived by applying a transformation to the Boolean formula.

In general, for any Boolean formulas $F$, $G$, $H$, the *distributive law* expressed as

$$F \wedge (G \vee H) = (F \wedge G) \vee (F \wedge H),$$
$$F \vee (G \wedge H) = (F \vee G) \wedge (F \vee H)$$

holds. It can easily be checked that these equations hold by substituting all the combinations of 0's and 1's for $F$, $G$, and $H$. Relating $\wedge$ to $\times$ and $\vee$ to $+$, the distributive law for the first equation is the same as that for the usual arithmetic equation. Furthermore, the second equation claims that the distributive law holds even if the $\vee$'s and $\wedge$'s in the first equation are interchanged. Similarly, the distributive law in which operations $\vee$ and $\wedge$ are applied from the right-hand side also holds as follows:

$$(G \vee H) \wedge F = (G \wedge F) \vee (H \wedge F),$$
$$(G \wedge H) \vee F = (G \vee F) \wedge (H \vee F).$$

Furthermore, for any Boolean formulas $F$ and $G$,

$$\overline{F \vee G} = \overline{F} \wedge \overline{G},$$
$$\overline{F \wedge G} = \overline{F} \vee \overline{G}$$

hold, which is called *De Morgan's law*. The first equation holds because the negation of "$F$ or $G$ are true" is equivalent to saying "both $F$ and $G$ are false," while the second equation holds because the negation of "both $F$ and $G$ are true" is equivalent to saying "$F$ is false or $G$ is false." These relations are verified by substituting all the combinations of 0's and 1's for $F$ and $G$ in the formulas. Replacing Boolean variables $F$ and $G$ by variables $A$ and $B$ for sets, respectively, and replacing the disjunction $\vee$, the conjunction $\wedge$, and the negation $\neg$ by the set operations of the union $\cup$, the intersection $\cap$, and the complement $\bar{\phantom{x}}$, respectively, we can obtain De Morgan's law for sets:

$$\overline{A \cup B} = \overline{A} \cap \overline{B},$$
$$\overline{A \cap B} = \overline{A} \cup \overline{B}.$$

By applying De Morgan's law and the distributive law to $\overline{(x_1 \wedge x_2) \vee (\overline{x}_1 \wedge \overline{x}_2)}$, we can transform the formula to obtain the following:

$$
\begin{aligned}
\overline{(x_1 \wedge x_2) \vee (\overline{x}_1 \wedge \overline{x}_2)} &= \overline{(x_1 \wedge x_2)} \wedge \overline{(\overline{x}_1 \wedge \overline{x}_2)} \\
&= (\overline{x}_1 \vee \overline{x}_2) \wedge (\overline{\overline{x}}_1 \vee \overline{\overline{x}}_2) \\
&= (\overline{x}_1 \vee \overline{x}_2) \wedge (x_1 \vee x_2) \\
&= (\overline{x}_1 \wedge x_1) \vee (\overline{x}_1 \wedge x_2) \vee (x_1 \wedge \overline{x}_2) \vee (\overline{x}_2 \wedge x_2) \\
&= (\overline{x}_1 \wedge x_2) \vee (x_1 \wedge \overline{x}_2).
\end{aligned}
$$

The equation derived says that the original formula is equivalent to saying "$x_1 = 0$ and $x_2 = 1$, or $x_1 = 1$ and $x_2 = 0$." In transforming the formula, we used equations such as $\overline{\overline{x}} = x$ and $x \wedge \overline{x} = 0$ in addition to De Morgan's law and the distributive law.

In expressing Boolean formulas, the operation symbol $\wedge$ is often omitted throughout this book. For example, "just two of $x_1$, $x_2$, $x_3$ are 1's" is expressed as $x_1 x_2 \overline{x}_3 \vee x_1 \overline{x}_2 x_3 \vee \overline{x}_1 x_2 x_3$, and "at least two of $x_1$, $x_2$, $x_3$ are 1's" is expressed as $x_1 x_2 \vee x_2 x_3 \vee x_3 x_1$.

## 2.6  Propositions and Proofs

A *proposition* is what claims an assertion whose validity is clearly determined. In particular, when a proposition includes variables, its validity is determined after substituting Boolean values for the variables. For propositions composed of various propositions, we shall explain the procedures to determine whether a composed proposition is true or false.

| Table 2.5 Illustration explaining $P(n) \Rightarrow Q(n)$ and $P(n) \Rightarrow R(n)$ | $n$ | $P(n)$: $n$ is divided by 6 | $Q(n)$: $n$ is divided by 3 | $R(n)$: $n$ is divided by 2 |
|---|---|---|---|---|
| | 1 | 0 | 0 | 0 |
| | 2 | 0 | 0 | 1 |
| | 3 | 0 | 1 | 0 |
| | 4 | 0 | 0 | 1 |
| | 5 | 0 | 0 | 0 |
| | 6 | 1 | 1 | 1 |
| | 7 | 0 | 0 | 0 |
| | 8 | 0 | 0 | 1 |
| | 9 | 0 | 1 | 0 |
| | 10 | 0 | 0 | 0 |
| | 11 | 0 | 0 | 1 |
| | 12 | 1 | 1 | 1 |
| | 13 | 0 | 0 | 0 |
| | 14 | 0 | 0 | 1 |
| | 15 | 0 | 1 | 0 |
| | ⋮ | ⋮ | ⋮ | ⋮ |

Given propositions $P$ and $P'$, we have a proposition "$P$ implies $P'$" which is usually denoted by $P \Rightarrow P'$. Similarly, "$P$ implies $P'$, and $P'$ implies $P$" is denoted by $P \Leftrightarrow P'$. When $P \Leftrightarrow P'$, $P$ and $P'$ are said to be *equivalent* with each other. Furthermore, the proposition "both $P$ and $P'$ are true" is denoted by $P \wedge P'$, "either $P$ or $P'$ is true" as $P \vee P'$, and "$P$ is false" as $\overline{P}$. In this way, new propositions are obtained by applying $\Rightarrow$, $\Leftrightarrow$, $\vee$, $\wedge$, and $^{-}$, etc., to variables.

Let $n$ be a natural number. We consider the propositions $P(n)$, $Q(n)$, and $R(n)$ in terms of parameter $n$ as follows:

$P(n)$:   $n$ is divided by 6,

$Q(n)$:   $n$ is divided by 3,

$R(n)$:   $n$ is divided by 2.

If $n$ is divided by 6, it is also divided by 3 and 2. Hence, clearly $P \Rightarrow Q$ and $P \Rightarrow R$ hold. This fact is also easily seen from Table 2.5, which shows whether $P(n)$, $Q(n)$, and $R(n)$ hold or not by means of 1 or 0. From Fig. 2.11, which schematically shows the relationship between $P(n)$ and $Q(n)$, we can see that not only $P \Rightarrow Q$ holds, but also $P \Rightarrow Q$ is equivalent to $\overline{Q} \Rightarrow \overline{P}$. This is because "for any $n$, $P(n) = 1$ implies $Q(n) = 1$" implies "for any $n$, $Q(n) = 0$ implies $P(n) = 0$." Notice that the reverse of this implication, namely, "for any $n$, $Q(n) = 0$ implies $P(n) = 0$" implies "for any $n$, $P(n) = 1$ implies $Q(n) = 1$," also holds.

$$\overline{Q} \Rightarrow \overline{P}$$

**Fig. 2.11** Explanation of
$P(n) \Rightarrow Q(n)$ and
$\overline{Q(n)} \Rightarrow \overline{P(n)}$

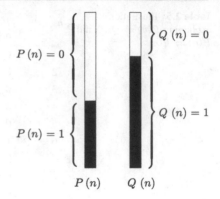

is called the *contraposition* of $P \Rightarrow Q$. In general, a proposition is equivalent to its contraposition.

Let $P$ and $Q$ denote conditions on the same collection of variables, and let $D(P)$ and $D(Q)$ denote the sets of the values of the variables that satisfy $P$ and $Q$, respectively. Since $P \Rightarrow Q$ means that values of the variables that satisfy $P$ also satisfy $Q$, $P \Rightarrow Q$ is equivalent to $D(P) \subseteq D(Q)$. Furthermore, since $P \Leftrightarrow Q$ is equivalent to "$P \Rightarrow Q$ and $Q \Rightarrow P$," $P \Leftrightarrow Q$ is equivalent to $D(P) = D(Q)$. We can examine these relations by using the condition $P(n)$, $Q(n)$, and $R(n)$ described above. As illustrated in Fig. 2.5, we can see $P(n)$ is equivalent to $Q(n) \wedge R(n)$. Furthermore, we have the following equations:

$$D(P \wedge Q) = D(P) \cap D(Q), \qquad D(P \vee Q) = D(P) \cup D(Q),$$
$$D(\overline{P}) = \overline{D(P)}.$$

Next, we describe how to make an argument to prove propositions. We begin with a proposition of the form $P \Rightarrow Q$ and explain how to prove that type of proposition. We discuss two types of arguments to prove such propositions: proof by contraposition and proof by contradiction.

*Proof by contraposition* is the method of proving $P \Rightarrow Q$ by deriving $\overline{Q} \Rightarrow \overline{P}$. The reason why we can argue this way is based on the fact that $P \Rightarrow Q$ is equivalent to $\overline{Q} \Rightarrow \overline{P}$. On the other hand, *proof by contradiction* is the method of proving $P \Rightarrow Q$ by showing that, if we assume $P$ and $\overline{Q}$, then a contradiction follows. The fact that "$P$ and $\overline{Q}$" leads logically to a contradiction means that "$P = 1$ and $Q = 0$" never happens. Therefore, the permissible value of $(P, Q)$ is either one of $(1, 1)$, $(0, 1)$, or $(0, 0)$, thereby proving that $P \Rightarrow Q$ holds, as is easily seen in Fig. 2.11. Note that if we can show that the assumption "$P$ and $\overline{Q}$" leads to a contradiction, we can consequently conclude $\overline{Q} \Rightarrow \overline{P}$. Furthermore, when the proposition that we want to prove simply takes the form $Q$ rather than $P \Rightarrow Q$, we can prove the proposition by verifying that the assumption $\overline{Q}$ leads to a contradiction, thereby proving $Q$. Comparing these two proof methods to prove $P \Rightarrow Q$, we can see that, in the case of proof by contraposition, we show that assuming $\overline{Q}$ leads to $\overline{P}$, whereas, in the case of proof by contradiction, we derive assuming that not only $\overline{Q}$ but also $P$

leads to a contradiction. But since the difference between these two proof methods is somewhat subtle, you may verify how they work in the following examples.

*Example 2.1* Let $x$, $y$, and $z$ be real numbers and let $P$ and $Q$ be conditions described as follows:

> $P$:   $x + y + z \geq 0$,
>
> $Q$:   at least one of $x, y, z$ is greater than or equal to 0.

We prove $P \Rightarrow Q$ by deriving $\overline{Q} \Rightarrow \overline{P}$ by *proof by contraposition*.

*Proof* Assume the negation of $Q$, namely, all $x$, $y$, and $z$ are less than 0. Then clearly we have $x + y + z < 0$, thereby $\overline{P}$ is derived. Thus, $\overline{Q} \Rightarrow \overline{P}$ holds.  □

Note that, in the example above, the reason why proof by contraposition makes it easy to prove is because, as compared to the condition "at least one of $x$, $y$, and $z$ is more than or equal to 0," the condition "all of $x$, $y$, and $z$ are less than 0" is easy to deal with, each of $x$, $y$, and $z$ being referred to independently.

*Example 2.2* Let $P$ be as follows:

> $P$:   there exist an infinite number of prime numbers.

In order to prove $P$ based on *proof by contradiction*, we shall derive a contradiction by supposing $\overline{P}$.

*Proof* Assume the negation $\overline{P}$ of $P$, namely, "the number of prime numbers is finite." So, let the primes be $p_1, p_2, \ldots, p_m$. Then the natural number $p_1 p_2 \cdots p_m + 1$ is not divided by any of the prime numbers $p_1, p_2, \ldots, p_m$ because the residue of the division is always 1. Therefore, that natural number is a prime number. This contradicts the fact that all of the prime numbers are listed as $p_1, p_2, \ldots, p_m$. Since either $P$ or $\overline{P}$ holds and $\overline{P}$ is negated by the contradiction, $P$ holds.  □

*Example 2.3* Let $i$, $j$, and $k$ be natural numbers and let $P$ and $Q$ be as follows:

> $P$:   $i^2 + j^2 = k^2$,
>
> $Q$:   at least one of $i, j, k$ is even.

We shall show that assuming $P$ and $\overline{Q}$ leads to a contradiction, thereby verifying $P \Rightarrow Q$ by proof by contradiction.

*Proof* Assume the negation of $Q$, namely, "any of $i$, $j$, and $k$ is odd." On the other hand, the square of an odd number, expressed as $2m + 1$, is given by

$$(2m + 1)^2 = 4(m^2 + m) + 1.$$

Hence the square of any odd number is odd. Since $\overline{Q}$ implies that any of $i^2$, $j^2$, and $k^2$ is odd, $i^2 + j^2$ is even, while $k^2$ is odd. Thus, this contradicts the assumption that $i^2 + j^2 = k^2$, which proves that $P \Rightarrow Q$.                                                    □

As the next method of proof, we discuss mathematical induction. The proposition that we want to prove is denoted by $P(n)$, which is described in terms of a positive integer $n$. Mathematical induction is the method used to prove that $P(n)$ holds for all $n$. In order to grasp the method intuitively, we consider an infinite sequence of domino tiles. We can intuitively accept the following: if we can verify the two statements, namely, "the first domino tile falls down" and "for any positive integer $n$ if the $n$th domino tile falls down, then the $(n + 1)$th domino tile falls down," then we can conclude that all of the domino tiles fall down. If we correspond the statement "the $n$th domino tile falls down" to the proposition "$P(n)$ holds," then what we conclude that the domino argument claims that $P(n)$ holds for all $n$. This is exactly what we want to derive. So all we need to do is to show the following statements (1) and (2), which correspond to the domino counterparts. Let

(1) $P(1)$ holds,
(2) For any positive integer $n$, if $P(n)$ holds, then $P(n + 1)$ also holds

*Mathematical induction* is the method for proving that $P(n)$ holds for all positive integer $n$'s by verifying that (1) and (2) above hold.

*Example 2.4* Let $P(n)$ express "$n^3 - n$ is divided by 3." By mathematical induction we shall prove that $P(n)$ holds for any $n \geq 1$.

*Proof* First, $P(1)$ holds since $1^3 - 1 = 0$ is divided by 3. Next, we shall derive that if $P(n)$ holds, $P(n + 1)$ also holds. We have

$$(n + 1)^3 - (n + 1) = \left(n^3 + 3n^2 + 3n + 1\right) - (n + 1)$$
$$= \left(n^3 - n\right) + 3n^2 + 3n.$$

Hence, if $n^3 - n$ is divided by 3 (that is, if $P(n)$ holds), then $(n + 1)^3 - (n + 1)$ is also divided by 3 (that is, $P(n + 1)$ holds).                                                    □

## 2.7    Descriptions of Algorithms

An *algorithm* is a procedure that can be automatically executed to solve a problem. Employing an example, we shall explain how to describe algorithms throughout this book.

What we take as an example is the problem to ask whether or not, given a directed graph $G$ and its two nodes $s$ and $t$, there exists a path from $s$ to $t$. It is called the *reachability problem*. First, an idea for solving it will be explained.

**Fig. 2.12** Explanation of
behavior of algorithm
*REACH*

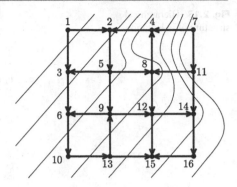

Figure 2.12 illustrates how an algorithm decides whether the node $t$ is reachable from node $1(= s)$ or not. The algorithm accumulates repeatedly the nodes that are reachable from node 1 to set $R$. It is shown in the figure that just as the wavefront travels, the algorithm repeatedly updates set $R$ consisting of nodes reachable from node 1: starting with $R = \{1\}$; add all the nodes reachable from node 1, letting $R = \{1, 2, 3\}$; further add the reachable nodes from $\{1, 2, 3\}$, setting $R = \{1, 2, 3, 5, 6\}$; similarly let $R = \{1, 2, 3, 5, 6, 8, 9, 10\}$ and so on until no further nodes are added to $R$. After all the reachable nodes are added to $R$ this way, the algorithm decides whether or not node $t$ is reachable from node $s$ depending on whether or not node $t$ belongs to $R$.

In general, steps of an algorithm are divided into a number of groups called stages, each performing certain small intended tasks. In this case the algorithm, denoted *REACH*, is divided into three stages, which are numbered **1**, **2** and **3** as illustrated below. In the algorithm, $R$ is used to denote the set of nodes that are found so far as the nodes reachable from node $s$, whereas $S$ is used to denote the set of nodes that belong to $R$ and are not yet checked to collect further nodes by going out from nodes in $S$ along edges.

In stage **1**, $R$ and $S$ are set to be the sets consisting of only the node $s$. In stage **2**, to collect further nodes not yet found to be reachable, if there exists a node $v$ in $S$, find all the new nodes $v'$ that are connected from $v$ by edges, and add them to both $R$ and $S$ ($R \leftarrow R \cup \{v'\}$, $S \leftarrow S \cup \{v'\}$), and finally the node $v$ is removed from $S$ ($S \leftarrow S - \{v\}$), where $\leftarrow$ means that what is expressed by its right-hand side is substituted into the left-hand side. So $S \leftarrow S \cup \{v'\}$, for example, simply means to add node $v'$ to the collection of nodes in $S$. Repeat the update of sets $R$ and $S$ in this way until no further nodes are added. When $S$ comes to be empty, hence no further nodes are being added, the algorithm proceeds to stage **3**. In stage **3**, the algorithm outputs YES if node $t$ belongs to $R$ and NO otherwise. This finishes the explanation of how the algorithm *REACH* works.

We present the algorithm *REACH* below. The description consists of the name of the algorithm, the input format, and finally the body of the algorithm consisting of the three stages.

**Fig. 2.13** Hierarchical
structure of indentation

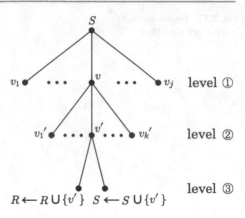

**Algorithm** *REACH*

Input: $\langle G, s, t \rangle$, where $s$ and $t$ are the nodes of the graph $G$ and the set of edges
        of $G$ is denoted by $E_G$.

**1.** $R \leftarrow \{s\}$, $S \leftarrow \{s\}$.
**2.** While $S \neq \emptyset$, do the following for each $v$ in $S$
    for each $(v, v')$ in $E_G$ with $v' \notin R$, do the following
        $R \leftarrow R \cup \{v'\}$,
        $S \leftarrow S \cup \{v'\}$.
    $S \leftarrow S - \{v\}$.
**3.** If $t \in R$, output YES.
    If $t \notin R$, output NO.

①
  ②
    ③

In the description of the algorithm, the first line shows that the name of the algorithm is *REACH*. The second line shows that the input is graph $G$ together with its two nodes $s$ and $t$, which are denoted as $\langle G, s, t \rangle$. In general, $\langle A \rangle$ denotes a description of entity $A$. For example, in the case that $A$ is a graph, as mentioned in Sect. 2.4, $\langle G \rangle$ may be a list of the nodes and a list of the edges of the graph, or alternatively, a list of the elements of the adjacency matrix of the graph. In particular, when the entity is a string $w$, an input is described as $w$ instead of $\langle w \rangle$ because in that case an algorithm can receive the string as input.

Taking *REACH* as an example of an algorithm, we explain how the *indentation* controls the execution of the algorithm. In the case of *REACH*, there are three levels ①, ②, and ③ of the indentation, as shown in the algorithm. In fact, the marks ①, ②, and ③ together with the associated vertical lines are just for explanation. When algorithms are written in practice, the levels of indentation are shown by just shifting the position of each line according to the corresponding level of the indentation.

Focusing exclusively on stage **2**, Fig. 2.13 illustrates how the indentation controls the execution of *REACH*. Stage **2** is executed as follows: at level ①, for each node $v$ in $S$, the one lower level ② (from the line beginning "for each $(v, v')$" to the line of $S \leftarrow S \cup \{v\}$) is executed repeatedly; at level ②, for each edge $(v, v')$, the one further lower level ③ (consisting of $R \leftarrow R \cup \{v'\}$, $S \leftarrow S \cup \{v'\}$) is executed repeatedly.

Figure 2.13 illustrates the moment when $v$ is chosen from $S = \{v_1, \ldots, v, \ldots, v_j\}$ in level ① and $v'$ is chosen from $\{v'_1, \ldots, v', \ldots, v'_k\}$ in level ②, where edges out of node $v$ are denoted by $(v, v'_1), \ldots, (v, v'), \ldots, (v, v'_k)$. Note that algorithm *REACH* works well no matter in what order you choose node $v$ and $v'$ mentioned above.

## 2.8   Problems

**2.1** Give the size of the power set of the set $\{1, 2, \ldots, n\}$ in terms of $n$.

**2.2** Let $m$ and $n$ be the sizes of sets $A$ and $B$, respectively. Give the number of functions $f : A \rightarrow B$.

**2.3**\*\*  Show that a partition $P$ is derived from a relation $R$ that satisfies the reflexive, symmetric, and transitive laws. To do so, show how $P$ is defined from $R$ and that $P$ defined so becomes a partition. Conversely, show that a relation $R$ that satisfies the reflexive, symmetric, and transitive laws is derived from a partition $P$. To do so, show how $R$ is defined from $P$ and that $R$ defined so satisfies the reflexive, symmetric, and transitive laws.

**2.4**\*\*  In the following, we give the proof of the statement that "a relation that satisfies the symmetric and transitive laws also satisfies the reflexive law." Find a flaw in the proof.

Suppose $xRy$ for any $x$ and $y$. From the symmetric law, $yRx$ holds. Therefore, since $xRy$ and $yRx$, we have $xRx$ from the transitive law. Thus, since $x$ is arbitrary, the reflexive law holds.

**2.5**\*\*  Show that, given an undirected graph $G$ with $n$ nodes, if there exists a path from $s$ to $t$ whose length is equal to or more than $n$, then there exists a path from $s$ to $t$ whose length is equal to or less than $n - 1$. Similarly, show that the same statement holds for a directed graph.

**2.6**\*\*  From De Morgan's law

$$\overline{F \vee G} = \overline{F} \wedge \overline{G},$$

$$\overline{F \wedge G} = \overline{F} \vee \overline{G},$$

derive the similar law

$$\overline{F \vee G \vee H} = \overline{F} \wedge \overline{G} \wedge \overline{H},$$

$$\overline{F \wedge G \wedge H} = \overline{F} \vee \overline{G} \vee \overline{H}.$$

Furthermore, generalize the above law to the case where the number of variables is arbitrary.

**2.7**\*\* In the following, we give the proof of "for any club, the birthplaces of the members of the club are all the same." Find a flaw in the proof.

Prove by induction on the number of members $n$ of the club.

The base of induction: For $n = 1$, clearly the statement holds.

Induction step: Suppose that the statement holds for $n \geq 1$. We will show that the statement holds when the number of members is $n + 1$. Given a club consisting of $n + 1$ members, make a group consisting of $n$ members, by removing one member in the club. From the hypothesis of induction, the birthplaces of all the members of the group are the same. Similarly, make another group consisting of $n$ members by removing another member from the club. Then the birthplaces of all the members of the second group are also the same by the hypothesis of induction. Therefore, all the members of the club with $n + 1$ members which is the union of the two groups are the same.

# Part II

# Automata and Languages

# Finite Automata

3

A finite automaton is a simple mathematical model to explain how a computer basically works. It is a model that receives as input a string of symbols and makes as output a decision of whether or not it accepts the string. A finite automaton is a model that has a finitely limited amount of memory. We shall explain how a finite automaton limits its behavior because of the finiteness of memory. A finite automaton serves as a core model that underpins the variety of computational models studied in this book.

## 3.1  Modeling by Automata

The automaton is a mathematical model which serves as a basis for many models that will be studied in what follows. After describing various practical problems in terms of the models, we give the definition of a finite automaton formally.

### Modeling of Problems

We consider three concrete problems and describe them in terms of finite automata.

*Example 3.1*  Figure 3.1, which will be called a state diagram, shows the behavior of a vending machine of juice for 150 yen. The nodes in the figure represent the states of the machine, and the edges represent transitions between the states. These states correspond to amounts of money that are fed just after the recent bottle of juice came out, and the edge labeled "50" represents the transition caused by inserting a coin of 50 yen. The figure illustrates that starting at the state 0 attached to the short arrow, a bottle of juice comes out every time the amount of coins inserted sums up to 150 yen. That is, the state 0 with the double circles is the state where a bottle of juice comes out. On the other hand, this diagram does not illustrate the vending machine exactly because, even though the diagram says an infinite number of bottles of juice are yielded by feeding an infinite number of coins, the real machine does not give

A. Maruoka, *Concise Guide to Computation Theory*,
DOI 10.1007/978-0-85729-535-4_3, © Springer-Verlag London Limited 2011

**Fig. 3.1** State diagram for a
vending machine

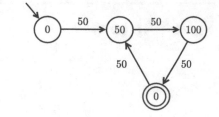

**Fig. 3.2** State diagram
explaining how to open and
close doors of a car

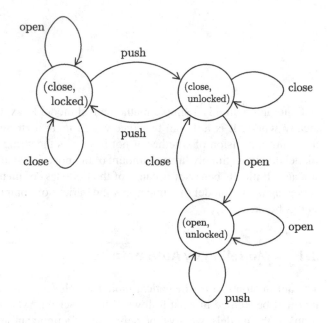

more bottles of juice than those that the machine holds to start with. In this sense
Fig. 3.1 is a model that abstracts away practical details.

*Example 3.2* We consider an example of a model for opening and closing a door of
a car by means of a remote-controlled switch. The states of the door are represented
in terms of a combination of conditions: whether it is opened or closed; whether
it is locked or unlocked. We consider three kinds of human actions: to "push" the
remote-controlled switch, and to "open" or "close" the door.

Figure 3.2 shows how the door works: what transition the states undergo according to the inputs of "push," "open," and "close." The switch of the remote controller,
which works as a toggle switch, changes the states from "locked" to "unlocked" and
vice versa, every time the switch is pushed. If the switch is pushed in the state (close,
locked), the door continues to be closed, but the second component of the state is
changed to be unlocked. That is, by the input "push" it goes from the state (close,
locked) to the state (close, unlocked). On the other hand, no change will occur if
we try to open or close the door in the state (close, locked). In other words, it goes

**Fig. 3.3** State diagram
explaining how to carry the
three objects

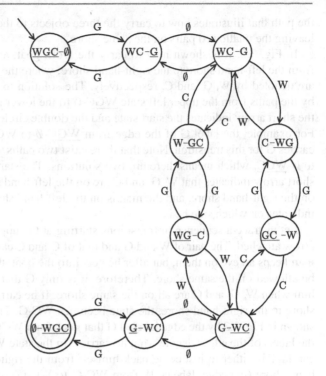

from the state (close, locked) to the same state (close, locked) by either the input
"open" or the input "close." The other state transitions can be similarly specified.
By the way, if we push the switch of the remote controller when the door is open
and unlocked, it is designed not to be locked (the author checked this fact with his
own car). So, by the input "push" it goes from the state (open, unlocked) to the same
state (open, unlocked).

*Example 3.3* In the example of the vending machine and that of the door of a car,
the system to be modeled is a visible concrete thing. In contrast, this example is
different from the previous ones in the sense that the object to be modeled is a
puzzle which is described as Fig. 3.3.

A man who has a wolf, a goat, and a cabbage is trying to cross a river to carry
them to the opposite shore by using a boat. In doing so he can carry at most one of
them on the boat. Furthermore, neither the pair of the wolf and the goat nor the pair
of the goat and the cabbage can be left on the same shore because one of them eats
the other. The problem is to show how all of them can be carried to the opposite
shore without violating the conditions.

The point of modeling the problem is how to define the state for this puzzle:
we formulate the state so that it indicates on which shore of the river the man, the
wolf, the goat, and the cabbage are placed. Once the state is defined appropriately,
the puzzle, which is seemingly difficult, can be automatically solved by drawing

the path that illustrates how to carry the three objects to the opposite shore without leaving the unallowed pairs on the same shore.

In Fig. 3.3, it is shown how to carry the wolf, goat, and cabbage successfully from the left-hand shore to the right-hand shore, where the wolf, goat, and cabbage are denoted by W, G, and C, respectively. The solution to the puzzle can be given by the paths from the upper left state WGC-Ø to the lower left state Ø-WGC, where the short arrow indicates the start state and the double circles indicate the final state. For example, the label G of the edge from WGC-Ø to WC-G shows that the man carries G for this transition. Note that there exist two paths which lead from WGC-Ø to Ø-WGC, which means there are two solutions. The start state WGC-Ø with the short arrow indicates that W, G, and C are on the left-hand shore and nothing exists on the right-hand shore, and the man is on the left-hand shore, where the underline indicates on which side he is.

Let us trace a sequence of transitions starting at the upper left state with a short arrow attached. The pair of W and G and that of G and C can stay together while the man keeps an eye on them, but after he gets into the boat, these pairs can no longer be allowed on the same shore. Therefore, it is only G that the man can carry with him when W, G, and C are all on the same shore. If he carries G from the left-hand shore to the right-hand shore, the state becomes WC-G. This way of carrying G is shown in Fig. 3.3 by the edge labeled G that goes from WGC-Ø to WC-G. Like this, the labels of the edges denote what he carries. In the state WC-G, the action that he can take is either to just come back himself from the right-hand shore to the left-hand shore (the edge, labeled Ø, from WC-G to WC-G) or to come back together with G from the right-hand shore to the left-hand shore (the edge, labeled G, from WC-G to WGC-Ø).

In this way, Fig. 3.3 is obtained by drawing all the edges of transitions that express the possible actions taken by the man. Since the state WGC-Ø denotes the state to start with and the state Ø-WGC denotes the state that we want to have at the end, the two paths from the state WGC-Ø to Ø-WGC in the figure show how the man carries the three successfully.

## Definition of a Finite Automaton

In the preceding subsection, three concrete examples are given in terms of the figures that describe their behavior. These figures are called the "state transition diagrams" or simply "state diagrams." A finite automaton is roughly what is described as a state diagram. Furthermore, the behavior of an automaton is expressed as the strings that the paths spell out, going from the start state with a short arrow to the final state with the double circles. In particular, the set of such strings is called the language accepted by the corresponding automaton. In this subsection an automaton together with the language that it accepts will be defined formally.

A state diagram helps us to figure out how the corresponding finite automaton behaves, but it cannot be precise in defining the notion. In Figs. 3.1 and 3.3, the states at which the automata start their transition are indicated by the short arrows, but this is not the case with Fig. 3.2. In Fig. 3.2, there are three edges, labeled "open," "close," and "push," respectively, going out of each state, whereas this is not the case with Fig. 3.3. So we shall need a precise definition of a finite automaton to tell us what is or is not a finite automaton. In this chapter, we define a deterministic finite automaton and a nondeterministic one, and then prove the theorem that claims that the two types of automata are equivalent, that is, given one type of automaton, we can construct the other type of automaton that behaves equivalently to the former. To prove such a theorem, we need to give precise definitions of both types of automata.

In general, a formal definition of something is a way of describing it precisely. As we can see below, the formal definition of a finite automaton is a list of objects that are needed to specify a concrete finite automaton precisely. If all of the objects that the formal definition requests are given, then a certain finite automaton will be specified precisely. Here, the "finite" part of finite automaton means that the number of states of the automaton is finite. Since in general a finite automaton stores information about input strings in terms of states, the fact that the number of states is finite means that so is the memory of a finite automaton.

**Definition 3.4** A *finite automaton* (*DFA* for short) is a 5-tuple $(Q, \Sigma, \delta, q_0, F)$, where
(1) $Q$ is a finite set of *states*.
(2) $\Sigma$ is a finite set of symbols, called the *alphabet*.
(3) $\delta : Q \times \Sigma \to Q$ is a *state transition function*.
(4) $q_0 \in Q$ is the *start state*.
(5) $F \subseteq Q$ is the set of *accept states*.
The finite set of input symbols is called an input alphabet or simply an alphabet. A state transition function is simply called a *transition function*.

Precisely, a finite automaton as defined above is called a *deterministic finite automaton* and is written as *DFA* for short, as will be explained later. Let us recall that the direct product $Q \times \Sigma$ in the definition is the set of all the pairs $(q, a)$'s of a state $q$ in $Q$ and an element $a$ in $\Sigma$.

*Example 3.5* Figure 3.4 shows a finite automaton that accepts all the binary sequences that have an even number of 1's. A binary sequence is one consisting of 0's and 1's. The start state is indicated by the short arrow attached and the accept state is shown by the double circles. In this example, the start state and the accept state are the same. Table 3.1 is the table, called a *state transition table*, which indicates the transition function of the finite automaton.

By describing the finite automaton according to Definition 3.4, we have $(\{q_0, q_1\}, \{0, 1\}, \delta, q_0, \{q_0\})$, where $\delta$ is given in Table 3.1. Furthermore, note that the finite automaton given by $(\{q_0, q_1\}, \{0, 1\}, \delta, q_0, \{q_1\})$ whose accept state is $q_1$ instead of $q_0$ accepts all the binary sequences that have an odd number of 1's.

In general, a *state transition diagram*, or *state diagram* for short, is used to show intuitively what the formal definition of a finite state machine describes. Nodes in a state diagram denote the states in $Q$, and an edge labeled $a$ directed from state

**Fig. 3.4** A finite automaton accepting strings that include an even number of 1's

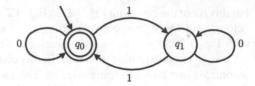

**Table 3.1** State diagram for the finite automaton given in Fig. 3.4

| State\Input | 0 | 1 |
|---|---|---|
| $q_0$ | $q_0$ | $q_1$ |
| $q_1$ | $q_1$ | $q_0$ |

$q$ to $q'$ means that $\delta(q, a) = q'$. In this way, the state diagram is considered to be a directed graph that specifies the transition function. Furthermore, as shown in Example 3.5, the start state is expressed as a state with a short arrow attached and accept states are expressed as double circles.

In general, the number of accept states is arbitrary, while that of the start states is one. This is because the set $F$ of accept states is specified by the condition $F \subseteq Q$, so any subset of $Q$ including $\emptyset$ can be specified as a set of accept states.

As mentioned above, although there are various ways to express a finite automaton, the formal definition of a finite automaton specifies clearly what is or is not a finite automaton. However, it is easier to understand intuitively how a finite automaton works when it is given in terms of a state diagram. For this reason, in the following, we use mostly a state diagram to specify a finite automaton.

In this book, we think of an automaton as something that decides whether an input string is accepted or not. Let us express a finite automaton $(Q, \Sigma, \delta, q_0, F)$ as $M$. The transition function $\delta$ is extended as follows so that whether an input string is accepted or not is easily described in terms of the extended transition function. For $q_1 \in Q$ and $a_1, \ldots, a_m \in \Sigma$, if

$$\delta(q_1, a_1) = q_1, \qquad \delta(q_2, a_2) = q_2, \qquad \ldots, \qquad \delta(q_m, a_m) = q_{m+1},$$

we define it as

$$\delta(q_1, a_1 a_2 \cdots a_m) = q_{m+1},$$

where $m \geq 0$. In particular, when $m = 0$, let $\delta(q_1, \varepsilon) = q_1$ for any $q_1$ in $Q$. In this way a function $\delta : Q \times \Sigma \to Q$ can be extended as

$$\delta : Q \times \Sigma^* \to Q.$$

Notice that the same symbol $\delta$ is used to represent a function $\delta : Q \times \Sigma \to Q$ as well as an extended transition function $\delta : Q \times \Sigma^* \to Q$. You may consult the context to figure out which meaning is intended. In what follows, if there exists a sequence of

**Fig. 3.5** Transitions driven
by a string $a_1 a_2 \cdots a_m$

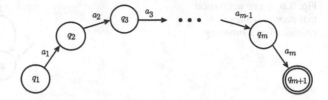

transitions as depicted in Fig. 3.5, it is expressed as

$$q_1 \xrightarrow{a_1} q_2 \xrightarrow{a_2} \cdots \xrightarrow{a_{m-1}} q_m \xrightarrow{a_m} q_{m+1}.$$

Moreover, $\delta(q_1, a_1 a_2 \cdots a_m) = q_{m+1}$ is alternatively expressed as

$$q_1 \xrightarrow{a_1 a_2 \cdots a_m} q_{m+1}.$$

In particular, $\delta(q_1, a_1) = q_2$ is often expressed as

$$q_1 \xrightarrow{a_1} q_2$$

as well.

Extending a transition function this way, we can define what a finite automaton accepts in terms of a transition function extended as follows. A finite automaton is said to accept a string $w$ in $\Sigma^*$ if $\delta(q_0, w) \in F$, where $\delta : Q \times \Sigma^* \to Q$ is the extended transition function. That is, starting at the start state $q_0$ and repeating transitions, a string that leads to an accept state is accepted. From the definition, in particular, if the start state is an accept state as well, the empty string $\varepsilon$ is accepted. A finite automaton $M$ accepts a language consisting of strings accepted by $M$. Let $L(M)$ denote the language accepted by $M$. Then, it is expressed as

$$L(M) = \{ w \in \Sigma^* \mid \delta(q_0, w) \in F \}.$$

Note that, if we follow the convention that the set of strings $w$ that satisfy condition $P(w)$ is denoted by $\{ w \mid P(w) \}$ as described in Sect. 2.1, then the language $L(M)$ above should be written as $\{ w \mid w \in \Sigma^*$ and $\delta(q_0, w) \in F \}$. But by convention we usually express the language in either way.

Let us note that, when we talk about what a finite automaton accepts, we have to make clear whether we are dealing with a string or a language. When, given a finite automaton, we talk about a language, there always exists a language that the finite automaton accepts. But when we talk about a string, there may be a case where the finite automaton accepts no strings. Even in such a case, we say the finite automaton accepts the language $\emptyset$ that does not contain any string. Furthermore, note that, in contrast to the cases of Examples 3.1, 3.2, and 3.3, a finite automaton that is given based on the formal definition has no interpretation of what its states, input, and state transition mean.

Next, several examples of finite automata are given. Through these examples, the readers are expected to understand intuitively what language a finite automaton accepts.

**Fig. 3.6** Finite automaton
that accepts strings that
contain 01 as a substring

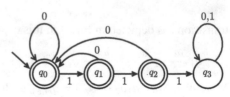

**Fig. 3.7** Finite automaton
that accepts strings that do
not contain 111 as a substring

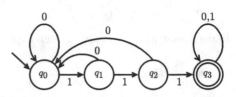

**Fig. 3.8** Finite automaton
that accepts strings that
contain 111 as a substring

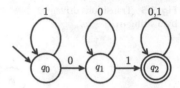

*Example 3.6* Consider the finite automaton shown in Fig. 3.6. In the figure, two
edges from $q_2$ to $q_2$, each labeled 0 and 1, are written as one edge labeled 0 and 1. As
can easily be seen, the automaton accepts the strings that include 01 as a substring,
that is, the strings that are expressed as $u01v$ for some $u, v$ in $\Sigma^*$.

*Example 3.7* What strings are accepted by the finite automaton given in Fig. 3.7?
At first it seems a little difficult to figure out what strings this automaton accepts.
How will the language accepted by the automaton be changed when we modify the
automaton, as illustrated in Fig. 3.8, by making state $q_3$ the accept state instead
of states $q_0$, $q_1$, and $q_2$? It will be easily seen that the automaton in Fig. 3.8 accepts the strings that contain 111 as a substring. Therefore, since the accept states
in Fig. 3.7 are exactly the non-accept state in Fig. 3.8, it follows that the automaton
in Fig. 3.7 accepts the strings that do not contain 111 as a substring. The relationship between the automaton in Fig. 3.7 and that of Fig. 3.8 can be generalized as
follows: given a finite automaton $M = (Q, \Sigma, \delta, q_0, F)$, construct a finite automaton $M' = (Q, \Sigma, \delta, q_0, Q - F)$ by making the accept states of $M$ the non-accept
states of $M'$, and vice versa. Then the language accepted by $M'$ is the complement
of $L(M)$, that is, $\Sigma^* - L(M)$.

*Example 3.8* The automaton in Example 3.5 accepts the strings that have an even
number of 1's. By letting the automaton keep track of not only the parity of the
number of 1's, but also the parity of the number of 0's, we can have the automaton,
given in Fig. 3.9, which accepts strings that have an even number of 1's and an even

**Fig. 3.9** Finite automaton that accepts strings that have an even number of 1's and an even number of 0's

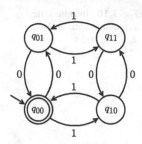

number of 0's. The four states of the automaton can be characterized in terms of the parity of the number of 1's and that of the number of 0's in input strings so far fed, which is described as follows:

$q_{00}$: both the number of 1's and the number of 0's are even.
$q_{01}$: the number of 1's is even and the number of 0's is odd.
$q_{10}$: the number of 1's is odd and the number of 0's is even.
$q_{11}$: both the number of 1's and the number of 0's are odd.

As described above, in general, to each state of an automaton there corresponds information that characterizes the strings that make the automaton move from the start state to that state. Such information is said to "be memorized as a state." Note that the number of states of a finite automaton is finite, and hence the number of anything that a finite automaton can memorize this way is finite. From this fact we can verify that a certain language cannot be accepted by any finite automaton as follows.

As will be explained in Sect. 3.4, there is a language that no finite automata can accept. Language $\{0^m 1^m \mid m \geq 0\} = \{\varepsilon, 01, 0011, \ldots\}$ is an example of such a language. In order to accept the language, it is necessary to memorize the length $m$ of a string at the time when $0^m$ has been read. This is because, in order to accept strings of the form $\{0^m 1^m\}$, it is necessary to decide whether or not the number of 1's that follow $0^m$ is exactly $m$. So there must exist an infinite number of states that memorize lengths $0, 1, 2, 3, \ldots$. Hence any automaton that has a finite number of states cannot accept the language. In this way, it is intuitively explained that the language $\{0^m 1^m \mid m \geq 0\}$ cannot be accepted by any finite automaton. A precise proof of this fact will be given in Sect. 3.4.

What happens if we remove the restriction that the number of states is finite? If we are allowed to use an infinite number of states, then it turns out that for any language $L \subseteq \Sigma^*$ we can construct an infinite state automaton, somehow defined, that accepts $L$. Letting $\Sigma = \{0, 1\}$, the state diagram of $M$ is given in Fig. 3.10. That is, we define the transition function $\delta$ as $\delta(q_w, a) = q_{wa}$ for any $w$ in $\Sigma^*$ and any $a$ in $\Sigma$. Then we can easily see that for any string $a_1 a_2 \cdots a_n$ in $\Sigma^*$ there exist transitions

$$q_\varepsilon \xrightarrow{a_1} q_{a_1} \xrightarrow{a_2} q_{a_1 a_2} \xrightarrow{a_3} \cdots \xrightarrow{a_n} q_{a_1 \cdots a_n}.$$

**Fig. 3.10** Infinite state
automaton with alphabet
$\Sigma = \{0, 1\}$

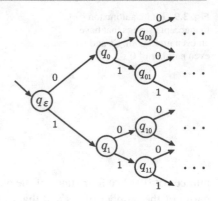

Obviously, to make the state accept a given language $L$, all we have to do is set the
start state to be $q_\varepsilon$ and set the accept states to be $F = \{q_w \mid w \in L\}$. One might say
this is only to specify language $L$ in terms of the collection of the accept states.

As stated so far, an automaton as defined in Definition 3.4 is deterministic in
the sense that the automaton goes to exactly one state for each present state and
each symbol fed as input. So the automaton defined in Definition 3.4 is called a
deterministic finite automaton when we need to explicitly say that the automaton
behaves deterministically. In the next section, by relaxing the condition of deterministic transition, we will introduce the notion of a nondeterministic finite automaton
and prove that the two notions are equivalent in terms of the power of accepting languages. So far we have mentioned nondeterministic behavior of finite automata, but
the notion will be extended to more powerful computational models such as a pushdown automaton and a Turing machine, which will be studied in Chaps. 5 and 6,
respectively.

## 3.2    Nondeterministic Finite Automaton

A finite automaton as studied so far is such that the next state is uniquely determined
from the present state and the symbol fed as input. A nondeterministic finite automaton is the same as a deterministic finite automaton except that, given a present state
and a input symbol, the automaton is allowed to move to more than one state. In this
section, the notion of a nondeterministic finite automaton is introduced, and then
it is shown that a nondeterministic finite automaton can be transformed to a deterministic finite automaton, which is equivalent to the former in the sense that both
automata accept the same language.

### Generalization to a Nondeterministic Finite Automaton

The notion of a nondeterministic finite automaton is very important and is repeatedly
used for several computational models other than a finite automaton. The readers are

**Fig. 3.11** Nondeterministic finite automaton that accepts strings that include 111 as a substring

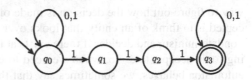

expected to fully understand how a nondeterministic model works through use of a simple computational model of a finite automaton.

First, let us begin with an example of a nondeterministic finite automaton.

*Example 3.9* As an example of a nondeterministic finite automaton, we consider the state diagram as shown in Fig. 3.11. The state diagram is not a finite automaton in the sense of Definition 3.4. This is because when the symbol 1 is given to the state $q_0$, the next state that the automaton moves to is not unique, that is, the next state may be $q_0$ or $q_1$. Furthermore, there is no edge with label 0 leaving state $q_1$. And so it is with state $q_2$. A nondeterministic finite automaton is one obtained by removing the restriction that, given a present state $q$ and input symbol $a$, the next state determined from $(q, a)$ is unique. How can we define a string that a nondeterministic finite automaton accepts?

In the case of the nondeterministic finite automaton given in Fig. 3.11, it accepts strings that contain 111 as a substring, that is, the strings expressed as $u111v$ for some strings $u$ and $v$. The correspondence between the form $u111v$ and the state diagram in Fig. 3.11 will be clear. That is, the part $u$ of $u111v$ corresponds to the transitions $q_0 \overset{0}{\to} q_0$ and $q_0 \overset{1}{\to} q_0$, and the part 111 corresponds to the transitions $q_0 \overset{1}{\to} q_1 \overset{1}{\to} q_2 \overset{1}{\to} q_3$, and the last part $v$ corresponds to the transitions $q_3 \overset{0}{\to} q_3$ and $q_3 \overset{1}{\to} q_3$. Through such correspondence, we have

$$q_0 \overset{u}{\to} q_0 \overset{111}{\longrightarrow} q_3 \overset{v}{\to} q_3,$$

thereby $q_0 \overset{u111v}{\longrightarrow} q_3$. For example, let us consider the sequence

$$0 \quad 1 \quad 1 \quad 0 \quad 1 \quad 0 \quad \overset{\checkmark}{1} \quad \overset{\checkmark}{1} \quad 1 \quad 1 \quad 0 \quad \overset{\checkmark}{1} \quad 1 \quad 1.$$

If the automaton is at state $q_0$, the transition $q_0 \overset{1}{\to} q_1$ is possible for any 1 in the sequence, but the accept state is reached only when $q_0 \overset{1}{\to} q_1$ takes place at any one of the three positions attached with $\checkmark$'s.

In general, a nondeterministic finite automaton accepts a string if the automaton can move from the start state to one of the accept states through an appropriately chosen path corresponding to the string in the state diagram, whereas a nondeterministic finite automaton does not accept a string if the automaton cannot go from the start state to any of the accept states no matter how a path corresponding to the string is chosen. By definition, a nondeterministic finite automaton cannot look ahead for symbols that will come as input except for the present symbol. But one way to figure out how the decision is made of whether or not an input string is accepted is to think of an entity that looks over all the possible moves caused by the nondeterminism and decides if there exists at least one sequence of transitions leading to an accept state. So, when we need to explain how a nondeterministic finite automaton behaves, we sometimes say that the automaton "guesses a sequence of transitions appropriately" that leads to an accept state, thereby guaranteeing that the string given as input is accepted.

The nondeterministic finite automaton in Fig. 3.11 can be transformed to an equivalent deterministic one, which is shown in Fig. 3.8. These automata accept the same language. As will be mentioned later, it can be proved that, in general, this equivalent transformation is always possible. By the way, which is easier to understand how Fig. 3.8 works or how Fig. 3.11 works? Once you figure out how a nondeterministic automaton works, a nondeterministic automaton is often easier to understand, compared to a deterministic automaton equivalent to it.

In the example above, the nondeterminism comes from the fact that there are more than one transition leaving a state. As is described in the following examples, there is another type of factor that, due to the empty string $\varepsilon$, causes nondeterminism. This is nondeterminism that comes from choosing either a transition attached with the empty string or a transition attached with an input symbol. In the former case, an input symbol is not read by an automaton, while in the latter case, a symbol is read.

*Example 3.10* Employing Fig. 3.12, we consider nondeterminism due to $\varepsilon$-*transition*, that is, transition by the empty string $\varepsilon$. In the figure, the sequence of the labels corresponding to the series of transitions

$$q_0 \xrightarrow{0} q_0 \xrightarrow{0} q_0 \xrightarrow{\varepsilon} q_1 \xrightarrow{1} q_1 \xrightarrow{\varepsilon} q_2 \xrightarrow{2} q_2$$

is $00\varepsilon 1\varepsilon 2$, which represents the string 0012. In this way, the empty string $\varepsilon$ is absorbed into the symbols adjacent to it; hence, the empty string does not appear in an input string unless the entire input string is $\varepsilon$.

The automaton shown in Fig. 3.12 accepts strings consisting of an arbitrary number of 0's, followed by an arbitrary number of 1's, and finally followed by an arbitrary number of 2's, where an arbitrary number can be zero. Let $a^i$ denote the string obtained by repeating $a$ $i$ times. In particular, when $i = 0$, $a^i$ means the empty string. Then a string that the automaton accepts is denoted by $0^i 1^j 2^k$, where $i \geq 0$, $j \geq 0$, and $k \geq 0$.

In general, when $\varepsilon$-transition is allowed, a nondeterministic finite automaton accepts a string $w$ if there exists at least one path going from the start state to an accept state among all the possible paths that spell out $w$. All the possible sequences of transitions that spell out the string 0012 or its prefixes are given as follows, and are illustrated in Fig. 3.13 as well:

$\varepsilon$:  $q_0$, $q_0 \xrightarrow{\varepsilon} q_1$ and $q_0 \xrightarrow{\varepsilon} q_1 \xrightarrow{\varepsilon} q_2$

0:  $q_0 \xrightarrow{0} q_0$, $q_0 \xrightarrow{0} q_0 \xrightarrow{\varepsilon} q_1$ and $q_0 \xrightarrow{0} q_0 \xrightarrow{\varepsilon} q_1 \xrightarrow{\varepsilon} q_2$

**Fig. 3.12** Nondeterministic finite automaton that accepts strings of the form $0^i 1^j 2^k$, where $i \geq 0$, $j \geq 0$, and $k \geq 0$

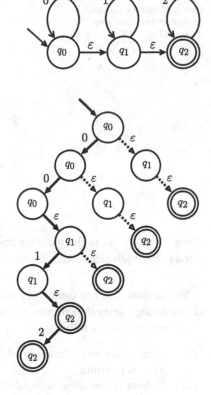

**Fig. 3.13** All the possible sequences of transitions that spell out the string 0012 or its prefixes

00:  $q_0 \xrightarrow{0} q_0 \xrightarrow{0} q_0, q_0 \xrightarrow{0} q_0 \xrightarrow{0} q_0 \xrightarrow{\varepsilon} q_1$ and $q_0 \xrightarrow{0} q_0 \xrightarrow{0} q_0 \xrightarrow{\varepsilon} q_1 \xrightarrow{\varepsilon} q_2$

001:  $q_0 \xrightarrow{0} q_0 \xrightarrow{0} q_0 \xrightarrow{\varepsilon} q_1 \xrightarrow{1} q_1$ and $q_0 \xrightarrow{0} q_0 \xrightarrow{0} q_0 \xrightarrow{\varepsilon} q_1 \xrightarrow{1} q_1 \xrightarrow{\varepsilon} q_2$

0012:  $q_0 \xrightarrow{0} q_0 \xrightarrow{0} q_0 \xrightarrow{\varepsilon} q_1 \xrightarrow{1} q_1 \xrightarrow{\varepsilon} q_2 \xrightarrow{2} q_2$

In the first four cases above, the whole of 0012 cannot be read by the automaton, but in the fifth transition, since string 0012 makes the automaton move from the start state $q_0$ to the accept state $q_2$, the string is accepted. In general, in deciding whether an input sequence is accepted or not, all the sequences of transitions are examined by inserting the empty string $\varepsilon$ in all the possible positions in the input sequence.

Now, as will be explained later, a nondeterministic finite automaton, in general, can be transformed to a deterministic finite automaton without changing the language to be accepted. Figure 3.14 shows a deterministic finite automaton equivalent to the automaton shown in Fig. 3.12. An edge labeled 0 is not allowed to leave state $q_1$ because 0 cannot follow 1 possibly already read, and similarly an edge labeled 0 or 1 is not allowed to leave state $q_2$ because neither 0 nor 1 can follow 2 possibly already read. On the other hand, in order to make the automaton in Fig. 3.14 deterministic, we have to recover these missing edges so that each state in Fig. 3.14 has three edges going out of it with labels 0, 1, and 2, respectively. Note that, although we add these missing edges in Fig. 3.14, any sequence of transitions going through any of these edges never leads to the accept state, hence adding no new string as a

**Fig. 3.14** Deterministic finite automaton equivalent to the nondeterministic one shown in Fig. 3.12

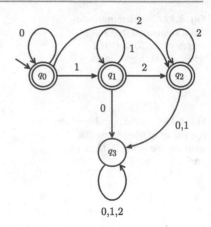

string accepted. As in the previous case, it is easier to understand Fig. 3.12, compared to the equivalent deterministic automaton given in Fig. 3.14.

We are now ready to summarize what was explained about nondeterministic behavior through several examples. For the factors that cause nondeterministic move there are two types as follows:

(1)   If there are more than one edge with the same label from a state, any of them is possible.

(2)   If there are an edge with label $\varepsilon$ and an edge with a symbol, both going out of a state, then either the transition via the edge with label $\varepsilon$ without reading any input symbol or the transition via an edge with a symbol label is possible.

The fact that a nondeterministic finite automaton accepts an input string $w$ means that "if nondeterministic moves of types (1) and (2) are chosen appropriately, there exists a path that spells out the sequence $w$ of labels, while going from the start state to an accept state." On the other hand, the fact that it does not accept an input string $w$ means that "no matter how nondeterministic moves of types (1) and (2) are chosen, there exists no path from the start state to any accept state that spells out the sequence $w$ of labels."

The automaton with no nondeterministic move allowed was defined in Definition 3.4. When we want to state explicitly that no nondeterministic move is allowed, we use the term *deterministic finite automaton*, sometimes abbreviated as DFA. A nondeterministic finite automaton is abbreviated as NFA. The formal definitions of both DFA and NFA are given as 5-tuples. Among the items of 5-tuples for these two types of finite automata the only item that differs between the two is the transition function. So a formal definition of NFA will be given after a transition function for NFA is explained.

In a state diagram for an NFA, when edges with label $a$ from state $q$ go to states $q_1, \ldots, q_m$, the corresponding transition function of the NFA is specified as $\delta(q, a) = \{q_1, \ldots, q_m\}$. In particular, if there is no edge with label $a$ from state $q$, it is defined as $\delta(q, a) = \emptyset$ where $a$ is an element of $\Sigma_\varepsilon$ with $\Sigma_\varepsilon = \Sigma \cup \{\varepsilon\}$. In this way, for any element $a$ in $\Sigma_\varepsilon$, $\delta(q, a)$ is defined to be a subset of the set of states $Q$. Hence, if the class set of all the subsets of $Q$ is denoted by $\mathcal{P}(Q)$, the state transition function of a nondeterministic state automaton is expressed as

$$\delta : Q \times \Sigma_\varepsilon \to \mathcal{P}(Q).$$

What should be noticed here is that an NFA is allowed to have nondeterministic moves, but it does not mean that the NFA *must* have nondeterministic moves. Hence, a deterministic finite automaton is automatically a nondeterministic finite automaton.

The formal definition for a nondeterministic finite automaton is given as follows.

**Definition 3.11** A *nondeterministic finite automaton* (NFA) is a 5-tuple $(Q, \Sigma, \delta, q_0, F)$, where

(1) $Q$ is a finite set of states.
(2) $\Sigma$ is a finite set of symbols, called the alphabet.
(3) $\delta : Q \times \Sigma_\varepsilon \to \mathcal{P}(Q)$ is a transition function.
(4) $q_0 \in Q$ is the start state.
(5) $F \subseteq Q$ is a set of accept states.

$\Sigma_\varepsilon$ has been defined as $\Sigma \cup \{\varepsilon\}$. By convention we think of $\Sigma_\varepsilon^*$ as the set of strings of symbols from $\Sigma_\varepsilon$, where $\varepsilon$ in the strings is treated exactly the same as symbols in $\Sigma$. For example, for $\Sigma = \{0, 1\}$, the strings of $\Sigma_\varepsilon^*$ such as $0\varepsilon 11\varepsilon$ or $\varepsilon 01\varepsilon\varepsilon 1$, etc., correspond to the string $011$ of $\Sigma^*$. The label of a path in a state diagram is the string obtained by concatenating the labels of the edges in the path, where $\varepsilon$ is treated exactly the same as the other symbols that never disappear. Hence, for an NFA, the label of a path is expressed as a string in $\Sigma_\varepsilon^*$.

An NFA $N$ is said to accept a string $w$ if there exists a string $a_1 a_2 \cdots a_m \in \Sigma_\varepsilon^*$ obtained by appropriately inserting $\varepsilon$'s between the symbols of $w$ (so by deleting the inserted $\varepsilon$'s from $a_1 a_2 \cdots a_m$ we have $w$) such that there exists a path with the label $a_1 a_2 \cdots a_m$ from the start state to an accept state. This is precisely defined as follows.

An NFA $N = (Q, \Sigma, \delta, q_0, F)$ accepts a string $w \in \Sigma^*$ if, for some $a_1, a_2, \ldots, a_m \in \Sigma_\varepsilon$ and some $q_1, \ldots, q_m \in Q$, the following three conditions hold:

(1) $w = a_1 a_2 \cdots a_m$
(2) $q_0 \xrightarrow{a_1} q_1 \xrightarrow{a_2} \cdots \xrightarrow{a_m} q_m$
(3) $q_m \in F$

where in the nondeterministic case we write $q \xrightarrow{a} q'$ if $q' \in \delta(q, a)$ for $a \in \Sigma_\varepsilon$.

In particular, if $m = 0$, the condition above implies that the start state is also an accept state, thereby $N$ accepts $\varepsilon$. String $w$ is treated as $a_{j_1} a_{j_2} \cdots a_{j_n}$ for $a_{j_1}, \ldots, a_{j_n} \in \Sigma$ if $a_1 a_2 \cdots a_m$ is expressed as $\varepsilon^{i_0} a_{j_1} \varepsilon^{i_1} a_{j_2} \cdots a_{j_n} \varepsilon^{i_n}$, which is obtained by inserting $\varepsilon$'s into $w$, where $j_1 = i_0 + 1, j_2 = i_0 + i_1 + 2, \ldots, j_n =$

$i_0+i_1+\cdots+i_{n-1}+n$. In particular, if $i=0$, $\varepsilon^i$ does not appear in the above expression for $w$. For example, when $w=011$ and $a_1a_2\cdots a_7=\varepsilon 0\varepsilon\varepsilon 11\varepsilon$, we have $i_0=1$, $i_1=2$, $i_2=0$, $i_3=1$, $j_1=i_0+1=2$, $j_2=i_0+i_1+2=5$, $j_3=i_0+i_1+i_2+3=6$.

Similarly as in a DFA, a language accepted by an NFA $N$, denoted by $L(N)$, is defined as

$$L(N)=\big\{w\in\Sigma^*\mid N\text{ accepts }w\big\}.$$

## Equivalence of Deterministic and Nondeterministic Automata

Since a DFA is also an NFA, a language accepted by a DFA is also accepted by an NFA. In this subsection, the converse of that will be proved: a language accepted by an NFA is also accepted by a DFA, thereby establishing the equivalence between DFA and NFA. To do this, for an arbitrary NFA $N=(Q,\Sigma,\delta,q_0,F)$, we construct a DFA $M$ equivalent to $N$: $M$ accepts the same language as the language that $N$ accepts. The DFA $M$, which we will construct, simulates the behavior of the NFA $N$. How can $M$ do this? The point of the construction is that we specify $M$ so that, when an input string $w$ is given, $M$ goes from the start state $q_0$ to the set of all the states that $N$ can go to from the start state $q_0$. In other words, if $P\subseteq Q$ consists of the states $q$'s such that $q_0\xrightarrow{w}q$ in $N$, we specify that $\{q_0\}\xrightarrow{w}P$ in $M$. The other point we should note here is that, since we take the subset $P$ of $Q$ as a state of $M$, $M$ can be thought of as going to a state uniquely specified. So when we talk about $M$, the set consisting of a single state $q_0$ is denoted by $\{q_0\}$ rather than $q_0$. So, if $N$ accepts $w$ because $q_0\xrightarrow{w}q$ in $N$ for an accept state $q$, then $\{q_0\}\xrightarrow{w}P$ in $N$ where $q\in P$. Thus, if we take such a subset $P$ that contains at least one accept state of $N$ as an accept state of $M$, then if $N$ accepts $w$ then $M$ also accepts $w$. What remains to be shown is how to specify each transition of $M$ so that what is described above holds. This will be explained shortly.

Next, based on these ideas, two examples for constructing a DFA equivalent to an NFA are given. In the first Example 3.12 NFA has no $\varepsilon$-transitions, while in the next Example 3.13 NFA has $\varepsilon$-transitions.

*Example 3.12* We take again NFA $N$ given in Fig. 3.11 which accepts strings of the form $u111v$ for some $u$ and $v$ and construct DFA $M$ equivalent to $N$.

To begin with, let $\{q_0\}$ be the start state. Since all the transitions in $N$ from the state $q_0$ are

$$q_0\xrightarrow{0}q_0,\qquad q_0\xrightarrow{1}q_0,$$

$$q_0\xrightarrow{1}q_1,$$

the transition function $\delta$ of $M$ is defined as

$$\delta\big(\{q_0\},0\big)=\{q_0\},\qquad \delta\big(\{q_0\},1\big)=\{q_0,q_1\},$$

**Fig. 3.15** Equivalent DFA obtained by transforming NFA in Fig. 3.8

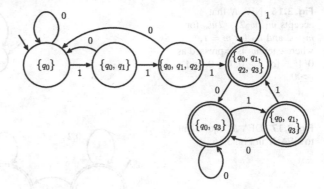

collecting all the related states. So, $\{q_0, q_1\}$ has newly become a state of $M$. Similarly, since

$$q_0 \xrightarrow{0} q_0, \qquad q_0 \xrightarrow{1} q_0,$$

$$q_0 \xrightarrow{1} q_1,$$

$$q_1 \xrightarrow{1} q_2,$$

it is defined as

$$\delta(\{q_0, q_1\}, 0) = \{q_0\}, \qquad \delta(\{q_0, q_1\}, 1) = \{q_0, q_1, q_2\}.$$

Proceeding step by step in this way, the state transition diagram shown in Fig. 3.15 is obtained by making $\{q_0\}$ the start state and letting subsets containing the accept state $q_3$ of $N$ be the accept states of $M$. It is easily seen that the automaton in Fig. 3.15 accepts strings that contain 111 as a substring.

Do we really need the three states $\{q_0, q_1, q_2, q_3\}$, $\{q_0, q_3\}$, and $\{q_0, q_1, q_3\}$ in the automaton in Fig. 3.15? Notice that by putting these states together to form a state and by redrawing the edges and renaming the states accordingly, we can obtain the DFA shown in Fig. 3.8, which accepts the same strings of the form $u111v$ for some $u$ and $v$. Clearly, in order to accept the strings of the form $u111v$, we don't need to distinguish these three states. This is because, once a string moves the automaton in Fig. 3.15 to state $\{q_0, q_1, q_2, q_3\}$, the automaton will accept an input string no matter what string follows after that. So, in general, by transforming an NFA according to the procedure described above, we can obtain an equivalent DFA, but the DFA obtained is not necessarily minimum in terms of the number of the states.

*Example 3.13* We consider the NFA $N$ with $\varepsilon$-transition shown in Fig. 3.16 and construct a DFA $M$ equivalent to it. Strings accepted by $N$ are in general expressed as

$$w_1 2 w_2 2 \cdots 2 w_m,$$

**Fig. 3.16** NFA $N$ that accepts $w_1 2 w_2 2 \cdots 2 w_m$ for $m \geq 2$ and $w_1$ for $m = 1$, where each $w_i$ is expressed as $0^j 1^k 2^l$ for $j \geq 0, k \geq 0$, and $l \geq 0$

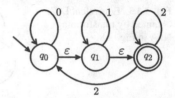

**Fig. 3.17** DFA $M$ equivalent to NFA $N$ in Fig. 3.16

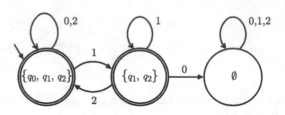

where each $w_i$ takes the form $0^j 1^k 2^l$ for some $j \geq 0, k \geq 0$, and $l \geq 0$. In particular, when $m = 1$ it takes the form $w_1$.

We will construct a DFA $M$ with no $\varepsilon$-transition that simulates $N$ with $\varepsilon$-transition. The idea of the construction is basically the same as that of the previous example, except that in this case we take into account the effect caused by $\varepsilon$-transition. The point of how to transform an NFA $N$ to an equivalent DFA $M$ is that when we make a collection of $N$'s states that correspond to a state of $M$, for each state $q$ and each symbol $a$, we collect not only the states $p$'s such that $q \xrightarrow{a} p$ in $N$, but also those $p$'s such that $q \xrightarrow{a\varepsilon^i} p$ in $N$ for $i \geq 1$. In the same spirit, the start state of $M$ is set to be the collection of the state $p$'s such that $q_0 \xrightarrow{\varepsilon^i} p$ in $N$ for $i \geq 0$, where $q_0$ is the start state of $N$. In our case, the collection turns out to be $\{q_0, q_1, q_2\}$. This is because we have in $N$

$$q_0 \xrightarrow{\varepsilon^0} q_0,$$

$$q_0 \xrightarrow{\varepsilon^1} q_1,$$

$$q_0 \xrightarrow{\varepsilon^2} q_2.$$

Note that by definition we left $q \xrightarrow{\varepsilon^0} q$ for any $q$. Thus, as shown in Fig. 3.17, $\{q_0, q_1, q_2\}$ becomes the start state in $M$.

Furthermore, we have $\delta_M(\{q_1, q_2\}, 2) = \{q_0, q_1, q_2\}$, as shown in Fig. 3.17. This is because

$$q_2 \xrightarrow{2} q_2,$$

$$q_2 \xrightarrow{2} q_0,$$

$$q_2 \xrightarrow{2} q_0 \xrightarrow{\varepsilon} q_1,$$

$$q_2 \xrightarrow{2} q_0 \xrightarrow{\varepsilon} q_1 \xrightarrow{\varepsilon} q_2$$

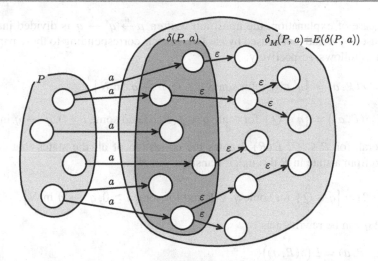

**Fig. 3.18** Given NFA $N$, how to specify transition function $\delta_M$ of an equivalent DFA $M$

and because there is no transition from state $q_1$ by a string of the form $2\varepsilon^i$. On the other hand, since any edge with label 0 leaves neither the state $q_1$ nor $q_2$, we specify $\delta_M(\{q_1, q_2\}, 0) = \emptyset$. Furthermore, we set $\delta_M(\emptyset, a) = \emptyset$ for any $a \in \{1, 2, 3\}$. This is simply because in this case there exists no state that $N$ starts with.

By the way, it is easy to see that $M$ beginning at the start state goes to state $\emptyset$ exactly when an input string contains 10 as a substring. Hence, $M$ accepts a string exactly when the string does not contain 10 as a substring. On the other hand, $M$ is equivalent to $N$ that accepts a string of the form $w_1 2 w_2 2 \cdots 2 w_m$. Thus it follows that the condition that string $w$ does not contain 10 as a substring is equivalent to the condition that string $w$ has the form of $w_1 2 w_2 2 \cdots 2 w_m$. In fact, in Problem 3.8**, it is proved that the two conditions are equivalent.

In general, given NFA $N = (Q, \Sigma, \delta, q_s, F)$ we shall describe how to construct a DFA $M$ equivalent to it.

First, we take the power set $\mathcal{P}(Q)$ of the set of states of $N$ as the set of states of $M$. Suppose that $M$ is in a state denoted by subset $P$ of $Q$. In Fig. 3.18 we illustrate schematically the subset of $Q$ that $M$ goes to from $P$ when $a$ is read as input. The subset is denoted by $\delta_M(P, a)$, where $\delta_M$ denotes the transition function of $M$. As you can see from Fig. 3.18, $\delta_M(P, a)$ consists of the states that $N$ goes to from a state $p \in P$ when symbol $a$ followed by any number (including zero) of $\varepsilon$ is read. Then $\delta_M : \mathcal{P}(Q) \times \Sigma \to \mathcal{P}(Q)$ is defined as

$$\delta_M(P, a) = \left\{ q \in Q \mid \text{for } p \in P, q' \in Q, \text{ and } i \geq 0, p \xrightarrow{a} q' \xrightarrow{\varepsilon^i} q \text{ in } N \right\}.$$

In this definition, since $i \geq 0$ in $q' \xrightarrow{\varepsilon^i} q$ is arbitrary, all the states $q$'s that can be reached from $q'$ by edges labeled $\varepsilon$ are put in $\delta_M(P, a)$. In particular, if $i = 0$, $p \xrightarrow{a} q' \xrightarrow{\varepsilon^i} q$ means $p \xrightarrow{a} q$.

For ease of explanation, the transition written $p \xrightarrow{a} q' \xrightarrow{\varepsilon^i} q$ is divided into the part caused by $a$ and that caused by $\varepsilon^i$. Two subsets corresponding to these parts are written as follows, respectively:

$$\delta(P, a) = \{q' \in Q \mid \text{for some } p \in P, p \xrightarrow{a} q' \text{ in } N\},$$

$$E\big(\delta(P, a)\big) = \{q \in Q \mid \text{for some } q' \in \delta(P, a) \text{ and some } i \geq 0, q' \xrightarrow{\varepsilon^i} q \text{ in } N\}.$$

In general, for $R \subseteq Q$, $E(R)$ denotes the collection of all the states that can be reached from a state in $R$ through $\varepsilon$-transitions in $N$:

$$E(R) = \{q \in Q \mid \text{for some } q' \in R \text{ and for some } i \geq 0, q' \xrightarrow{\varepsilon^i} q \text{ in } N\}.$$

Then, $\delta_M$ can be rewritten as

$$\delta_M(P, a) = E\big(\delta(P, a)\big).$$

Let the start state and the set of accept states of $M$ be denoted by $q_M$ and $F_M$, respectively. These are defined as follows:

$$q_M = E(\{q_s\}),$$
$$F_M = \{P \in P(Q) \mid F \cap P \neq \emptyset\}.$$

Having defined $M$ from NFA $N$ as above, we will show that $N$ and $M$ are equivalent with each other. To do so, we show that a string accepted by $N$ is also accepted by $M$ and vice versa.

First, we suppose that a string $a_1 a_2 \cdots a_n$ is accepted by $N$. Then, in the state diagram of $N$ there exists a path from the start state to an accept state that spells out a string written as $a_1 a_2 \cdots a_n$. If we write explicitly all the symbols including the empty symbol $\varepsilon$ on the path, we have a sequence

$$\varepsilon^{i_0} a_1 \varepsilon^{i_1} a_2 \varepsilon^{i_2} \cdots a_n \varepsilon^{i_n},$$

where $i_0 \geq 0, \ldots$, and $i_n \geq 0$ are the numbers of $\varepsilon$'s appearing consecutively on each part. Then, there are states $q_0, q_1', q_1, q_2', q_2, \ldots, q_n$ in $N$ as shown in Fig. 3.19 such that

$$q_s \xrightarrow{\varepsilon^{i_0}} q_0 \xrightarrow{a_1} q_1' \xrightarrow{\varepsilon^{i_1}} q_1 \xrightarrow{a_2} \cdots \xrightarrow{\varepsilon^{i_{n-1}}} q_{n-1} \xrightarrow{a_n} q_n' \xrightarrow{\varepsilon^{i_n}} q_n, \quad q_n \in F. \tag{3.1}$$

In the argument above, if we vary $i_0$ in all possible ways, we have the collection of the possible states $q_0$'s, which will be denoted by $P_0$. Similarly, if we vary $q_0$ in $P_0$ and $i_1$ together with the corresponding paths taken in all possible ways, we have the collection of the possible states $q_1$'s, which will be denoted by $P_1$. In a similar way, $P_2, P_3, \ldots, P_n$ are obtained, as is shown in Fig. 3.19. Then, from the definition of the transition function, it is clear that

$$P_0 \xrightarrow{a_1} P_1 \xrightarrow{a_2} \cdots \xrightarrow{a_n} P_n \tag{3.2}$$

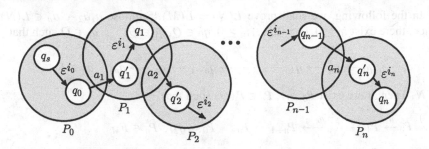

**Fig. 3.19** State transition of DFA $M$

holds in $M$. Furthermore, since $q_n \in P_n$ and $q_n \in F$, we have $F \cap P_n \neq \emptyset$, and hence $P_n$ is an accept state. Therefore, $M$ accepts $a_1 a_2 \cdots a_n$.

For the proof of the converse direction, we assume that a string $a_1 a_2 \cdots a_n$ is accepted by $M$, that is,

$$P_0 \xrightarrow{a_1} P_1 \xrightarrow{a_2} \cdots \xrightarrow{a_{n-1}} P_{n-1} \xrightarrow{a_n} P_n,$$

where $P_0$ is the start state and $P_n$ an accept state of $M$. Assuming that, what we want to derive is the existence of a path in $N$ from $q_s$ to $q_n$ shown in Fig. 3.19. The point of this derivation is to trace the path in the reverse direction. First, since $P_n$ is an accept state, there exists an accept state $q_n \in P_n$. From the definition of the transition in $M$, there exists $q_{n-1} \in P_{n-1}$ such that $q_{n-1} \xrightarrow{a_n \varepsilon^{i_n}} q_n$ in $N$ and $i_n \geq 0$, which guarantees that $q_n \in P_n$. Repeating a similar argument, we can conclude that there exist $q_n \in P_n, q_{n-1} \in P_{n-1}, \ldots, q_0 \in P_0$ such that

$$q_s \xrightarrow{\varepsilon^{i_0}} q_0 \xrightarrow{a_1 \varepsilon^{i_1}} q_1 \xrightarrow{a_2 \varepsilon^{i_2}} \cdots \xrightarrow{a_n \varepsilon^{i_n}} q_n, \quad q_n \in F$$

for $i_n \geq 0, \ldots, i_1 \geq 0, i_0 \geq 0$. Therefore, the string $a_1 a_2 \cdots a_n$ is accepted by $N$.

**Theorem 3.14** *For any NFA $N$, there exists a DFA that is equivalent to $N$.*

*Proof* Let $N = (Q, \Sigma, \delta, q_s, F)$ be an arbitrary NFA. We define an equivalent DFA $M = (\mathcal{P}(Q), \Sigma, \delta_M, q_M, F_M)$ as follows. First, $\delta_M : \mathcal{P}(Q) \times \Sigma_\varepsilon \to \mathcal{P}(Q)$, $q_M$ and $F_M$ are defined as follows:

$$\delta_M(P, a) = \big\{ q \in Q \mid \text{there exist } p \in P, p' \in Q \text{ and } i \geq 0$$

$$\text{such that } p \xrightarrow{a} p' \xrightarrow{\varepsilon^i} q \text{ in } N \big\},$$

where $P \in \mathcal{P}(Q)$ and $a \in \Sigma_\varepsilon$. In particular, we let $p' \xrightarrow{\varepsilon^0} q$ mean that $p' = q$. Furthermore,

$$q_M = \big\{ q \in Q \mid \text{there exists } i \geq 0 \text{ such that } q_s \xrightarrow{\varepsilon^i} q \text{ in } N \big\},$$
$$F_M = \{ P \in P(Q) \mid P \cap F \neq \emptyset \}.$$

In the following, we shall prove $L(N) = L(M)$. Suppose $a_1 a_2 \cdots a_n \in L(N)$. Then, there exist $i_0 \geq 0, i_1 \geq 0, \ldots, i_n \geq 0, q_0 \in Q, q_1 \in Q, \ldots, q_n \in Q$ such that

$$q_s \xrightarrow{\varepsilon^{i_0}} q_0 \xrightarrow{a_1} q_1' \xrightarrow{\varepsilon^{i_1}} q_1 \xrightarrow{a_2} \cdots \xrightarrow{\varepsilon^{i_{n-1}}} q_{n-1} \xrightarrow{a_n} q_n' \xrightarrow{\varepsilon^{i_n}} q_n, \quad q_n \in F$$

in $N$. Then, there exist $P_0, \ldots, P_n \in P(Q)$ such that

$$P_0 \xrightarrow{a_1} P_1 \xrightarrow{a_2} \cdots \xrightarrow{a_{n-1}} P_{n-1} \xrightarrow{a_n} P_n, \quad P_0 = q_M, \ P_n \in F_M$$

in $M$. Therefore, $a_1 a_2 \cdots a_n \in L(M)$.

Conversely, suppose that $a_1 a_2 \cdots a_n \in L(M)$. Then, there exist $P_0, \ldots, P_n \in P(Q)$ such that

$$P_0 \xrightarrow{a_1} P_1 \xrightarrow{a_2} \cdots \xrightarrow{a_{n-1}} P_{n-1} \xrightarrow{a_n} P_n, \quad P_0 = q_M, \ P_n \in F_M$$

in $M$. Furthermore, there exists $q_n \in P_n$ such that $q_n \in F$. Then, there exist $q_{n-1} \in Q, \ldots, q_1 \in Q$ and $q_0 \in Q$ such that

$$q_s \xrightarrow{\varepsilon^{i_0}} q_0 \xrightarrow{a_1 \varepsilon^{i_1}} q_1 \xrightarrow{a_2 \varepsilon^{i_2}} \cdots \xrightarrow{a_{n-1}} q_{n-1} \xrightarrow{a_n \varepsilon^{i_n}} q_n$$

in $N$ for $i_n \geq 0, \ldots, i_2 \geq 0, i_1 \geq 0$, and $i_0 \geq 0$. Therefore, $a_1 a_2 \cdots a_n \in L(N)$. $\qquad \square$

So far we have described how to construct a DFA $M$ that is equivalent to a given NFA $N$. Can we really construct an equivalent DFA by the method described? In defining $\delta_M(P, a)$ we have to decide if $p \xrightarrow{a} p' \xrightarrow{\varepsilon^i} q$ holds for some $i$. The difficulty concerning the decision is that there might exist a possibility that we have to check if $p' \xrightarrow{\varepsilon^i} q$ holds for infinitely many $i$'s. But actually this is not the case. In fact, we only have to check the condition for $i = 1, 2, \ldots, k - 1$, where $k$ is the number of states of $N$. This is because if there exists a path of length $m \geq k$ from $q$ to $q'$ in a state diagram,

$$q_0(=q) \to q_1 \to q_2 \to \cdots \to q_m(=q'),$$

then there exists a path of length less than or equal to $k - 1$ from $q$ to $q'$ in the state diagram. The argument above is illustrated in Fig. 3.20 for the typical case of $k = 8$. Since there are 8 kinds of states, at least two of the 9 states that appear on the path must be the same. Figure 3.20 shows that $q_2$ and $q_6$ are the same states. So in this case we have the path $q_0 \to q_1 \to q_2(=q_6) \to q_7 \to q_8$ of length 4. In short, if two states are connected through a long path, then the two states are connected through a short path, thereby we only have to check if two states are connected through a short path. This argument can be generalized as in Fig. 3.20.

The point of the argument above is that if there are $k$ different kinds of objects, then the same kind of object appears in the collection of more than $k$ objects. This type of argument is called the pigeonhole principle, which is described as follows. Let $A = \{a_1, \ldots, a_m\}$ denote a set of pigeons, and $B = \{b_1, \ldots, b_k\}$ a set of $k$ holes.

**Fig. 3.20** If there are 8 different states, the same state appears more than once in a path on which 9 states appear

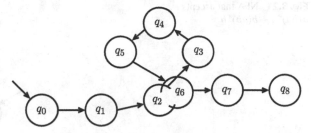

Suppose that $m > k$ and that each pigeon in $A$ is placed into a hole. The *pigeonhole principle* asserts that some hole must have more than one pigeon in it. In the above example, the states correspond to the holes, whereas the states that appear on the path correspond to the pigeons. Some arguments discussed later are also based on the pigeonhole principle.

## 3.3   Regular Expression

In this section, we introduce the regular expression as a way of expressing a language, and show that regular expressions and finite automata are equivalent in the power of expressing languages.

### Definition of a Regular Expression

Whether it is deterministic or nondeterministic, a finite automaton specifies a language by giving a way of judging whether an input is accepted or not. In contrast, a regular expression is one obtained by performing three kinds of operations repeatedly to symbols of an input alphabet, which turns out to express a language. So a regular expression expresses directly how a string of the corresponding language is composed from symbols.

Next, in order to explain intuitively the relation between a regular expression and the language expressed by it, we take the NFA shown in Fig. 3.21. The regular expression that expresses the language accepted by the automaton is written

$$a(ba)^*a + b(ab)^*b.$$

More precisely, the expression may be written as

$$a \cdot (b \cdot a)^* \cdot a + b \cdot (a \cdot b)^* \cdot b.$$

Note that the symbol for the operation "·" in the latter expression is omitted in the former expression. The language represented by the above expression will be clear from the state diagram shown in Fig. 3.21. Just as the expression, say, $5 \times 5^2 + 75$ means 200, the expression $a \cdot (b \cdot a)^* \cdot a + b \cdot (a \cdot b)^* \cdot b$ expresses the language

$$\{aa, abaa, ababaa, \ldots, bb, babb, bababb, \ldots\}.$$

**Fig. 3.21** NFA that accepts
$a(ba)^*a + b(ab)^*b$

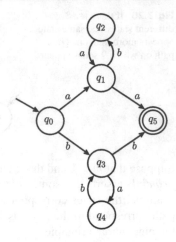

**Fig. 3.22** How to construct
$a(ba)^*a + b(ab)^*b$

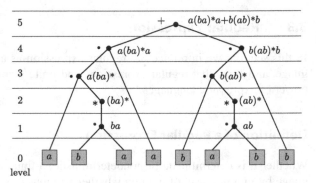

As shown in the above example, regular expressions have three kinds of operations expressed as "+", "·", "*", which, respectively, represent the union, the concatenation, and the concatenation applied repeatedly an arbitrary number of times. In particular, the operation "*" is called star. A regular expression is obtained by applying these operations to symbols repeatedly, which is precisely defined as follows.

**Definition 3.15** A *regular expression* over an alphabet $\Sigma = \{a_1, \ldots, a_k\}$ is defined as follows.
(1) Each of $a_1, \ldots, a_k$, $\varepsilon$, and $\emptyset$ is a regular expression.
(2) If $r$ and $s$ are regular expressions, so is each of $(r + s)$, $(r \cdot s)$, and $(r^*)$.

Figure 3.22 shows how the regular expression mentioned in the above example is built up according to the definition by assigning regular expressions to the corresponding nodes. For example, $b$ and $a$ are regular expressions from (1) of the definition. So, if they are considered to be $r$ and $s$, respectively, $(b \cdot a)$ is also a regular expression from (2). Similarly, if $(b \cdot a)$ is considered to be $r$, $((b \cdot a)^*)$ is a regular expression. Similarly, $(a \cdot ((b \cdot a)^*))$ and $((b \cdot ((a \cdot b)^*)) \cdot b)$, etc., turn out to be regular expressions. As shown in this example, parentheses in a regular expression may be omitted appropriately.

As is shown by the horizontal lines in Fig. 3.22, a regular expression has a level, which is shown by an integer in the figure. Let level($r$) denote the level of an expression $r$. Then, a level is defined along with the definition of regular expressions as follows. Let level($r$)=0 for the regular expression $r$ defined in (1) of Definition 3.15. Also, for the regular expressions $(r + s)$, $(r \cdot s)$, and $(r^*)$ defined in (2), let

$$\text{level}\big((r + s)\big) = \max\{\text{level}(r), \text{level}(s)\} + 1,$$

$$\text{level}\big((r \cdot s)\big) = \max\{\text{level}(r), \text{level}(s)\} + 1,$$

$$\text{level}\big((r^*)\big) = \text{level}(r) + 1,$$

where max$\{x, y\}$ denotes the largest of $x$ and $y$.

Definition 3.15 says that "if $r$ and $s$ are regular expressions, so is $(r + s)$." Therefore, to decide if $(r + s)$ is a regular expression, we have to go back to the same question for smaller regular expressions $r$ and $s$. So we can say that in this definition a notion for a larger object is defined in terms of the same notion for smaller objects. Such a way of defining a notion is called an *inductive definition*. Given an expression, in general, we have to go back to (2) of Definition 3.15 repeatedly until we come to (1), so that we finally can decide whether the expression is regular or not without referring to anything. This corresponds to breaking down toward lower levels repeatedly in Fig. 3.22 until we get to level 0. If we rely on Definition 3.15 precisely, the regular expression $a(ba)^*a + b(ab)^*b$ is written

$$\big(\big(\big((a \cdot (b \cdot a)^*)\big) \cdot a\big) + \big((b \cdot ((a \cdot b)^*)) \cdot b\big)\big)$$

in which the parentheses and the operation "$\cdot$" can be omitted appropriately as we now describe.

First, "$\cdot$" is usually omitted as in the case of mathematical expressions. Furthermore, parentheses can be omitted if the order in which the operations are applied is uniquely determined based on the precedence order between the operations:

$$+ < \cdot < *.$$

For example, according to the precedence order, $a + b \cdot c^*$ is interpreted as $a + (b \cdot (c^*))$. So if this is what is intended, then the parentheses may be omitted. In the expression $b \cdot c^*$, the operation "$*$" claims that $b \cdot (c^*)$, whereas the operation "$\cdot$" claims that $(b \cdot c)^*$, but it is interpreted as the former since "$*$" precedes "$\cdot$" in the precedence order. On the other hand, if we assumed the precedence order $* < \cdot < +$, then $a + b \cdot c^*$ would be interpreted as $((a + b) \cdot c)^*$.

You might require that a regular expression and the language that it represents should be treated as different things. To distinguish the two, we introduce the notation $L(r)$, which denotes the language that a regular expression $r$ represents. $L(r)$ can be defined as in the case of Definition 3.15 as follows:

---

(1) For $a_i$ with $1 \leq i \leq k$,

$$L(a_i) = \{a_i\}, \qquad L(\varepsilon) = \{\varepsilon\}, \qquad L(\emptyset) = \emptyset.$$

(2) If $r$ and $s$ are regular expressions,

$$L(r + s) = L(r) \cup L(s),$$
$$L(r \cdot s) = \{ww' \mid w \in L(r), w' \in L(s)\},$$
$$L(r^*) = \{w_1 w_2 \cdots w_k \mid \text{if } k = 0, \varepsilon \text{ and if } k \geq 1,$$
$$w_i \in L(r) \text{ for } 1 \leq i \leq k\}.$$

---

Among the above equations the first equation in (2) shows simply union of sets, and in the second equation $w$ and $w'$ range over all the strings in $L(r)$ and those in $L(s)$, respectively. Similarly, in the third equation, for an arbitrary $k$, $w_1, w_2, \ldots, w_k$ range over all the strings in $L(r)$. Hence, using the operation "$\cdot$", $L(r^*)$ can also be expressed as follows:

$$r^* = \{\varepsilon\} \cup L(r) \cup L(r) \cdot L(r) \cup \cdots,$$

where in general, the concatenation of languages $L_1$ and $L_2$ is defined as

$$L_1 \cdot L_2 = \{w_1 \cdot w_2 \mid w_1 \in L_1, w_2 \in L_2\}.$$

In the following, for convenience of explanation, a regular expression is taken to express either the expression itself or the language represented by it, depending on the context. For example, $r + s$ and $r \cdot s$ may be interpreted to represent $L(r) \cup L(s)$ and $L(r) \cdot L(s)$, respectively.

In the following, a couple of examples of regular expressions are given. Two regular expressions $r$ and $s$ are called *equivalent* if they express the same language, which is denoted by $r = s$.

*Example 3.16* As mentioned above, in general, $r \cdot s = \{ww' \mid w \in r, w' \in s\}$, which means $L(r \cdot s) = \{ww' \mid w \in L(r), w' \in L(s)\}$. In particular, since no string can be chosen from the language $\emptyset$, we have

$$\emptyset \cdot s = \emptyset.$$

Similarly, since $\varepsilon \cdot r = \{ww' \mid w \in \varepsilon, w' \in r\} = \{\varepsilon w' \mid w' \in r\}$, we have

$$\varepsilon \cdot r = r.$$

Furthermore, by replacing $r$ in $r^* = \varepsilon \cup r \cup r^2 \cup \cdots$ with $\emptyset$ and $\varepsilon$, we have

$$\emptyset^* = \varepsilon \cup \emptyset \cup \emptyset^2 \cup \cdots = \varepsilon,$$
$$\varepsilon^* = \varepsilon \cup \varepsilon \cup \varepsilon^2 \cup \cdots = \varepsilon,$$

respectively. We have $r^* = \varepsilon + rr^*$ as follows:

$$r^* = \varepsilon \cup r \cup r^2 \cup r^3 \cup \cdots = \varepsilon \cup r \cdot \left(\varepsilon \cup r \cup r^2 \cdots\right)$$
$$= \varepsilon \cup r \cdot r^*.$$

Note that $rsrsrsr$ is written either $r(sr)(sr)(sr)$ or $(rs)(rs)(rs)r$. As this simple example suggests, we have, in general, $r(sr)^* = (rs)^*r$, which is derived as follows:

$$r(sr)^* = r \cup r(sr) \cup r(sr)(sr) \cup \cdots$$
$$= r \cup (rs)r \cup (rs)(rs)r \cup \cdots$$
$$= \left(\varepsilon \cup (rs) \cup (rs)(rs) \cup \cdots\right)r$$
$$= (rs)^*r.$$

*Example 3.17* Some more examples are given as follows:

$$(0+1)^*110(0+1)^* = \left\{w \in \{0,1\}^* \mid w \text{ includes } 110 \text{ as a substring}\right\},$$

$$(0+1)(0+1)(0+1)(0+1)^*$$
$$= \left\{w \in \{0,1\}^* \mid \text{the length of } w \text{ is greater than or equal to } 3\right\},$$

$$0 + 1 + 0(0+1)^*0 + 1(0+1)^*1$$
$$= \left\{w \in \{0,1\}^* \mid \text{the first and the last symbols of } w \text{ are the same}\right\},$$

$$0^*1^*2^* = \left\{0^i 1^j 2^k \mid i \geq 0, j \geq 0, k \geq 0\right\}.$$

## Equivalence of Regular Expression and Finite Automaton

In this subsection, we shall show that the language represented by a regular expression is accepted by a finite automaton and, conversely, the language accepted by a finite automaton is represented by a regular expression. First, we consider the regular expression $1(0+1)^*1$ and construct an NFA that accepts the language represented by the expression. It is clear that the NFA is given as shown in Fig. 3.23. The NFA has been constructed according to Definition 3.15, relying on (1) and (2) of the definition. We give another equivalent state diagram in Fig. 3.24, which makes it easy to figure out how the NFA was constructed from the regular expression.

The following theorem is proved by constructing an NFA that accepts the language represented by a given regular expression, according to Definition 3.15.

**Theorem 3.18** *For any regular expression, there exists an NFA that accepts the language represented by the regular expression.*

*Proof* First, the proposition $P(m)$ for $m \geq 0$ is defined as follows:

**Fig. 3.23** NFA accepting
$1(0+1)*1$

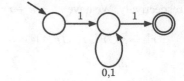

**Fig. 3.24** Construction of
NFA accepting $1(0+1)*1$

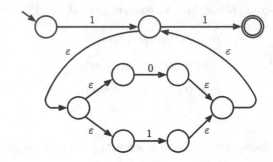

**Fig. 3.25** NFA accepting $\varepsilon$,
$\emptyset$, and $a$, respectively

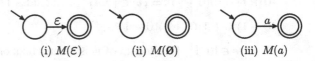

(i) $M(\varepsilon)$      (ii) $M(\emptyset)$      (iii) $M(a)$

$P(m)$: For a regular expression $r$ in which the number of the operations appearing
in it is no more than $m$, there exists an NFA with only one accept state that accepts
the language represented by $r$.

In the following, it is proved by mathematical induction on $m$ that for any $m \geq 0$
$P(m)$ holds. It is clear that this suffices for the proof of the theorem. Let $M(r)$
denote an NFA that accepts the language represented by the regular expression $r$.

*Proof of $P(0)$* For the regular expressions in which no operation symbols appear,
$M(\varepsilon)$, $M(\emptyset)$, and $M(a)$ are constructed as shown in Fig. 3.25.

*Proof of the fact that $P(m-1)$ implies $P(m)$* Let $m \geq 1$. A regular expression in
which the operation symbols appear $m$ times for $m \geq 1$ is expressed as either $r + s$,
$r \cdot s$, or $r^*$. Then, the number of operation symbols that appear in $r$ and that in $s$ are
less than or equal to $m-1$ because the operations $+$, $\cdot$, and $*$ in these expressions are
already counted one among the $m$ operation symbol expressions. Therefore, from
the assumption $P(m-1)$, $M(r)$ and $M(s)$ can be constructed where the start and
accept states of $M(r)$ are denoted by $q_s$ and $q_f$, respectively, and similarly they are
denoted by $p_s$ and $p_f$ for $M(s)$.

(i) Construction of $M(r+s)$ $M(r+s)$ is constructed by connecting $M(r)$ and
   $M(s)$ in parallel as shown in Fig. 3.26. The two rectangles enclosed by the dashed
   lines in the figure represent the state diagrams of $M(r)$ and $M(s)$, respectively.
   Details of the diagrams are not shown except for the start and accept states. Note

**Fig. 3.26** Construction of $M(r+s)$

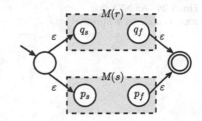
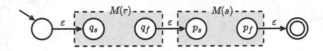

**Fig. 3.27** Construction of $M(r \cdot s)$

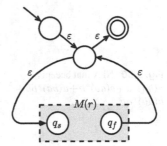

**Fig. 3.28** Construction of $M(r^*)$

that in Fig. 3.26 the two new states are introduced and are specified as the start state and the accept state of $M(r+s)$.

(ii) Construction of $M(r \cdot s)$  $M(r \cdot s)$ is constructed by connecting $M(r)$ and $M(s)$ in serial. Figure 3.27 illustrates the state diagram of $M(r \cdot s)$ in a similar way to the above example, where the two new states are introduced as the start state and the accept state.

(iii) Construction of $M(r^*)$  $M(r^*)$ is constructed by introducing a state that serves as both the start state and the accept state. As shown in Fig. 3.28, $M(r)$ connected with the state added forms a loop so that we can go through $M(r)$ an arbitrary number of times.

It is clear that each of the state diagrams constructed above has one start state and one accept state and that $M(r+s)$, $M(r \cdot s)$, and $M(r^*)$ accept $r+s$, $r \cdot s$, and $r^*$, respectively, thus completing the proof.  □

In the following, we shall show that a language accepted by a finite automaton is represented by a regular expression. To do so, we show that a finite automaton can be transformed to the regular expression that represents the language accepted by the finite automaton.

To explain such a transformation intuitively, we first examine an easy case, and then proceed to explain a somewhat complicated case. As an easy case, we take the state diagram of the nondeterministic automaton shown in Fig. 3.29. Notice that the state diagram is nested in the sense that each subdiagram at a higher level con-

**Fig. 3.29** An NFA that
accepts $a(ba)^*a + b(ab)^*b$

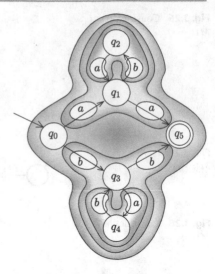

**Fig. 3.30** NFA that accepts
$a(ba)^*a + b(ab)^*b + a(ba)^*bb + b(ab)^*aa$

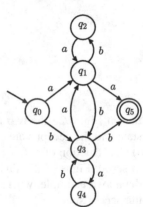

tains subdiagrams at the lower level, where the level of a subdiagram is that of the corresponding regular expression as shown in Fig. 3.21. Obviously, parallel connection, serial connection, and loop connection correspond to the three operations $+$, $\cdot$, and $*$, respectively. In the case of a state diagram with such a nested structure, it is easy to see that we can obtain an equivalent regular expression based on the diagram. This is because the nested structure immediately indicates how the equivalent regular expression is constructed along the line of Definition 3.15. In fact, the regular expression that corresponds to the state diagram in Fig. 3.29 is given by $a(ba)^*a + b(ab)^*b$.

However, this is not always the case. In fact, there exist state diagrams that cannot be decomposed into such nested structures. We will proceed to a complicated case of a state diagram that does not have a nested structure. As an example, we take the state diagram of Fig. 3.30 and show that the state diagram can be transformed to an equivalent regular expression. Thus it is easy to see that the state diagram in Fig. 3.30, which does not have a nested structure, can be transformed to an equivalent state diagram that has a nested structure. After having discussed how to transform the diagrams of the examples equivalently, what remains to be explored is how to derive a regular expression equivalent to an NFA that does not have a nested structure in general.

The NFA shown in Fig. 3.30 is an example that does not have a nested structure. The edges that correspond to $q_3 \xrightarrow{a} q_1$ and $q_1 \xrightarrow{b} q_3$ in the figure break the nested structure of the diagram. Because of these two edges we cannot derive an equivalent regular expression in the way described above for the nested structure case. In the particular case of Fig. 3.30, by tracing the paths from the start state to the accept state in Fig. 3.30, we can conclude that a string that the diagram accepts takes one of four types of forms, namely, $awa$, $bwb$, $awb$, and $bwa$ for appropriately chosen $w$. Clearly this is because there are only two edges leaving state $q_0$, namely, $q_0 \xrightarrow{a} q_1$ and $q_0 \xrightarrow{b} q_3$, and only two edges entering state $q_5$, namely, $q_1 \xrightarrow{a} q_5$ and $q_3 \xrightarrow{b} q_5$. The languages consisting of strings with the four types of forms $awa$, $bwb$, $awb$, and $bwa$ are written $a(ba)^*a$, $b(ab)^*b$, $a(ba)^*bb$, and $b(ab)^*aa$. Thus the language that the diagram in Fig. 3.30 accepts is given by taking the union of the four languages as follows:

$$a(ba)^*a + b(ab)^*b + a(ba)^*bb + b(ab)^*aa.$$

Instead of examining each particular state diagram this way, we develop a general procedure to obtain the equivalent regular expression from an NFA. We explain the procedure as follows.

First we generalize state diagrams so that not only a symbol in an alphabet but also a regular expression is allowed to be assigned as a label of an edge in the state diagrams. The idea is that, if a regular expression $r$ is assigned to an edge from state $q$ to state $q'$, then any string $w$ in $L(r)$ can move the automaton from $q$ to $q'$. After having generalized state diagrams this way, we use a procedure that transforms a state diagram by repeatedly removing a state one by one without changing the language accepted by the state diagrams until we obtain a state diagram that has only one transition $q_s \xrightarrow{r} q_f$ from the start state to the accept state with a label of regular expression $r$. The regular expression $r$ that we seek represents the language accepted by the state diagram given first. The point of the transformation is that, after a state is removed from a diagram, edges with regular expressions appropriately chosen are added so that all the transitions that go through the removed state are compensated for.

In Fig. 3.31 the left-hand side shows the part of Fig. 3.30 related to state $q_1$, whereas the right-hand side shows the corresponding part that compensates for all the transitions caused by removing state $q_1$. It is easy to see that if the four edges with the regular expressions are added as in the right-hand part of Fig. 3.31, then those edges compensate for all the transitions that pass through state $q_1$. Note that the two edges associated with $q_1 \xrightarrow{b} q_2 \xrightarrow{a} q_1$ in the left-hand part of the figure are put together as the edge $q_1 \xrightarrow{ba} q_1$ in the right-hand part. When the state diagram

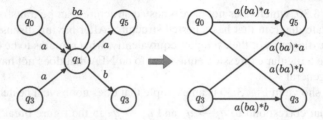

**Fig. 3.31** Edges to compensate for removed state $q_1$

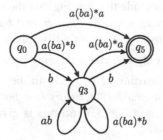

**Fig. 3.32** State diagram after $q_1$ and $q_2$ are removed from the diagram in Fig. 3.30

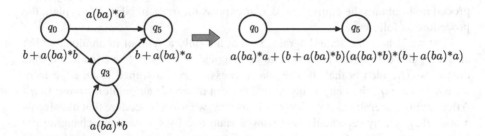

**Fig. 3.33** Transformation caused by removing state $q_3$

in Fig. 3.30 is given, we first delete state $q_1$ and then delete $q_2$ according to the transformation shown in Fig. 3.31. After deleting $q_1$ and $q_2$ in this way, we have the diagram in Fig. 3.32.

By removing further the state $q_3$ from the state diagram of Fig. 3.32, we have the transformation shown in Fig. 3.33. Note that $q_0 \overset{b}{\to} q_3$ and $q_0 \xrightarrow{a(ba)^*b} q_3$ are put together to make $q_0 \xrightarrow{b+a(ba)^*b} q_3$. Similarly, we have $q_3 \xrightarrow{b+a(ba)^*a} q_5$ and $q_3 \xrightarrow{a(ba)^*b} q_3$. Finally we have the regular expression $r$ of $q_0 \overset{r}{\to} q_5$ which is given as

$$a(ba)^*a + \big(b + a(ba)^*b\big)\big(a(ba)^*b\big)^*\big(b + a(ba)^*a\big).$$

We shall simplify this regular expression by using the equality and the inclusion relation shown as follows:

$$a(ba)^* = ab(ab)^*, \qquad \left(a(ba)^*b\right)^* = \left(ab(ab)^*\right)^* = (ab)^*,$$

$$a(ba)^*a = (ab)^*aa, \qquad b(ab)^*b = (ba)^*bb,$$

$$ab(ab)^*aa \subseteq (ab)^*aa = a(ba)^*a.$$

Recall that $r = s$ means that $L(r) = L(s)$ and that $r \subseteq s$ means that $L(r) \subseteq L(s)$, where $L(r)$ denotes the language that regular expression $r$ represents. Using these relations and the distributive law, the regular expression shown in Fig. 3.33 is simplified as follows:

$$a(ba)^*a + \left(b + a(ba)^*b\right)\left(a(ba)^*b\right)^*\left(b + a(ba)^*a\right)$$
$$= a(ba)^*a + \left(b + ab(ab)^*\right)(ab)^*\left(b + (ab)^*aa\right)$$
$$= a(ba)^*a + b(ab)^*b + b(ab)^*aa + ab(ab)^*b + ab(ab)^*aa$$
$$= a(ba)^*a + b(ab)^*b + b(ab)^*aa + a(ba)^*bb.$$

Using the example shown in Fig. 3.30, we have explained the procedure for obtaining an equivalent regular expression from an NFA.

We now generalize the argument so far. Figure 3.34 shows the generalization of the transformation shown in Fig. 3.31 to remove a state. Let $q_{rm}$ denote the state to be removed and let it be neither the start state nor the accept state. Furthermore, it is assumed that $k$ edges go to state $q_{rm}$ and $l$ edges go out of $q_{rm}$, where there may be a state that appears both in $\{q_1, \ldots, q_k\}$ and in $\{p_1, \ldots, p_l\}$. Then remove state $q_{rm}$ together with the edges incident to the state, and add instead the edge $(q_i, p_j)$ for every pair of $q_i$ in $\{q_1, \ldots, q_k\}$ and $p_j$ in $\{p_1, \ldots, p_l\}$, where the edge $(q_i, p_j)$ is attached with the regular expression $r_i s^* t_j$. It is clear that the transitions deleted by removing state $q_{rm}$ are compensated for by introducing these new edges. If there exists an edge from $q_i$ to $p_j$ attached with regular expression $u_{ij}$ in the original state diagram, regular expression $r_i s^* t_j$ in Fig. 3.34 should be replaced with regular expression $u_{ij} + r_i s^* t_j$ for every pair of states $q_i$ and $p_j$.

We are now ready to prove the statement that a language accepted by an NFA is represented by a regular expression. But, before proceeding to the proof, we shall make some comments about a generalized NFA. A generalized NFA is defined to be the same as an NFA except that every edge other than edges going to the start state and those going out of the accept state is assigned with a regular expression rather than a symbol. Generalizing the statement above, we shall prove the statement obtained by replacing "NFA" in the statement with "generalized NFA." In a generalized NFA, $q \xrightarrow{r} p$ for a regular expression $r$ means that the automaton goes from state $q$ to state $p$ when a string in $r$ is fed. So when there exists no edge from $q$ to $p$, we consider that the edge from $q$ to $p$ is assigned with $\emptyset$, i.e., $q \xrightarrow{\emptyset} p$. Therefore, in Fig. 3.34, $\{q_1, \ldots, q_k\}$ consists of all the states except for $q_{rm}$ and the accept state, whereas $\{p_1, \ldots, p_l\}$ consists of all the states except for $q_{rm}$ and the start state.

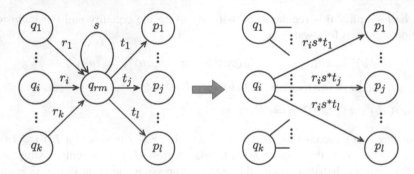

**Fig. 3.34** Transformation to remove state $q_{\mathrm{rm}}$

By the way, what is explained in Fig. 3.34 can also be applied to transformation for a generalized NFA. For example, $q_i \xrightarrow{r_i} q_{\mathrm{rm}}$, $q_{\mathrm{rm}} \xrightarrow{s} q_{\mathrm{rm}}$, and $q_{\mathrm{rm}} \xrightarrow{t_j} q_j$ are replaced by $q_i \xrightarrow{r_i s^* t_j} p_j$. If there is no transition, say, from $q_i$ to $q_{\mathrm{rm}}$, that is, $r_i = \emptyset$, then we have $r_i s^* t_j = \emptyset$ which is equivalent to saying that there is no transition from $q_i$ to $q_j$ via state $q_{\mathrm{rm}}$. On the other hand, if $s = \emptyset$, that is, there is no transition from $q_{\mathrm{rm}}$ to $q_{\mathrm{rm}}$, we have $r_i s^* t_j = r_i \varepsilon t_j = r_i t_j$ since $s^* = \varepsilon$ from the definition. In this way, whether some of the related edges exist or not, the general form of $r_i s^* t_j$ is appropriate for compensating for the removed state $q_{\mathrm{rm}}$.

In a generalized NFA, a regular expression is assigned to every edge except for edges going to the start state and those going out of the accept state. The regular expression assigned to a pair of states $(q, q')$ is denoted by $R(q, q')$. The set of regular expressions is denoted by *REG*.

**Definition 3.19** A *generalized nondeterministic finite automaton* (GNFA) is a 5-tuple $(Q, \Sigma, R, q_s, q_f)$, where $Q$ denotes a set of states, $\Sigma$ an alphabet, and $q_s$ the start state, and $q_f$ is the single accept state. The state transition is defined by a transition function $R : (Q - \{q_f\}) \times (Q - \{q_s\}) \to REG$, where it is supposed that there are no edges going out of the accept state and no edges going to the start state. A GNFA $(Q, \Sigma, R, q_s, q_f)$ accepts a string $w$ if there exist $m \geq 0$, $q_0, q_1, \ldots, q_m \in Q$ and $w_1, \ldots, w_m \in \Sigma^*$ such that

$$w = w_1 \cdots w_m, \quad w_1 \in R(q_0, q_1), \ w_2 \in R(q_1, q_2), \ \ldots, \ w_m \in R(q_{m-1}, q_m),$$

$$q_0 = q_s, \ q_m = q_f.$$

As in the case of a finite automaton, when $w \in R(q, p)$ holds, we express it as $q \xrightarrow{w} p$. Furthermore, when $q_0 \xrightarrow{w_1} q_1 \xrightarrow{w_2} q_2 \to \cdots \xrightarrow{w_m} q_m$ holds, we express it as $q_0 \xrightarrow{w_1 w_2 \cdots w_m} q_m$.

In the definition above, it is supposed that the number of accept states is one and that there are no edges going out of the accept state and no edges going to the start state. This assumption is just used to make the proof of the following Theorem 3.20

easy. We can make the assumption without loss of generality because, when given a GNFA that does not satisfy the assumption, we can transform it to one that satisfies the assumption by introducing a new start state $q'_s$ and a new accept state $q'_f$ and adding transitions $q'_s \xrightarrow{\varepsilon} q_s$ and $q_f \xrightarrow{\varepsilon} q'_f$ for all of the accept states $q_f$ of the GNFA. It is clear that the NFA constructed in this way satisfies the assumption mentioned above and accepts the same language as the one accepted by the GNFA given first.

**Theorem 3.20** *If a language is accepted by a generalized nondeterministic finite automaton (GNFA), then it is described by a regular expression.*

Since an NFA is just a GNFA restricted in such a way that a regular expression assigned to each edge is a symbol in the alphabet or $\varepsilon$, the following corollary is obtained immediately from the above theorem.

**Corollary 3.21** *If a language is accepted by a nondeterministic finite automaton, then it is described by a regular expression.*

*Proof of Theorem 3.20* We will describe formally how to repeatedly transform a GNFA to a GNFA with one fewer state that accepts the same language as the one that the original GNFA accepts until we obtain a GNFA with the two states, namely, the start state and the accept state, thereby obtaining a regular expression on the edge that describes the language accepted by the original GNFA.

Let us denote a given GNFA by $M = (Q, \Sigma, R, q_s, q_f)$, and the number of states of $M$ by $m$. We apply the algorithm that transforms a GNFA to an equivalent GNFA with one fewer state repeatedly, starting with $M$ which is denoted by $M_m$. We assume $m \geq 2$. The sequence of GNFAs obtained this way are denoted by $M_m, M_{m-1}, \ldots, M_2$. If we are able to prove that the algorithm transforms a GNFA to a GNFA, with one fewer state, that accepts the same language as the old one, the proof will be completed. This is true because since the algorithm removes a state that is neither the start state nor the accept state, $M_2$, which we finally obtain, consists of the start state and the accept state so that the regular expression assigned with the edge from the start state to the accept state represents the language that the original GNFA accepts.

Next, we give the algorithm, denoted *CONVERT*, that transforms a GNFA with more than two states to an equivalent GNFA with one fewer states.

Algorithm *CONVERT*:
1. Choose a state different from the start and the accept state and denote it by $q_{\text{rm}}$.
2. For all $(q, p) \in (Q - \{q_{\text{rm}}, q_f\}) \times (Q - \{q_{\text{rm}}, q_s\})$ relabel the regular expressions denoted in terms of $R$ to the ones denoted in terms of $R'$ as follows:
$$R'(q, p) \leftarrow R(q, p) + R(q, q_{\text{rm}})R(q_{\text{rm}}, q_{\text{rm}})^* R(q_{\text{rm}}, p).$$

Figure 3.35 illustrates how the regular expressions labeled with the edge are altered in step 2 by the algorithm. Assume that the algorithm transforms a GNFA denoted by $M = (Q, \Sigma, R, q_s, q_f)$ to the GNFA denoted by $M' =$

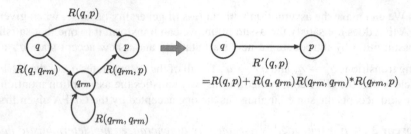

**Fig. 3.35** Algorithm *CONVERT* removes state $q_{\text{rm}}$ and alters labels of regular expressions appropriately

$(Q - \{q_{\text{rm}}\}, \Sigma, R', q_s, q_f)$. Since state $q_{\text{rm}}$ to be removed is neither the start state $q_s$ nor the accept state $q_f$, and $M$ satisfies the conditions for a GNFA given in Definition 3.19, $M'$ also satisfies the conditions, where the conditions for the GNFA are that there exists no edge coming in to $q_s$ and no edge going out of $q_f$, and that there exist the single start state and the single accept state that is different from the start state.

Next, we shall verify that $M$ and $M'$ accept the same language. First, suppose that a string $w$ is accepted by $M$. Then, there are states $q_0, q_1, \ldots, q_m \in Q$ and $w_1, \ldots, w_m \in \Sigma^*$ such that

$$w = w_1 w_2 \cdots w_{m-1} w_m,$$

$$q_0 \xrightarrow{w_1} q_1 \xrightarrow{w_2} \cdots \xrightarrow{w_{m-1}} q_{m-1} \xrightarrow{w_m} q_m, \quad q_0 = q_s, \ q_m = q_f,$$

where $w_i \in R(q_{i-1}, q_i)$ for $1 \le i \le m$. If $q_i \ne q_{\text{rm}}$ and $q_{i+1} \ne q_{\text{rm}}$, since

$$w_i \in R(q_i, q_{i+1}) \subseteq R'(q_i, q_{i+1}),$$

we have $q_i \xrightarrow{w_i} q_{i+1}$ in $M'$.

On the other hand, if consecutive $q_{\text{rm}}$ appearing in the sequence are surrounded by $q_i$ and $q_j$ that are different from $q_{\text{rm}}$ shown as follows:

$$q_i \xrightarrow{w_{i+1}} q_{\text{rm}} \xrightarrow{w_{i+2}} q_{\text{rm}} \xrightarrow{w_{i+3}} \cdots \xrightarrow{w_{j-1}} q_{\text{rm}} \xrightarrow{w_j} q_j,$$

then we have

$$w_{i+1} \cdots w_j \in R(q_i, q_j) + R(q_i, q_{\text{rm}}) R(q_{\text{rm}}, q_{\text{rm}})^* R(q_{\text{rm}}, q_j) = R'(q_i, q_j).$$

Hence, we have

$$q_i \xrightarrow{w_{i+1} \cdots w_j} q_j$$

in $M'$. Therefore, since we can argue in that way no matter where the $q_{\text{rm}}$ appears, we have

$$q_0 \xrightarrow{w_1 w_2 \cdots w_m} q_m$$

in $M'$. Thus, we can conclude that $w = w_1 w_2 \cdots w_m$ is accepted by $M'$.

Next, in order to prove the reverse direction, suppose that string $w$ is accepted by $M'$. Then, in $M'$, there exist states $q_0, q_1, \ldots, q_m \in Q - \{q_{rm}\}$ and $w_1, \ldots, w_m \in \Sigma^*$ such that

$$q_0 \xrightarrow{w_1} q_1 \xrightarrow{w_2} \cdots \xrightarrow{w_{m-1}} q_{m-1} \xrightarrow{w_m} q_m,$$

$$w = w_1 w_2 \cdots w_{m-1} w_m,$$

$$w_1 \in R'(q_0, q_1), \ w_2 \in R'(q_1, q_2), \ \ldots, \ w_m \in R'(q_{m-1}, q_m),$$

$$q_0 = q_s, \ q_m = q_f,$$

where $w_i \in R'(q_{i-1}, q_i)$ for $1 \le i \le m$. So we have

$$w_i \in R'(q_{i-1}, q_i) = R(q_{i-1}, q_i) + R(q_{i-1}, q_{rm}) R(q_{rm}, q_{rm})^* R(q_{rm}, q_i).$$

If $w_i \in R(q_{i-1}, q_i)$, we have

$$q_{i-1} \xrightarrow{w_i} q_i$$

in $M$. On the other hand, if $w_i \in R(q_{i-1}, q_{rm}) R(q_{rm}, q_{rm})^* R(q_{rm}, q_i)$, then there exists a sequence of transitions in $M$

$$q_{i-1} \to q_{rm} \to q_{rm} \to \cdots \to q_{rm} \to q_i$$

that spells out $w_i$. Stringing these transitions together, we have

$$q_0 \xrightarrow{w_1 w_2 \cdots w_m} q_m$$

in $M$. Thus, since $q_0$ is the start state and $q_m$ is the accept state, we conclude that $M$ accepts $w = w_1 w_2 \cdots w_m$, completing the proof. □

It has been shown that the three conditions specifying a language are equivalent with each other: (1) a language is accepted by a deterministic finite automaton; (2) a language is accepted by a nondeterministic finite automaton; (3) a language is represented as a regular expression. A language that satisfies one of these conditions is called a *regular language*, and the set of regular languages is called the *class of regular languages*.

## 3.4 Properties of Regular Languages

We present a precise version of the following statement about regular languages: if a string in a regular language is somehow altered by repeating a certain substring of the string, then the resulting string turns out to be also contained in the original language. Based on this property, a language can be shown not to be regular by verifying that the language does not have that property. Furthermore, it will be shown that the class of regular languages has a property that the class is closed under various set operations.

## Limit of Language Accepting Power of a Finite Automaton

There is a language that cannot be accepted by any finite automaton. Language $\{0^n 1^n \mid n \geq 1\}$ is an example of such a language. We intuitively figure out that the language $\{0^n 1^n \mid n \geq 1\}$ cannot be accepted by any finite automaton as follows. Suppose that there is a deterministic finite automaton that accepts the language. Then it will be shown that, beginning with the start state, string $0^n$ takes the automaton to different states for different $n$'s. Hence we have to prepare different states for $n = 1, 2, 3, \ldots$. But this is impossible because the automaton has only a finite number of states.

Now, we verify that

$$q_s \xrightarrow{0^n} q, \qquad q_s \xrightarrow{0^m} q', \quad \text{and} \quad n \neq m$$

imply that $q \neq q'$. This is because, if $q = q'$, then as shown in Fig. 3.36 we have

$$q_s \xrightarrow{0^n} q \xrightarrow{1^n} q'', \qquad q_s \xrightarrow{0^m} q' \xrightarrow{1^n} q''$$

for a certain state $q''$. If $q''$ is an accept state, then, since $n \neq m$, the automaton accepts $0^m 1^n \notin \{0^n 1^n \mid n \geq 1\}$. This is a contradiction. On the other hand, if $q''$ is not an accept state, then the automaton does not accept $0^n 1^n \in \{0^n 1^n \mid n \geq 1\}$. This is also a contradiction. In this way, the contradiction is derived for either case. Therefore, if $q_s \xrightarrow{0^n} q$ and $q_s \xrightarrow{0^m} q'$ for $n \neq m$, then $q$ and $q'$ are different from each other.

We shall show that all regular languages have a certain property, which we call the repeat condition. So if we can show that a language does not satisfy the repeat condition, then we can conclude that the language is not regular. Employing the repeat condition, we can verify that a language is not regular in a more systematic way as compared to the arguments above to derive that $\{0^n 1^n \mid n \geq 1\}$ is not regular.

The repeat condition is roughly stated as follows. Let $L$ denote a language. $L$ is said to satisfy the repeat condition if the following condition is satisfied: for any sufficiently long string $w$ in $L$, $w$ can be separated into three parts $w = xyz$ in such a way that all the strings $xz, xyz, xyyz, xyyyz, \ldots$ belong to the original language $L$ as well. Any regular language turns out to satisfy the repeat condition, whose proof is intuitively described as follows. Let $w$ be a sufficiently long string in a regular language $L$. Let $w$ be of length $n$ and denoted by $a_1 a_2 \cdots a_n$. Then, in a deterministic finite automaton that accepts $L$, we have

$$q_0 \xrightarrow{a_1} q_1 \xrightarrow{a_2} q_2 \xrightarrow{a_3} \cdots \xrightarrow{a_n} q_n,$$

**Fig. 3.36** An illustration to show that $\{0^n 1^n \mid n \geq 1\}$ is not accepted by any finite automaton

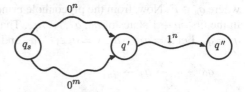

**Fig. 3.37** An illustration of the repeat condition

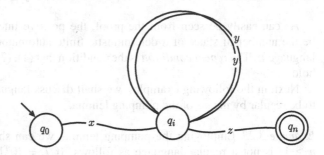

where $q_0$ is the start state and $q_n$ an accept state. Now, if the length $n$ of $w$ is greater than or equal to the number of states of the automaton, then at least one state appears twice in $q_0, q_1, \ldots, q_n$. Let such a state be $q_i$ and let $y$ be the substring of $w$ surrounded by two $q_i$'s, thereby the whole string is denoted by $w = xyz$. The trace of transitions in Fig. 3.37, denoted by $zyz$, is one that spells out $w$. Then, as shown in the figure, $xyyz$ is also accepted. In this way we can figure out intuitively that the regular language satisfies the repeat condition, which says that $xy^i z$ belongs to the regular language for any $i \geq 0$.

The argument above is stated as the following theorem, which is traditionally called the *pumping lemma*. In the theorem, $|w|$ denotes the length of string $w$.

**Theorem 3.22** (Pumping lemma)  *For a regular language $L$ over $\Sigma$ there is a positive integer $m$ that satisfies the following condition: any string $w$ in $L$ whose length is greater than or equal to $m$ is expressed as $w = xyz$ for some $x, y, z \in \Sigma^*$ and satisfies the following three conditions:*
(1) *For any $i \geq 0$, $xy^i z \in L$*
(2) $|y| \geq 1$
(3) $|xy| \leq m$

*Proof*  Let $M = (Q, \Sigma, \delta, q_0, F)$ be a deterministic finite automaton that accepts $L$, where the number of states of $M$ is denoted by $m$. Let $a_1 a_2 \cdots a_n$ be a string in $L$ whose length is greater than or equal to $m$. Then we have the sequence of transitions to accept $a_1 a_2 \cdots a_n$

$$q_0 \xrightarrow{a_1} q_1 \xrightarrow{a_2} q_2 \xrightarrow{a_3} \cdots \xrightarrow{a_n} q_n,$$

where $q_n \in F$. Now, from the pigeonhole principle, some state appears at least twice in the first $m + 1$ states of $q_0, q_1, \ldots, q_n$. That is, there are $0 \le i < j \le m$ such that $q_i = q_j$. Let $x = a_1 \cdots a_i$, $y = a_{i+1} \cdots a_j$ and $z = a_{j+1} \cdots a_n$. Since

$$q_0 \xrightarrow{x} q_i \xrightarrow{y} q_j \xrightarrow{z} q_n,$$

and $q_i = q_j$, $M$ accepts $xy^i z$ for any $i \ge 0$. Figure 3.37 illustrates the case of $i = 2$. Furthermore, since $i \ne j$, we have $|y| > 0$ and $|xy| = j \le m$.                                    □

As can easily be seen from the proof, the positive integer $m$ is considered to be the number of states of a deterministic finite automaton that accepts a regular language $L$. The *repeat condition* is the condition that (1), (2), and (3) of the theorem hold.

Next, in the following examples, we shall discuss languages that are shown not to be regular by means of the pumping lemma.

*Example 3.23* Employing the pumping lemma, we can show that the set $\{0^n 1^n \mid n \ge 1\}$ is not a regular language as follows. If $L = \{0^n 1^n \mid n \ge 1\}$ is a regular language, there exists $m$ that satisfies the conditions of the pumping lemma. Let $w = 0^m 1^m$. From the pumping lemma, $w$ is expressed as $xyz$ for some $x, y, z$, and by the condition that $|xy| \le m$, $xy$ is in the former part $0^m$ of $w$. Therefore, $xz$, which is obtained by removing $y$ from $xyz = 0^m 1^m$, is written $xz = 0^{m-|y|} 1^m \in L$. However, since $|y| \ge 1$, and hence $m - |y| \ne m$, this is a contradiction.

*Example 3.24* We consider a language $L$ consisting of strings that have an equal number of 0's and 1's. That is, $L = \{w \in \{0, 1\}^* \mid N_0(w) = N_1(w)\}$, where $N_0(w)$ and $N_1(w)$ denote the number of 0's in $w$ and that of 1's, respectively. Taking $0^m 1^m$ as string $w$, we can show that the language $L$ is not a regular language in a similar way to the previous example. Let us note that we have to choose an appropriate string $w$ so that a contradiction is easily derived. For example, if string $(01)^m$ is chosen in this case, a contradiction cannot be found by an argument similar to the previous example.

*Example 3.25* In order to prove that the language $\{ww \mid w \in \{0, 1\}^*\}$ is not regular, we take $0^m 10^m 1$, which leads to a contradiction in a similar way to Example 3.23.

## Closure Properties of the Class of Regular Languages

Let a binary operation, denoted by $\square$, be defined on set $U$. As shown in Fig. 3.38, it is said that a subset $S$ of $U$ is closed under the operation $\square$ if for any $a$ and $b$

$$a \in S \quad \text{and} \quad b \in S \quad \Rightarrow \quad a \square b \in S.$$

In other words, $S$ is closed under an operation if the resultant of the operation to any elements of $S$ does not go out of $S$. For example, the set of even numbers is closed under addition, whereas the set of odd numbers is not closed under addition.

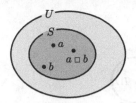
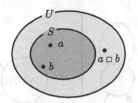

(i) $S$ is closed under $\square$:
for any $a \in S$ and $b \in S$, $a \square b \in S$

(ii) $S$ is not closed under $\square$:
there exist $a \in S$ and $b \in S$ such that $a \square b \notin S$

**Fig. 3.38** Closure property under operation $\square$ on $S$

The closure property of an $r$-ary operation is defined similarly to that for a binary operation.

In this subsection, we take the power set of $\Sigma^*$ as $U$ and the class of regular languages as $S$:

$$U = \{L \mid L \subseteq \Sigma^*\},$$

$$S = \{L \mid L \text{ is accepted by a finite automaton}\}.$$

We are now ready to discuss the closure properties of the class of regular languages under various operations such as union $\cup$, intersection $\cap$, complement $^-$, concatenation $\cdot$, and star $*$.

It follows immediately from the definition of a regular expression that the class of regular languages is closed under union, concatenation, and star operations. The condition that the class of regular languages is closed under an operation, say, union $+$, is stated as follows:

$r$ and $s$ are regular expressions $\quad \Rightarrow \quad r + s$ is a regular expression.

This is just (2) of the definition of regular expressions. Similarly, we can derive that the class of regular languages is closed under concatenation and star operations.

In the following, we shall show that the class of regular languages is also closed under intersection $\cap$ and complement $^-$, where the operation of complement is with regard to $\Sigma^*$ and hence the complement of $L \subseteq \Sigma^*$, denoted by $\overline{L}$, is such that $\overline{L} = \Sigma^* - L$. The condition for the closure property concerning intersection becomes the following:

$L$ and $L'$ are accepted by finite automata

$\Rightarrow \quad L \cap L'$ is accepted by a finite automaton.

To prove this closure property, let $M$ and $M'$ be finite automata that accept $L$ and $L'$, respectively. Employing these automata, a finite automaton that accepts $L \cap L'$ is constructed as follows. The automaton to be constructed is called a *Cartesian prod-*

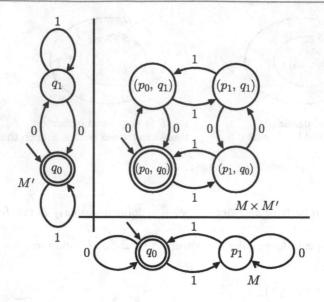

**Fig. 3.39** Cartesian product automaton that accepts strings consisting of an even number of 1's and an even number of 0's

*uct automaton*, which is denoted by $M \times M'$. A state of $M \times M'$ is denoted by a pair of a state of $M$ and that of $M'$, denoted by $(p, q)$, where $p$ and $q$ keep track of a state of $M$ and a state of $M'$, respectively. That is, $(p, q) \xrightarrow{a} (p', q')$ in $M \times M'$ means that $p \xrightarrow{a} p'$ in $M$ and $q \xrightarrow{a} q'$ in $M'$. Furthermore, the start state is $(q_0, q_0')$, where $q_0$ and $q_0'$ are the start states of $M$ and $M'$, respectively, and $(p, q)$ is an accept state of $M \times M'$ if $p$ and $q$ are accept states of $M$ and $M'$, respectively. Then, it is easy to see that $M \times M'$ accepts $L \cap L'$.

*Example 3.26* Let $N_0(w)$ and $N_1(w)$ denote the numbers of 0's and 1's that appear in $w \in \{0, 1\}^*$, respectively. Let $L = \{w \in \{0, 1\}^* \mid N_1(w)$ is even$\}$ and $L' = \{w \in \{0, 1\}^* \mid N_0(w)$ is even$\}$. Then, we have

$$L \cap L' = \{w \in \{0, 1\}^* \mid \text{both } N_1(w) \text{ and } N_0(w) \text{ are even}\}.$$

Figure 3.39 shows finite automata $M$ and $M'$ that accept $L$ and $L'$, respectively, and the Cartesian product automaton $M \times M'$ as well. If we interpret the horizontal moves on the state diagram of $M \times M'$ as transitions of $M$, and the vertical moves as transitions of $M'$, we can easily see that the Cartesian product automaton $M \times M'$ accepts the strings that both $M$ and $M'$ accept.

**Theorem 3.27** *The class of regular languages is closed under union $\cup$, intersection $\cap$, complement $^-$, and star $*$.*

*Proof* It is clear from the definition of a regular expression that the class of regular languages is closed under union, concatenation, and star. Let $L$ be a regular language and let $M = (Q, \Sigma, \delta, q_0, F)$ be a deterministic finite automaton that accepts $L$. Then, we consider the deterministic finite automaton $\overline{M} = (Q, \Sigma, \delta, q_0, Q - F)$ by exchanging the roles of the accept states and non-accept states of $M$. Clearly $\overline{M}$ accepts the complement $\overline{L} = \Sigma^* - L$ of $L$. Thus we can conclude that the class of regular languages is closed under the operation of complement.

Next, we shall prove that the class is closed under the operation of intersection. Let $M_1 = (Q_1, \Sigma, \delta_1, q_{s1}, F_1)$ and $M_2 = (Q_2, \Sigma, \delta_2, q_{s2}, F_2)$ be deterministic finite automata that accept regular languages $L_1$ and $L_2$, respectively. The transition function of the Cartesian product automaton

$$M_1 \times M_2 = (Q_1 \times Q_2, \Sigma, \delta, (q_{s1}, q_{s2}), F_1 \times F_2)$$

is defined as

$$\delta((q_1, q_2), a) = (\delta_1(q_1, a), \delta_2(q_2, a)).$$

Clearly $M$ accepts $L_1 \cap L_2$. Thus $L_1 \cap L_2$ is a regular language, completing the proof. $\qquad\square$

Here is another way of proving that the class of regular languages is closed under the operation of intersection. First, let us note that for regular languages $L_1$ and $L_2$ their intersection is expressed by using De Morgan's law as

$$L_1 \cap L_2 = \overline{\overline{L_1} \cup \overline{L_2}}.$$

On the other hand, since the class of regular languages is closed under union and complement, as already shown, it is also closed under intersection, which is expressed as above in terms of union and complement.

So far we have given two proofs to verify that the class of regular languages is closed under intersection: one based on the Cartesian product automaton and the other based on De Morgan's law. Compared to the latter, the former seems to help us figure out how intersection of regular languages is accepted by a finite automaton.

## 3.5 Problems

**3.1** Describe concisely the strings accepted by the automata defined by the following state diagrams.

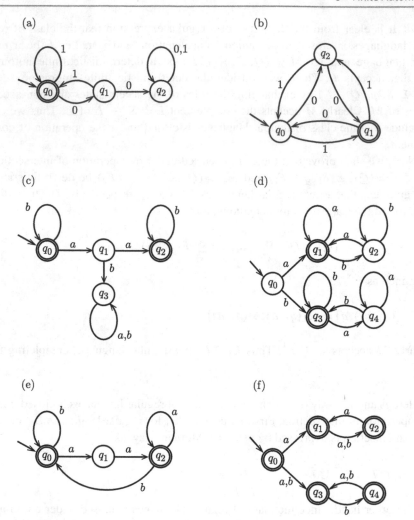

**3.2** Show deterministic finite automata that accept the following languages. Let the alphabet be {0, 1}.

(1)   {$w$ | $w$ contains 101 as a substring}
(2)*  {$w$ | $w$ contains a 1 in the third position from the end}
(3)   {$w$ | every 1 in $w$ is preceded and is followed by at least one 0}

**3.3** Show state diagrams of simple nondeterministic finite automata that accept the languages of (1) and that of (2) of Problem 3.2.

**3.4** Transform the NFA in Problem 3.3 that accepts the language of (2) in Problem 3.2 to an equivalent DFA.

**3.5**
(1) Informally describe the language accepted by the NFA shown below.
(2) Transform the NFA to an equivalent DFA.

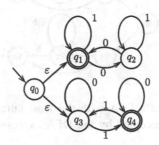

**3.6** Let $\{0, 1\}$ be the alphabet. Give regular expressions that represent the following languages.
(1)  The language consisting of strings that contain 101 as a substring.
(2)  The language consisting of strings such that every consecutive 1 is followed by 0's of length at least two.
(3)  The language consisting of strings that begin with 00 and end with 11.
(4)  The language consisting of strings that begin with 01 and end with 01.
(5)  The language consisting of strings whose lengths are multiples of 2 or 3.
(6)* The language consisting of strings in which 101 never appears as a substring.

**3.7** Let $\{0, 1\}$ be the alphabet. Answer the following questions.
(1)*  Let a string of even length be cut into pieces so that the length of each piece is 2. Then, each piece is 00, 11, 01, or 10. Give the condition in terms of the numbers of 00's, 11's, 01's, and 10's that is equivalent to the condition that both the number of 0's and the number of 1's in a string are even.
(2)** Noticing that the two conditions in (1) are equivalent with each other, give a regular expression that represents the language consisting of strings that contain an even number of 0's and that of 1's.
(3)*  Give a regular expression by transforming the automaton shown in Example 3.8 in the way described in Sect. 3.3 and compare it with the regular expression obtained in (2) above.

**3.8**** Verify directly, but differently from Example 3.13, that for a string $w$ in $\{0, 1, 2\}^*$, the condition that "$w$ is expressed as $w_1 2 w_2 2 \cdots 2 w_m$, where $w_i \in 0^* 1^* 2^*$ for each $1 \le i \le m$" is equivalent to the condition that "10 does not appear as a substring in $w$", where, when $m = 1$, $w_1 2 w_2 2 \cdots 2 w_m$ becomes $w_1$.

**3.9**** Show that if a DFA accepts a string whose length is greater than or equal to the number of states, then the number of strings accepted by the automaton is infinite.

**3.10** Give a state diagram of a DFA that is equivalent to the NFA given below.

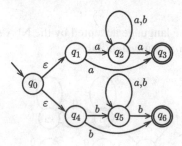

**3.11** Give a state diagram of a DFA with the least number of states that accepts the same language as that accepted by the DFA expressed by the following state diagram.

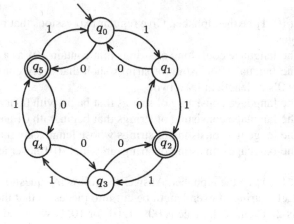

# Context-Free Languages

# 4

Among what are called languages, there are artificial languages such as programming languages and natural languages such as English, German, French, and Japanese, etc. The languages that we study in this book belong to the former group, and in this chapter we study context-free languages in the former group. A regular language studied in the previous chapter is defined to be the language that a finite automaton *accepts*. In this chapter, we introduce a *context-free grammar*, and define a *context-free language* to be the one that a context-free grammar *generates*. The class of context-free languages includes the class of regular languages and, furthermore, contains nonregular languages like $\{0^n 1^n \mid n \geq 0\}$. So the generating power of a context-free grammar exceeds the accepting power of a finite automaton. Corresponding to the pumping lemma for regular languages, we introduce a pumping lemma for context-free languages, which shows that a language does not belong to the class of context-free languages. Context-free grammar are also used to describe practical programming languages.

## 4.1 Context-Free Grammar

Sentences of natural languages can be thought of as sequences of words that are placed according to certain rules. No matter whether it's English or Japanese, there exist rules, characteristic to each language, that specify how words are placed to form a sentence. The linguist Noam Chomsky proposed to express the rules as rewriting rules. For example, we can consider the following rewriting rules that describe a couple of English sentences

⟨sentence⟩ → ⟨noun phrase⟩ ⟨verb phrase⟩,

⟨verb phrase⟩ → ⟨verb⟩ ⟨noun phrase⟩,

⟨noun phrase⟩ → ⟨article⟩ ⟨noun⟩,

⟨article⟩ → a,　　⟨article⟩ → the,

A. Maruoka, *Concise Guide to Computation Theory*,
DOI 10.1007/978-0-85729-535-4_4, © Springer-Verlag London Limited 2011

**Fig. 4.1** Example of a parse
tree

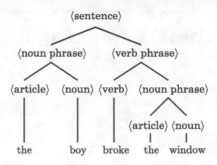

$\langle\text{noun}\rangle \to \text{boy}, \qquad \langle\text{noun}\rangle \to \text{girl},$

$\langle\text{noun}\rangle \to \text{window}, \qquad \langle\text{noun}\rangle \to \text{bed},$

$\langle\text{verb}\rangle \to \text{broke}, \qquad \langle\text{verb}\rangle \to \text{made}.$

In general, $x \to y$, which is called a *substitution rule* or simply a *rule*, means
that $x$ can be replaced by $y$. The collection of substitution rules produces sentences:
starting at $\langle\text{sentence}\rangle$, we apply the substitution rules repeatedly until we obtain a
sentence. For example, the sentence "the boy broke the window" is generated as
follows:

$\langle\text{sentence}\rangle \Rightarrow \langle\text{noun phrase}\rangle\langle\text{verb phrase}\rangle$

$\qquad \Rightarrow \langle\text{article}\rangle\langle\text{noun}\rangle \ \langle\text{verb phrase}\rangle$

$\qquad \Rightarrow \text{the} \ \langle\text{noun}\rangle \ \langle\text{verb phrase}\rangle$

$\qquad \Rightarrow \text{the boy} \ \langle\text{verb phrase}\rangle$

$\qquad \Rightarrow \text{the boy} \ \langle\text{verb}\rangle \ \langle\text{noun phrase}\rangle$

$\qquad \Rightarrow \text{the boy broke} \ \langle\text{noun phrase}\rangle$

$\qquad \Rightarrow \text{the boy broke} \ \langle\text{article}\rangle \ \langle\text{noun}\rangle$

$\qquad \Rightarrow \text{the boy broke the} \ \langle\text{noun}\rangle$

$\qquad \Rightarrow \text{the boy broke the window}.$

Such a sequence of substitutions is called a derivation. Figure 4.1 depicts how
the substitutions are performed. This kind of graph is called a *parse tree*.

The grammatical terms written inside brackets in the substitution rules are taken
to be individual symbols, which will usually be represented by capital letters. In
the course of applying substitution rules, any rule can be applied provided that the
symbol on the left-hand side of the rule appears in the resulting sequence. Context
is defined to be a sequence fragment that surrounds the symbol. The grammar consisting of these types of substitution rules is called a context-free grammar because
we are allowed to apply substitution rules independently of context. A context-free

grammar is compared to a context-sensitive grammar, which we will discuss shortly. In the case of a context-sensitive grammar, we are allowed to apply substitution rules depending on context. For example, let's consider $A \rightarrow w$ in a context-free grammar and $AB \rightarrow wB$ in a context-sensitive grammar. Clearly, in the case of the context-free grammar we are always allowed to replace symbol $A$ with string $w$, whereas in the case of the context-sensitive grammar we are allowed to do so only when symbol $B$ appears immediately to the right of symbol $A$. As this example shows, for a context-free grammar, the number of symbols on the left-hand side of each substitution rule is restricted to be one, while, for a context-sensitive grammar, that number is allowed to be arbitrary.

Any grammar consisting of substitution rules of such context-free type cannot seem to generate all of the correct English sentences, no matter how we choose the substitution rules. Furthermore, for correct English sentences, we may need to exclude meaningless sentences such as "the window broke the boy," which can be generated by using the substitution rules described above. Among programming languages, however, there is a language such that a context-free grammar specified appropriately generates all the correct programs of the programming language. Such a context-free grammar can be used not only to define a programming language concisely but also to design a compiler for the programming language. Furthermore, although context-free grammars do not seem to have enough control mechanisms on how a string is replaced by substitution, there is a specified context-free grammar that can approximate a natural language in the sense that the grammar generates a fragment of the natural language. Once such context-free grammars are specified, it turns out to be easy to handle natural languages by means of context-free grammars, which therefore become a basis of processing the natural language.

We are ready to describe the formal definition of a context-free grammar. To define a context-free grammar we specify an alphabet consisting of symbols that appear in the strings of substitution rules. For the example above, each string surrounded by brackets "⟨" and "⟩", such as ⟨sentence⟩ and ⟨noun phrase⟩, is considered as a symbol. The remaining symbols in the example are a, t, h, e, b, o, y, etc. These symbols make up the alphabet. One may think of an invisible blank as a symbol, denoted by " ", so that the words do not run together. There are two types of symbols: one, called a *nonterminal symbol*, that is eventually replaced by another string; and the other, called a *terminal symbol*, that is never replaced. In the example above, the strings surrounded by brackets are nonterminal symbols, whereas a, t, h, e, b, o, y, etc. and " " are terminal symbols. A substitution rule is in general expressed as $A \rightarrow x$, where $A$ is a nonterminal symbol and $x$ is a string of terminal and nonterminal symbols. We specify a symbol of the nonterminal symbols as the *start symbol*, from which a derivation starts. Starting at the start symbol, we apply substitution repeatedly until the string obtained this way has no nonterminal symbol. The context-free grammar *generates* such a string. In the course of a derivation, any rule is allowed to be applied to a string when the symbol on the left-hand side of the rule appears in the string. In this sense there exists nondeterministic choice in each of the steps of a derivation.

**Fig. 4.2** Parse tree for string
000111

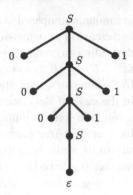

*Example 4.1* Let substitution rules be given by

$$S \rightarrow 0S1, \qquad S \rightarrow \varepsilon.$$

We write several rules with the same left-hand side symbol into a single line: in our case the rules $S \rightarrow 0S1$, $S \rightarrow \varepsilon$ are written as $S \rightarrow 0S1 \mid \varepsilon$. String 000111, for example, is generated as follows:

$$S \Rightarrow 0S1 \Rightarrow 00S11 \Rightarrow 000S111 \Rightarrow 000111.$$

Note that, by convention, when we write a substitution rule we use $\rightarrow$, whereas when we describe a derivation we use $\Rightarrow$. Figure 4.2 shows the corresponding parse tree. In this case, the only nonterminal symbol is $S$, the terminal symbols are 0 and 1, and the language generated is written $\{0^n1^n \mid n \geq 0\}$. The root of the parse tree, which is the node at the top of the parse tree, is assigned the start symbol.

A context-free grammar is formally defined as follows.

**Definition 4.2** A *context-free grammar* (CFG) is a 4-tuple $(V, \Sigma, P, S)$, where
(1)  $V$ is a finite set of *nonterminals* (also called *variables*).
(2)  $\Sigma$ is a finite set of *terminals*, called the alphabet, where $\Sigma$ is disjoint from $V$.
(3)  $P$ is a finite set of *substitution rules*, each of which has the form of

$$A \rightarrow x,$$

for $A \in V$ and $x \in (V \cup \Sigma)^*$. A substitution rule is also simply called a *rule*.
(4)  $S \in V$ is the *start symbol*.

For strings $x$ and $y$, $x \Rightarrow y$ means that string $x$ can be replaced by string $y$: more precisely, there exists a rule $A \rightarrow w$ and $u, v \in (V \cup \Sigma)^*$ such that

$$x = uAv, \qquad y = uwv.$$

Furthermore, the transitive closure of the relation $\Rightarrow$ is denoted by $\overset{*}{\Rightarrow}$. That is, $x \overset{*}{\Rightarrow} y$ is defined as follows:

$$x \overset{*}{\Rightarrow} y \quad \Leftrightarrow \quad \begin{array}{ll} (1) & x = y, \text{ or} \\ (2) & \text{there exist } k \geq 1 \text{ and } x_1, \ldots, x_{k-1} \in (V \cup \Sigma)^* \text{ such that} \\ & x_0 \Rightarrow x_1 \Rightarrow \cdots \Rightarrow x_k, \\ & x_0 = x, \ x_k = y. \end{array}$$

If $x \overset{*}{\Rightarrow} y$, it is said that $y$ is *derived* from $x$. And $x_0 \Rightarrow x_1 \Rightarrow \cdots \Rightarrow x_k$ is called a *derivation* of $x_k$ from $x_0$. Grammar $G$ *generates* a string $w$ when $w$ consists of terminal symbols and is derived from the start symbol. Grammar $G$ *generates* language $L$ when $L$ consists of all the strings that are generated by $G$. The language that $G$ generates is denoted by $L(G)$. That is,

$$L(G) = \left\{ w \in \Sigma^* \mid S \overset{*}{\Rightarrow} w \right\}.$$

As in the case of a finite automaton, when we talk about what is generated by a context-free grammar, we have to make clear whether it is a string or a language. But when it is clear from the context, we do not mention that it is a string or a language. A string in $(V \cup \Sigma)^*$ that is derived from the start symbol is called a sentential form. That is, a *sentential form* is a string $u \in (V \cup \Sigma)^*$ such that $S \overset{*}{\Rightarrow} u$. In particular, when we want to make explicit the context-free grammar that we assume, $\Rightarrow$ and $\overset{*}{\Rightarrow}$ are denoted by $\Rightarrow_G$ and $\overset{*}{\Rightarrow}_G$, respectively. If two grammars $G_1$ and $G_2$ generate the same language, i.e., $L(G_1) = L(G_2)$, then $G_1$ and $G_2$ are *equivalent*. As a finite automaton is specified by a state diagram, a context-free grammar is specified by listing substitution rules. The set of nonterminals is specified by the set of symbols appearing on the left-hand side of the rules, and the set of terminals is specified by the remaining symbols that appear in the rules. By convention, the start symbol is the nonterminal that appears on the left-hand side of the first rule in the list.

*Example 4.3* The context-free grammar $G = (\{S\}, \{(,)\}, \{S \rightarrow (S) \mid SS \mid \varepsilon\}, S)$ generates well-nested strings of parentheses. Figure 4.3 shows the parse tree for the string $((()) ())$. Note that, if we want to precisely define the well-nested strings of parentheses, we need something like substitution rules. In fact, the grammar $G$ can be interpreted to give an inductive definition of the well-nested strings of parentheses:
(1) $\varepsilon$ is a correct string of parentheses.
(2) If $S$ and $S'$ are correct strings of parentheses, $(S)$ and $SS'$ are correct strings of parentheses.

A *leftmost derivation* is one such that a substitution rule is applied to the leftmost nonterminal in every sentential form of the derivation. Figure 4.3 gives a parse tree for string $((()) ())$ which illustrates the leftmost derivation by attaching natural

**Fig. 4.3** Parse tree for
$(((\,)\,)(\,))$

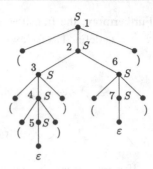

numbers with nonterminals in the order that substitutions are applied. In this case, the leftmost derivation is given as follows:

$$S \Rightarrow (S) \Rightarrow (SS) \Rightarrow ((S)S)$$

$$\Rightarrow (((S))S) \Rightarrow (((\,))S)$$

$$\Rightarrow (((\,))(S)) \Rightarrow (((\,))(\,)).$$

In general, a parse tree has different derivations. But if we restrict ourselves to leftmost derivations, there exists only one derivation corresponding to a parse tree.

*Example 4.4* Consider the context-free grammar $G = (\Gamma, \Sigma, P, E)$ that generates arithmetic expressions, where

$$\Gamma = \{E, T, F\},$$
$$\Sigma = \{x, +, \times, (\,,)\,\},$$
$$P = \{E \rightarrow E + T \mid T, \ T \rightarrow T \times F \mid F, \ F \rightarrow (E) \mid x\}.$$

Nonterminals $E$, $T$, and $F$ correspond to arithmetic expressions, terms, and factors, respectively. Figures 4.4 and 4.5 show parse trees for $x + x \times x$ and $(x + x) \times x$, respectively.

Typically, an arithmetic expression is a sum of terms, and a term is a product of factors. For example, the arithmetic expression $x + x \times x \times x$ has terms $x$ and $x \times x \times x$, and these terms have a factor $x$. Furthermore, since there is rule $F \rightarrow (E)$, expression $E$ surrounded by parentheses $(\,,)$ turns out to be factor $F$. For example, surrounding $x + xx$ by parentheses gives $(x + xx)$ which is considered a factor. Replacing the last $x$ in $x + xxx$ with $(x + xx)$ gives $x + xx(x + xx)$, which is also an expression.

In general, rule $A \rightarrow w$ can be interpreted as follows: $A$ is defined to be $w$, where $A$ is a nonterminal and $w$ is a string of nonterminal and terminal symbols. Therefore, the rules of this example say that an expression is represented in terms of a term, a term is represented in terms of a factor, and finally a factor is represented in terms of an expression. The referring relation is represented in Fig. 4.6, in which there is

**Fig. 4.4** Parse tree for
$x + x \times x$

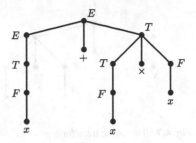

**Fig. 4.5** Parse tree for
$(x + x) \times x$

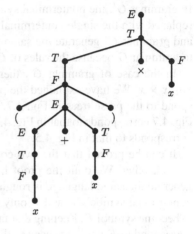

**Fig. 4.6** Referring relation
among $E$, $T$, and $F$

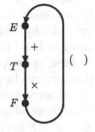

a circular referring. Note that the definition is valid because an expression refers to a smaller expression, thereby avoiding referring an infinite number of times.

*Example 4.5* We consider the context-free grammar $G' = (\Gamma', \Sigma, P', E)$ that is obtained by slightly modifying the grammar in Example 4.4, where

$$\Gamma' = \{E\}, \qquad \Sigma = \{x, +, \times, (, )\},$$
$$P' = \{E \to E + E \mid E \times E \mid (E) \mid x\}.$$

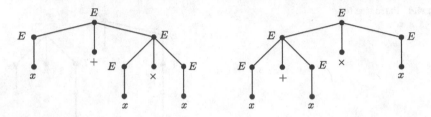

**Fig. 4.7** Two parse trees for $x + x \times x$

In grammar $G'$, the nonterminal symbols $E$, $T$, and $F$ in the previous example are replaced with the single nonterminal symbol $E$. As will be shown later, grammar $G$ and grammar $G'$ generate the same language. Is grammar $G'$ preferable compared to grammar $G$ because the rules of $G'$ are simpler?

In the case of grammar $G'$, there are two parse trees, shown in Fig. 4.7, for $x + x \times x$. We have explained the parse trees, based on grammar $G$, which correspond to the parse trees in Fig. 4.7. That is, the parse tree on the left-hand side of Fig. 4.7 corresponds to that in Fig. 4.4, whereas the parse tree on the right-hand side corresponds to that in Fig. 4.5.

It can be proved that the two context-free grammars $G$ and $G'$ are equivalent to each other. We omit the proof here, but the idea of why these two grammars generate the same language is roughly explained as follows. The point is that in $G$ nonterminal symbols $E$ and $T$ only appear as transitional symbols, which turn out to become symbol $F$. Keeping that fact in mind, if we observe that $E$, $T$, and $F$ in $G$ correspond to $E$ in $G'$, we can see that $E \to E + T$, $T \to T \times F$, and $F \to (E)$ in $G$ and $E \to E + E$, $E \to E \times E$, and $E \to (E)$ in $G'$ are equivalent collectively to each other. Finally, note that in the case of $G'$ there exists a string that can be generated by different parse trees as shown in Fig. 4.7, whereas in the case of $G$ any string in $L(G)$ can be generated by a unique parse tree. Roughly speaking, this is because, in each stage of generating terms and their factors in $G$, the structure of the corresponding parse tree is such that first the terms are generated and then their factors are generated. On the other hand, as we can see in Fig. 4.7, that structure of the corresponding parse tree in $G'$ can be formed by applying rules $E \to E + E$ and $E \to E \times E$ interchangeably.

Suppose that we are given a context-free grammar. As with grammar $G'$, if there exists a string that is generated by at least two different parse trees, then the grammar is said to be an *ambiguous* grammar. On the other hand, if any string is generated by at most one parse tree, then the grammar is said to be an *unambiguous* grammar.

As an application of context-free grammar, we discuss describing programming languages in terms of the grammars. A context-free grammar is said to generate a programming language if the grammar generates exactly all the strings that represent the programs written by the programming language. The context-free grammar generating a program language is required to be such that a parse tree of a program corresponds to how the program is executed. As an example of this, take the parse tree in Fig. 4.4 which says that operation × is executed first, and then operation + is executed. Furthermore, suppose we are given a programming language and a context-free grammar that generates the language. If we have an algorithm that, given a program as input, produces a parse tree as output, then we can use the algorithm to produce the codes that represent how the program is executed. In that case, the structure of the parse tree should reflect the order in which the codes are executed to do what the original program intends. Thus the context-free grammar that generates a programming language should be unambiguous so that a parse tree based on the grammar reflects on how a program is executed, which is supposed to be uniquely determined by the program.

## 4.2 Generating and Accepting

A finite automaton and a context-free grammar specify a language, but the manner in which they decide whether or not each string belongs to the language is different. In the case of an automaton, when a string is given as input, the automaton decides whether it is accepted or rejected. In the case of a grammar, choosing each rule to be applied, the grammar generates a string that belongs to the language specified by the grammar. Thus we can say that, in order to specify a language, an automaton behaves passively, while a grammar behaves actively by itself.

Before explaining more about context-free languages, we briefly discuss various models of automata and grammars other than those we have studied so far. As models of automata, there are a finite automaton, a pushdown automaton, a linear bounded automaton, and a Turing machine; as models of grammars, there are a regular grammar, a context-free grammar, a context-sensitive grammar, and a phrase structure grammar. In what follows, ↔ means that models are equivalent in expressive ability. For example, the power with which pushdown automata accept languages is equivalent to the power with which context-free grammars generate languages. More specifically, this means that, given a language accepted by a pushdown automaton, we can construct a context-free grammar that accepts the language, and *vice versa*, which will be verified in Chap. 5. The lower an item is in the following list, the more power it has to accept or generate languages. In other words, the smaller the type number, the larger the corresponding class of languages. The four classes of languages listed below constitute the *Chomsky hierarchy*,

finite automaton ↔ regular grammar   (type 3),

pushdown automaton ↔ context-free grammar   (type 2),

linear bounded automaton ↔ context-sensitive grammar   (type 1),

Turing machine ↔ phrase structure grammar   (type 0).

In this section, after a regular grammar is defined, it will be verified that finite automata and regular grammars are equivalent.

**Fig. 4.8** Finite automaton
that accepts strings that
contain even number of 1's

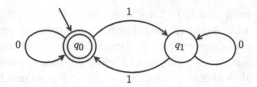

   In this book, we omit the details of what a *linear bounded automaton* is, except
to say that it is a somehow restricted nondeterministic single-tape Turing machine,
which we shall study in Chap. 6: the machine is never allowed to use cells outside
of the region of cells on the tape where the input is first placed. We briefly explain
a context-sensitive grammar and a phrase structure grammar as follows. First of
all, a context-free grammar is generalized to obtain a context-sensitive grammar,
which is in turn generalized to obtain a phrase structure grammar. Suppose that $V$,
$\Sigma$, and $S$ are as in the case of a context-free grammar, namely, as in Definition 4.2.
A context-sensitive grammar is such that the substitution rules take the forms $S \to \varepsilon$
or $\alpha_1 A \alpha_2 \to \alpha_1 \beta \alpha_2$, where $\alpha_1$, $\alpha_2$, and $\beta$ are in $(V \cup \Sigma)^*$, and $A$ is in $V$, $\beta \neq \varepsilon$, and
$S$ does not appear in $\beta$. You can interpret this form as $A$ is replaced by $\beta$ provided
that $A$ appears in the context of $\alpha_1$ and $\alpha_2$. A *phrase structure grammar* is such
that the substitution rules take the form $\alpha \to \beta$, where $\alpha$ and $\beta$ are in $(V \cup \Sigma)^*$, and
$\alpha \neq \varepsilon$. You can easily generalize the derivation, denoted by $\Rightarrow$, for these generalized
grammars: for any $x, y \in (V \cup \Sigma)^*$ let $xuy \Rightarrow xvy$ provided that $u \to v$ is a rule.
The corresponding languages generated by these grammars are defined in a similar
way to the case of a context-free grammar.

*Example 4.6* We shall show a regular grammar that is equivalent to the finite automaton given by the state diagram in Fig. 4.8 (the same as that in Fig. 3.4). The
formal definition of a regular grammar will be given shortly.
   The regular grammar $G$ is given as $G = (\{A, B\}, \{0, 1\}, P, A)$, where the set $P$
of substitution rules is given as follows:

$$P = \{A \to 0A, A \to 1B, B \to 0B, B \to 1A, A \to \varepsilon\}.$$

Then it can be easily seen that $G$ is equivalent to the state diagram in Fig. 4.8. For
example, a sequence of transitions by which 101 is accepted is given as

$$q_0 \xrightarrow{1} q_1 \xrightarrow{0} q_1 \xrightarrow{1} q_0,$$

which corresponds to the derivation

$$A \Rightarrow 1B \Rightarrow 10B \Rightarrow 101A \Rightarrow 101.$$

From the example, it is seen that although transition of an automaton and derivation
of a regular grammar are different from each other in form, the regular grammar
simulates the behavior of the finite automaton in essentially the same way. The

point of the simulation is that the grammar keeps track of the present state of the automaton in terms of the corresponding nonterminal. Suppose that states $q_0$ and $q_1$ correspond to nonterminals $A$ and $B$, respectively. Then for transition $q_0 \overset{a}{\to} q_1$ we prepare rule $A \to aB$ so that, when transition $q_0 \overset{a}{\to} q_1$ occurs, derivation proceeds as

$$wA \Rightarrow waB,$$

where $w$ is the string that is fed to the automaton so far. In addition, since state $q_0$ associated with nonterminal $A$ is the accept state, we add rule $A \to \varepsilon$ so that we have $wA \Rightarrow w$ for any string $w$.

Next, we give the formal definition of a regular grammar which is obtained from a context-free grammar by imposing a further restriction on the substitution rules.

**Definition 4.7** A regular grammar is a 4-tuple $(V, \Sigma, P, S)$, where
(1) $V$ is a finite set of nonterminals.
(2) $\Sigma$ is a finite set of terminals, called the alphabet, where $\Sigma$ is disjoint from $V$.
(3) $P$ is a finite set of substitution rules, each of which takes one of the following forms:

$$A \to aB, \qquad A \to a, \qquad A \to \varepsilon,$$

where $A, B \in V$ and $a \in \Sigma$.
(4) $S \in V$ is the start symbol.

Next, we describe how to construct an equivalent regular grammar from a deterministic finite automaton as follows:

---

(1) Given a finite automaton, make nonterminals so that a nonterminal corresponds to a state of the automaton in a one-to-one fashion.
(2) Let the nonterminal symbol corresponding to the start state be the start symbol.
(3) Let $A_q$ and $A_p$ be nonterminals corresponding to states $q$ and $p$, respectively. When $\delta(q, a) = p$, add $A_q \to aA_p$ as a rule.
(4) When $A$ corresponds to an accept state, add $A \to \varepsilon$ as a rule.

---

The regular grammar constructed from a finite automaton as described above generates the language that the finite automaton accepts. This is because for each transition there corresponds a rule so that the regular grammar can simulate the finite automaton.

Conversely, an equivalent nondeterministic finite automaton can be constructed from a regular grammar as follows:

(1)  Given a regular grammar, let each nonterminal correspond to a state denoted
     by the same symbol as the nonterminal. In addition to those states, add a
     state denoted $Q_\varepsilon$.
(2)  Let the state corresponding to the start symbol be the start state.
(3)  When $A \to aB$ is a rule, let

$$B \in \delta(A, a),$$

while when $A \to a$ is a rule, let

$$Q_\varepsilon \in \delta(A, a).$$

(4)  When $A \to \varepsilon$ is a rule, let $A$ be an accept state. In addition to those, let $Q_\varepsilon$
     be the accept state.

As in the case of simulating a finite automaton by the regular grammar, it is easy
to see that the nondeterministic finite automaton obtained above accepts the same
language as that generated by the regular grammar.

From what we have discussed so far, we have the following theorem.

**Theorem 4.8**  *The following two conditions about language L are equivalent.*
(1)  *There is a finite automaton that accepts L.*
(2)  *There is a regular grammar that generates L.*

## 4.3    Chomsky Normal Form

If any substitution rule of a context-free grammar takes the form of either $A \to BC$
or $A \to a$, it is said that the grammar is in Chomsky normal form, where $A$, $B$,
and $C$ are nonterminals and $a$ is a terminal. As shown in the following, in general,
a context-free grammar can be transformed to a grammar in Chomsky normal form
without changing the language to be generated. So, it can be assumed that a context-
free language is generated by rules of the form $A \to BC$ or $A \to a$. This fact will
be used in Sect. 4.5 when we design an algorithm which, provided that a context-
free grammar is given, decides whether or not a string is generated by the grammar.
This fact will also be used when we prove in Sect. 5.2 that context-free grammars
and pushdown automata are equivalent in the power of specifying languages, which
we mentioned when we explained Chomsky hierarchy in the previous section. The
definition of a pushdown automaton will be given in the next chapter.

**Definition 4.9**  A context-free grammar is in *Chomsky normal form* if every rule is
of either of the following forms:
(1)  $A \to BC$
(2)  $A \to a$

where $a$ is a terminal, $A$ is a nonterminal, and $B$ and $C$ are nonterminals other than the start symbol. Moreover, the $\varepsilon$ rule $S \to \varepsilon$ is permitted only for the start symbol $S$.

A rule of type $A \to \varepsilon$ is called the $\varepsilon$ *rule* and a rule of type $A \to B$ is called a *renaming rule*. A rule $A \to u_1 u_2 \cdots u_k$ in which the length of the right-hand side $k$ is more than or equal to 3 is called a *long rule*. Next, it will be shown that any context-free grammar can be transformed successively into Chomsky normal form without changing the language to be generated. This transformation is done by removing the three types of rules mentioned above.

We delete the first two types of rules and then delete the third type. First we shall explain the idea for deleting the three types of rules. The basic idea to delete rules is the same for $\varepsilon$ rules and renaming rules, so we discuss only the first type, namely, the $\varepsilon$ rule. Since in Chomsky normal form $S \to \varepsilon$ is permitted for the start symbol $S$, the $\varepsilon$ rules we must delete are the ones with nonterminals on the left-hand side other than the start symbol. Clearly such an $\varepsilon$ rule is applied after some other rule. For example, suppose that we have rules $A \to BCDCE$ and $C \to \varepsilon$. Then, as far as rule $A \to BCDCE$ is concerned, we can delete $C \to \varepsilon$ if we add $A \to BDE \mid BCDE \mid BDCE$ and leave $A \to BCDCE$ as it is. Of course, to complete the deletion of $C \to \varepsilon$, we must do this for all the rules that have symbol $C$ on their right-hand side. Similarly, the renaming rules can be deleted by adding this type of "short-cut" rule appropriately. On the other hand, the idea to delete a long rule is just to generate a long sequence by successively applying rules of the type $A \to BC$ appropriately chosen. For example, we can replace $A \to BCDE$ with $A \to BX$, $X \to CY$, and $Y \to DE$. Of course, we need to consider the case where the right-hand side of a long rule consists of both nonterminals and terminals, which will be shown in the proof of the next theorem.

**Theorem 4.10** *Any context-free language is generated by a context-free grammar in Chomsky normal form.*

*Proof* Let $G = (\Gamma, \Sigma, P, S)$ be a context-free grammar that generates a context-free language $L$. We transform $G$ to an equivalent context-free grammar in Chomsky normal form by deleting the $\varepsilon$ rules, the renaming rules, and finally the long rules from the rules of $G$. First, we add the rule $S_0 \to S$ and consider the context-free grammar $G' = (\Gamma \cup \{S_0\}, \Sigma, P \cup \{S_0 \to S\}, S_0)$ with the new start symbol $S_0$. Observe that this change does not change the language to be generated and that the start symbol $S_0$ never appears on the right-hand side of any rule.

*Deletion of $\varepsilon$ rules* For the case that there is a rule of type $A \to \varepsilon$ for nonterminal $A$ other than $S_0$, we delete it and add instead all the rules obtained by deleting $A$, in all the possible ways, from the right-hand side of every rule that contains $A$ on its right-hand side. This is called deletion of $\varepsilon$ rules. Note that since $S_0$ never appears on any right-hand side of rules by definition, the deletion of $S_0 \to \varepsilon$ will never be applied. Once the deletion of $A \to \varepsilon$ is applied for all the rules in the grammar, the $\varepsilon$ rule $A \to \varepsilon$ will never be taken up again in the later deletion process.

Let us consider an example of the deletion of $A \to \varepsilon$ that is applied for rule $D \to u_1 A u_2 A u_3$, where $A$ does not appear in any of the strings $u_1$, $u_2$, and $u_3$ in $(\Gamma \cup \Sigma)^*$. Then the rules added in accordance with the deletion of $A \to \varepsilon$ are $D \to u_1 A u_2 A u_3 \mid u_1 u_2 A u_3 \mid u_1 A u_2 u_3 \mid u_1 u_2 u_3$. In all of the deletion process the deletion is applied with respect to every $\varepsilon$ rule once. In the example above, if $D \to u_1 u_2 u_3$ becomes an $\varepsilon$ rule, that is, $u_1 u_2 u_3 = \varepsilon$, then $D \to \varepsilon$ is newly added to the grammar only if the deletion of $D \to \varepsilon$ was not applied so far. After the deletion of all the $\varepsilon$ rules for all the rules is done, the only $\varepsilon$ rule that may possibly be left is $S_0 \to \varepsilon$.

*Deletion of renaming rules* For each renaming rule $A \to B$, the deletion of $A \to B$ is applied for all the rules that take the form of $B \to w$, where $w \in (\Gamma \cup \Sigma)^*$. By the deletion, we eliminate the rule $A \to B$ and add all the rules that are expressed as $A \to w$ provided that $A \to w$ was not deleted so far. For each renaming rule this type of deletion and addition will be applied concerning all the remaining rules, including ones that are added while applying other deletions.

*Deletion of long rules* We delete long rule $A \to u_1 u_2 \cdots u_k$ and instead add rules $A \to u_1 B_2$, $B_2 \to u_2 B_3, \ldots, B_{k-2} \to u_{k-2} B_{k-1}$, and $B_{k-1} \to u_{k-1} u_k$, where $k \geq 3$ and $B_2, \ldots, B_{k-1}$ are newly introduced nonterminals. Then the transformation by $A \to u_1 u_2 \cdots u_k$ can be realized by the derivation $A \Rightarrow u_1 B_2 \Rightarrow u_1 u_2 B_3 \Rightarrow \cdots \Rightarrow u_1 \cdots u_{k-3} B_{k-2} \Rightarrow u_1 u_2 \cdots u_k$, where each $u_i$ is a nonterminal or a terminal. This deletion is applied to all the long rules. The nonterminals newly introduced this way are made to be different from each other so that the derivation that does not correspond to the original long rules never happens.

The length of the right-hand side of each rule obtained this way is less than or equal to 2. Among those rules there might be left some forbidden rules which we must eliminate. The only possible rule with a right-hand side of length 0 is $S_0 \to \varepsilon$ which is allowed to exist in Chomsky normal form. The rule whose length of the right-hand side is 1 takes the form of $A \to a$, which is also of the type permitted as a Chomsky normal form. Finally, let the rule whose length of the right-hand side is 2 be expressed as $A \to u_1 u_2$, where $u_1$ and $u_2$ are a nonterminal or a terminal. If both $u_1$ and $u_2$ are nonterminals, the rule is of the type permitted in Chomsky normal form. If both $u_1$ and $u_2$ are terminals, then introducing new nonterminals $U_1$ and $U_2$, we delete $A \to u_1 u_2$ and add instead $A \to U_1 U_2$, $U_1 \to u_1$, and $U_2 \to u_2$. If $u_1$ is a terminal and $u_2$ is a nonterminal which is denoted by $U_2$, then introducing new nonterminal $U_1$, we delete $A \to u_1 U_2$ and add instead $A \to U_1 U_2$ and $U_1 \to u_1$. A similar modification is done for the case where $u_1$ is a nonterminal and $u_2$ is a terminal. The nonterminals introduced in this way are made different from each other.

It is clear that the grammar $G'$ obtained by modifying the grammar $G$ as described above is equivalent to $G$ and in Chomsky normal form, completing the proof.                                                                                                      □

*Example 4.11* Let a context-free grammar be denoted by $G = (\{S\}, \{a, b\}, P, S)$, where

$$S \to aS \mid aSbS \mid \varepsilon.$$

We transform the grammar $G$ to an equivalent context-free grammar which is in Chomsky normal form in the way described above.

First, introducing the start symbol $S_0$, we modify the grammar as follows:

$$S_0 \to S,$$

$$S \to aS \mid aSbS \mid \varepsilon.$$

Next, we delete $S \to \varepsilon$ by applying what is described above to the three rules: By applying to $S_0 \to S$ we have $S_0 \to S \mid \varepsilon$. By applying to $S \to aS$ we have $S \to aS \mid a$. By applying to $S \to aSbS$ we have $S \to aSbS \mid abS \mid aSb \mid ab$. Putting them together, we have

$$S_0 \to S \mid \varepsilon,$$

$$S \to aS \mid a \mid aSbS \mid abS \mid aSb \mid ab.$$

Next, to delete $S_0 \to S$, we add $S_0 \to w$ for each rule of the form $S \to w$, thereby obtaining rules as follows:

$$S_0 \to aS \mid a \mid aSbS \mid abS \mid aSb \mid ab \mid \varepsilon,$$

$$S \to aS \mid a \mid aSbS \mid abS \mid aSb \mid ab.$$

Finally we delete long rules. Although the newly introduced symbols are made to be different from each other in the proof of Theorem 4.10, in this example we allow some of them to be assigned the same symbol as long as the modified rules can simulate the original rules. For example, deleting $S \to aSbS$, we add instead $S \to X_a A_2$, $A_2 \to SA_3$, $A_3 \to X_b S$, $X_a \to a$, and $X_b \to b$. In this way we have the equivalent rules in Chomsky normal form as follows:

$$S_0 \to X_a S \mid a \mid X_a A_2 \mid X_a B_2 \mid X_a C_2 \mid X_a X_b \mid \varepsilon,$$

$$S \to X_a S \mid a \mid X_a A_2 \mid X_a B_2 \mid X_a C_2 \mid X_a X_b,$$

$$A_2 \to SA_3, \qquad A_3 \to X_b S,$$

$$B_2 \to X_b S,$$

$$C_2 \to SX_b,$$

$$X_a \to a,$$

$$X_b \to b.$$

By the way, the grammar $G$ generates strings such that the number of $a$'s is more than or equal to the number of $b$'s in any substring starting from the leftmost symbol. For example, $ab, aab, abab \in L(G)$, and $baa, ababba, aabbb \notin L(G)$, and so on. To get an idea of how such a string is generated by the grammar, we introduce a graph, shown in Fig. 4.9, which illustrates the condition that a generated string

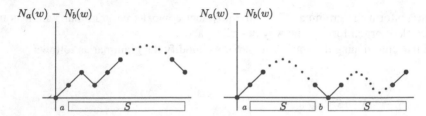

**Fig. 4.9**  Two typical examples of strings generated by first applying rules $S \to aS$ and $S \to aSbS$, respectively

satisfies. In the graph the horizontal axis corresponds to a string and the vertical axis represents $N_a(w) - N_b(w)$, where $N_a(w)$ and $N_b(w)$ are the number of $a$'s and $b$'s that appear in $w$, respectively. The condition that $w \in L(G)$ turns out to be equivalent to "the graph that corresponds to the string $w$ begins at the origin and stays in the region above the horizontal axis." For graphs associated with a string $w$ in $L(G)$ there are two cases: the case where the graph touches the horizontal axis only at the origin, and the case where the graph touches the horizontal axis other than at the origin. These cases tell us which of the rules $S \to aS$ and $S \to aSbS$ is applied first to generate the string. Observe that this is true for any substring of the string that is derived from symbol $S$. For the observation to be valid, we must shift the initial point of such a substring to the origin.

## 4.4  Limitation on Generating Power of Context-Free Grammars

As described in the previous section, a regular grammar can be thought of as a context-free grammar with further restriction on substitution rules. Thus a regular grammar is automatically a context-free grammar. Hence the class of regular languages is included in the class of context-free languages. According to various grammars in the Chomsky hierarchy, corresponding languages are introduced: a language generated by a phrase structure, context-sensitive, context-free, or regular grammar is a *phrase structure*, *context-sensitive*, *context-free*, or *regular language*, respectively. In particular, a phrase structure language is usually called *recursively enumerable*, or *Turing-recognizable*. Since in a Chomsky hierarchy one type of grammar is obtained by imposing restrictions on substitution rules of another, we have an inclusion relation between classes of languages as in the case of regular languages and context-free languages described above. This inclusion relation is shown schematically in Fig. 4.10.

Figure 4.10 shows two languages which belong to one class, but not the other. That is, language $\{a^n b^n \mid n \geq 0\}$ is context-free, but not regular, and language $\{a^n b^n c^n \mid n \geq 0\}$ is context-sensitive, but not context-free. In Sect. 3.4 it was shown that the pumping lemma for regular languages verifies that the language $\{a^n b^n \mid n \geq 0\}$ is not accepted by any finite automaton. In this section a similar,

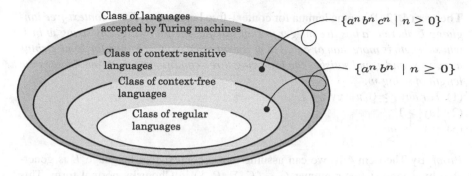

Fig. 4.10 Inclusion relation among classes of languages in Chomsky hierarchy

Fig. 4.11 Repeated structure illustrated by means of parse trees

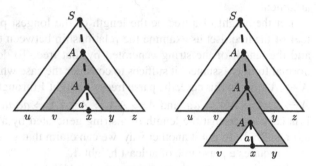

but more complex, pumping lemma for context-free languages is presented, which is used to show that the language $\{a^n b^n c^n \mid n \geq 0\}$ is not generated by any context-free grammar. Roughly speaking, the pumping lemma for context-free languages claims that, if we somehow repeat some parts of a long string in a context-free language, then the resulting string still remains in the language. Figure 4.11 shows how this happens by examining parse trees. As this figure shows, since a long string has a long path from the root with the start symbol $S$ to a leaf with a terminal, say, $a$ in the parse tree, there exists a nonterminal, denoted by $A$ in the figure, that appears at least twice on the path. Thus the string is divided into five parts, denoted by $u$, $v$, $x$, $y$, and $z$ in the figure, so that the string obtained by repeating the second and the fourth parts still remains in the language. In the right-hand part of Fig. 4.11 the second and the fourth parts are repeated twice. Obviously, even if these parts are repeated an arbitrary number of times, the statement above still holds. More precisely, since $A \overset{*}{\Rightarrow} uxy$ holds as is shown in Fig. 4.11, we have

$$S \overset{*}{\Rightarrow} uvAyz \overset{*}{\Rightarrow} uvvxyyz \overset{*}{\Rightarrow} uv^i xy^i z$$

for any $i \geq 0$.

The pumping lemma for context-free languages is stated as Theorem 4.12 below, which claims what is described above.

**Theorem 4.12** (Pumping lemma for context-free languages) *For a context-free language $L$, there is a positive integer $m$ that satisfies the following: any string $w$ in $L$ whose length is more than or equal to $m$ is expressed as $w = uvxyz$ for some strings $u, v, x, y$, and $z$ and satisfies the following three conditions, where $|w|$ denotes the length of string $w$:*

(1) *For any $i \geq 0$, $uv^i xy^i z \in L$*

(2) *$|vy| \geq 1$*

(3) *$|vxy| \leq m$*

*Proof* By Theorem 4.10 we can assume that a context-free language $L$ is generated by a context-free grammar $G = (\Gamma, \Sigma, P, S)$ in Chomsky normal form. This is just used for ease of explanation. For the statement concerning a general context-free grammar, we need to choose an integer appropriately for $m$ in the following argument.

Let the height of a tree be the length of the longest path from the root to any leaf of the tree. Let us examine the relationship between the height of a parse tree and the length of the string generated by that tree. To do so, since the Chomsky normal form is assumed, it suffices to consider the case where the rules are given by $A \to AA \mid a$. In this case, the parse trees of height 1 through 3 are generated through $A \Rightarrow a$, $A \Rightarrow AA \overset{*}{\Rightarrow} aa$, and $A \Rightarrow AA \overset{*}{\Rightarrow} AAAA \overset{*}{\Rightarrow} aaaa$, respectively. In general, it is easy to see that the length of a string generated by a parse tree of height $h$ is at most $2^{h-1}$. To put it another way, we can claim that generating a string of length $2^{h-1}$ requires a parse tree of at least height $h$.

Replacing $h$ in the above statement with $|\Gamma| + 1$, it follows that, if a parse tree generates a string of length at least $2^{|\Gamma|}$, then the height of the parse tree is at least $|\Gamma| + 1$. So set $m = 2^{|\Gamma|}$. Then we can conclude that any parse tree that generates a string of length at least $m$ has a path from the root to a leaf of length at least $|\Gamma| + 1$. Note that such a path of length at least $|\Gamma| + 1$ consists of one node with a terminal at a leaf and at least $|\Gamma| + 1$ nodes with nonterminals. Since $G$ has $|\Gamma|$ nonterminals, some nonterminal, say, $A$ as shown in Fig. 4.12, appears at least twice on this path. For (3) in the theorem to hold, let $A$ be chosen from the lowest $|\Gamma| + 1$ nonterminals on the path. Figure 4.12 is a more detailed version of the left-hand side of Fig. 4.11. From Fig. 4.12 we can see that the region with hatched gray in Fig. 4.11, which corresponds to strings $v$ and $y$, connects to the remaining parts of the parse tree through the nodes assigned with $A$. So the region can be patched as shown in the right-hand side of Fig. 4.11.

We are now ready to verify the three conditions of the theorem.

Condition (1): For an arbitrary number $i$ we can construct a parse tree that generates $uv^i xy^i z$ by patching the region corresponding to $v$ and $y$ $i$ times. For $i = 0$, we only have to eliminate the region.

Condition (2): From the construction of the proof of Theorem 4.12 the only possible $\varepsilon$ rule of $G$ is $S \to \varepsilon$ and the start symbol $S$ does not appear on any right-hand side of any rule; hence $S$ never appears in the tree except for the root. Let $A \to BC$ be the rule that is applied to the upper $A$ in the left-hand side tree of Fig. 4.11. Then clearly at least one of $v$ and $y$ contains substring $s$ such that $X \overset{*}{\Rightarrow} s$ for $X \in \{B, C\}$

**Fig. 4.12** Path from root $S$ to leaf $a$ in parse tree

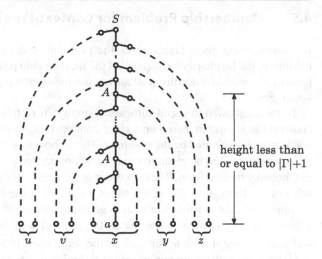

height less than or equal to $|\Gamma|+1$

and $s \in \Sigma^*$ such that $|s| \geq 1$. (In the case of Fig. 4.12, $y$ is such string.) This is because $G$ is in Chomsky normal form. Notice that $B$ or $C$, but not both, might be the lower $A$. This completes the proof of condition (2).

Condition (3): Since $A$ is chosen from the lowest $|\Gamma|+1$ nonterminals on the path, the height of the upper $A$ in Fig. 4.12 is at most $|\Gamma|+1$. Hence the length of the string $vxy$ that the upper $A$ generates is at most $2^{(|\Gamma|+1)-1} = 2^{|\Gamma|} = m$.

$\square$

*Example 4.13* Let language $L$ be

$$L = \left\{ a^n b^n c^n \mid n \geq 0 \right\}.$$

We shall show that the language $L$ is not a context-free language using Theorem 4.12. To do so, suppose that $L$ is a context-free language. We show that this leads to a contradiction. Let $a^n b^n c^n$ be a string whose length is more than or equal to $m$, where $m$ is the integer specified in the theorem. Then, from Theorem 4.12, the following hold.

(1) $a^n b^n c^n$ is expressed as $uvxyz$.

(2) $uv^2xy^2z$ belongs to $L$. That is, for some $l$, $uv^2xy^2z$ is expressed as $a^l b^l c^l$.

Since we have $|vy| \geq 1$ from condition (2) of Theorem 4.12, at least one of $v$ and $y$ is not the empty string. So we assume without loss of generality that string $v$ is not the empty string. In any string in $L$ there exist exactly two points where symbol changes from $a$ to $b$ and from $b$ to $c$ occur. So the substring $v$ cannot contain $ab$ or $bc$ with such a point because, if it does contain it, the number of such points in $wv^2xy^2z \in L$ increases, which is a contradiction. Hence $v$ is contained entirely in one of $a^n$, $b^n$, and $c^n$. In a similar way we can conclude that $y$ is contained entirely in one of $a^n$, $b^n$, and $c^n$. So when string $uvxyz$, expressed as $a^n b^n c^n$, is replaced with the longer string $uv^2xy^2z$ expressed as $a^l b^l c^l$, at most two of the numbers of $a$'s, $b$'s, and $c$'s are increased, which is a contradiction.

## 4.5    Membership Problem for Context-Free Languages

In general, for a given language $L$, the problem of deciding whether a string $w$ belongs to the language or not is called the membership problem. In this section, we present an efficient algorithm that solves the membership problem for context-free languages.

If we are allowed to spend sufficiently enough time to solve the problem we can construct an algorithm, based on a given context-free grammar, to decide whether or not a string is generated by the grammar. Throughout this section, a context-free language is assumed to be given in terms of a context-free grammar $G = (\Gamma, \Sigma, P, S)$ in Chomsky normal form. Then, in order to generate a string of length $n$, we apply rules of the form $A \to BC$ $n - 1$ times and rules of the form $A \to a$ $n$ times. So we apply rules $2n - 1$ times in total to generate a string of length $n$. Hence, given a string $w$ of length $n$, we enumerate all the sequences of $2n - 1$ rules $r_1 r_2 \cdots r_{2n-1}$ and check to see if $S \overset{*}{\Rightarrow} w$ for each of the sequences. If $S \overset{*}{\Rightarrow} w$ for at least one of these sequences, the grammar generates $w$, and does not generate $w$ otherwise. This is a typical example of a brute-force search. When the number of rules in the grammar is denoted by $m$, then the total number of sequences of rules of length $2n - 1$ is given by $m^{2n-1}$, which is an exponential function in the length $n$ of an input string and becomes enormously large as $n$ becomes large. Hence this way of solving the membership problem cannot be practical.

We shall describe the CYK (Cocke–Younger–Kasami) algorithm which efficiently solves the membership problem, based on *dynamic programming*. Dynamic programming is a powerful technique for designing an efficient algorithm, which, in brief, repeats the following process: divide a given problem into smaller problems, solve the smaller problems in advance, memorize the answers for them, and finally combine these answers for the small problems to compute the answer for the original problem. In order to decide whether $S \overset{*}{\Rightarrow} a_1 \cdots a_n$ holds or not, we check whether $A \overset{*}{\Rightarrow} a_i \cdots a_j$ or not for every nonterminal $A$ and for every subsequence $a_i \cdots a_j$ of $a_1 \cdots a_n$, and memorize the set of $A$'s such that $A \overset{*}{\Rightarrow} a_i \cdots a_j$ as the content of variable $X_{ij}$. Since Chomsky normal form is assumed, $A \overset{*}{\Rightarrow} a_i a_{i+1} \cdots a_j$ holds if $a_i \cdots a_j$ is divided into two substrings $a_i \cdots a_k$ and $a_{k+1} \cdots a_j$ in such a way that $B \overset{*}{\Rightarrow} a_i \cdots a_k$ and $C \overset{*}{\Rightarrow} a_{k+1} \cdots a_j$ holds for some nonterminals $B$, $C$ and for some rule $A \to BC$. In other words, to decide whether or not $A \in X_{ij}$, we only have to check whether or not there exists an integer $k$ such that $i \leq k \leq j - 1$ and nonterminals $B$ and $C$ such that $B \in X_{ik}$, $C \in X_{k+1,j}$, and $A \to BC \in P$, where $P$ denotes the set of rules. To compute the membership problem this way, we place each $X_{ij}$ as illustrated in Fig. 4.13 for $n = 6$. The variables $X_{11}, \ldots, X_{nn}$ in row 1 of the table contain nonterminals that generate $a_1, \ldots, a_n$, respectively. The variables in $X_{12}, \ldots, X_{n-1,n}$ in row 2 contain nonterminals that generate $a_1 a_2, a_2 a_3, \ldots, a_{n-1} a_n$, respectively. Similarly, nonterminals in rows numbered 1, 2, 3, ... are specified. The algorithm to solve the membership problem somehow computes the nonterminals of the variables from the bottom row toward the top row and decides that $S \overset{*}{\Rightarrow} a_1 \cdots a_n$ holds if the variable $X_{1n}$ contains the start symbol $S$.

| Row No./String | $a_1$ | $a_2$ | $a_3$ | $a_4$ | $a_5$ | $a_6$ |
|---|---|---|---|---|---|---|
| 6 | $X_{16}$ | | | | | |
| 5 | $X_{15}$ | $X_{26}$ | | | | |
| 4 | $X_{14}$ | $X_{25}$ | $X_{36}$ | | | |
| 3 | $X_{13}$ | $X_{24}$ | $X_{35}$ | $X_{46}$ | | |
| 2 | $X_{12}$ | $X_{23}$ | $X_{34}$ | $X_{45}$ | $X_{56}$ | |
| 1 | $X_{11}$ | $X_{22}$ | $X_{33}$ | $X_{44}$ | $X_{55}$ | $X_{66}$ |

**Fig. 4.13** Derivation table for substrings of $a_1a_2a_3a_4a_5a_6$

**Fig. 4.14** Pairs of $X_{2,k}$ and $X_{k+1,6}$ to compute $X_{26}$ where $2 \leq k \leq 5$

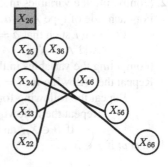

So the problem left is how to compute the nonterminals of each variable $X_{ij}$, which will now be explained.

First, let element $X_{ii}$ be a set of all $A$ such that $A \to a_i \in P$. Suppose that nonterminals in the variables that are placed in row 1 up to row $l - 1$ have been computed so far. We shall describe how to compute nonterminals in $X_{ij}$'s that are placed on row $l$, where $n \geq l = j - i + 1 \geq 2$. What we want to compute is $A$, which by definition consists of nonterminals such that $A \overset{*}{\Rightarrow} a_i \cdots a_j$. Clearly we have $A \overset{*}{\Rightarrow} a_i \cdots a_j$ if there exists an integer $k$ such that $i \leq k \leq j - 1$ and nonterminals $B$ and $C$ such that

(1)  $B \in X_{ik}$

(2)  $C \in X_{k+1,j}$

(3)  $A \to BC \in P$

Note that, since $k - i + 1 \leq j - 1 - i + 1 = j - i < l$ and $j - (k+1) + 1 \leq j - (i+1) + 1 = j - i < l$, both $X_{ik}$ and $X_{k+1,j}$ are placed in row $1, \ldots,$ and row $l - 1$. Since nonterminals in $X_{ik}$ and those in $X_{k+1,j}$ are already computed, we can check the conditions (1), (2), and (3). Figure 4.14 illustrates how to compute $X_{26}$ from the four pairs $(X_{22}, X_{36})$, $(X_{23}, X_{46})$, $(X_{24}, X_{56})$, and $(X_{25}, X_{66})$.

Assume that $X_{11}, \ldots, X_{nn}$ in the bottom row are already computed. Then from the bottom row toward the top row of the table, we proceed by adding nonterminals to variable $X_{ij}$ in the $(j - 1 + 1)$th row as follows:

For each rule $A \to BC$ and each $i \leq k \leq j - 1$, if $B \in X_{ik}$ and $C \in X_{k+1,j}$, then add $A$ to $X_{ij}$.

We are now ready to give the algorithm that solves the membership problem for context-free languages.

**Algorithm** *MEM*
Input: $a_1 a_2 \cdots a_n$
**1.** *Accept* if $a_1 a_2 \cdots a_n = \varepsilon$ and $S \to \varepsilon$ is a rule.
**2.** (computing the variables in the first row)
    For each rule of type $A \to a$.
        For each $i$ such that $a = a_i$.
            Add $A$ to $X_{ii}$.
**3.** (computing the variables in the $l$th row where $l \geq 2$)
    Repeat the following for $l = 2, \ldots, n$.
        Repeat the following for each $X_{ij}$ in the $l$th row such that $j - i + 1 = l$.
            Repeat the following for each rule $A \to BC$ and each $i \leq k \leq j - 1$.
                If $B \in X_{ik}$ and $C \in X_{k+1,j}$, then add $A$ to $X_{ij}$.
**4.** *Accept* if $S \in X_{1n}$.

We shall estimate the number of steps of the algorithm by evaluating the order of the function that gives the steps in terms of input length $n$. The order of a function will be explained in Sect. 8.1. For example, when we say a function $f(n)$ is $O(n^3)$, it means that, when $k$ is appropriately chosen, $f(n) \leq kn^3$ holds for sufficiently large $n$. Since a context-free language is given in terms of a context-free grammar $G$ that is fixed, the number of rules of $G$ is considered to be fixed independent of input length $n$. So, we consider only stages **2** and **3** in the algorithm, where the number of steps required depends on input length $n$. In stage **2**, for each rule $A \to a$, it is required to check if $a = a_i$ for $a_1, \ldots, a_n$. So the order of the steps required in stage **2** is $O(n)$. In stage **3**, $l$ varies over $2, \ldots, n$, $i$ varies over $1, \ldots, n - l + 1$, and finally $k$ varies over $i, \ldots, i + l - 2$. Since the size of each range is $O(n)$, the number of steps required in stage **3** is $O(n^3)$, which in turn gives the steps required by the entire algorithm. This is a great improvement over the number of steps, given by $O(m^{2n-1})$, of the algorithm based on brute-force search described at the beginning of this section.

*Example 4.14* We try to apply the algorithm described above to a context-free grammar given as follows:

$$S \to AB \mid AC \mid AA,$$
$$A \to CB \mid a,$$
$$B \to AC \mid b,$$
$$C \to CC \mid b.$$

| 5 | $\{S, A, B\}$ | | | | |
|---|---|---|---|---|---|
| 4 | $\{S, A\}$ | $\{S, A, B\}$ | | | |
| 3 | $\{S\}$ | $\{A\}$ | $\{S, B\}$ | | |
| 2 | $\{A, C\}$ | $\emptyset$ | $\{S, B\}$ | $\{A, C\}$ | |
| 1 | $\{B, C\}$ | $\{B, C\}$ | $\{A\}$ | $\{B, C\}$ | $\{B, C\}$ |
| Row No./Column No. | 1 | 2 | 3 | 4 | 5 |
| String | $b$ | $b$ | $a$ | $b$ | $b$ |

**Fig. 4.15** Derivation table for substrings of *bbabb*

Figure 4.15 shows the derivation table for substrings of *bbabb*. For example, let us see how $X_{25}$ is computed. Variable $X_{25}$ comes from three pairs $(X_{22}, X_{35})$, $(X_{23}, X_{45})$, and $(X_{24}, X_{55})$, which correspond to $(\{B, C\}, \{S, B\})$, $(\emptyset, \{A, C\})$, and $(\{A\}, \{B, C\})$, respectively. Since we have

$$\{B, C\} \cdot \{S, B\} = \{BS, BB, CS, CB\},$$
$$\emptyset \cdot \{A, C\} = \emptyset,$$
$$\{A\} \cdot \{B, C\} = \{AB, AC\},$$

and the rules such that these strings of length 2 appear on the right-hand side of the rules are

$$S \to AB, \quad S \to AC, \quad A \to CB \quad \text{and} \quad B \to AC,$$

we have $X_{25} = \{S, A, B\}$. Other elements are similarly obtained. Since $X_{15} = \{S, A, B\}$, which includes the start symbol $S$, it is concluded that the string *bbabb* is generated by the context-free grammar.

## 4.6 Problems

**4.1** Let the context-free grammar be given whose rules are

$$S \to SbS \mid ScS \mid a.$$

Give all the parse trees that generate string *abaca*. Furthermore, give the leftmost derivation corresponding to each of these parse trees.

**4.2*** Let languages $L_1$ and $L_2$ be generated by context-free grammars $G_1 = (V_1, \Sigma, P_1, S_1)$ and $G_2 = (V_2, \Sigma, P_2, S_2)$, respectively. Give a context-free grammar that generates the language $L_1 \cup L_2$ in terms of $G_1$ and $G_2$.

**4.3** Let $(w_1 w_2 \cdots w_n)^R = w_n \cdots w_2 w_1$ for string $w_1 w_2 \cdots w_n$ of length $n$. Give a context-free grammar that generates $\{w \in \{a, b\}^* \mid w = w^R\}$.

**4.4**\*\* Let $w^R$ be defined as in Problem 4.3. Answer the following questions.

(1) Prove that for a string $w = w_1 w_2 \cdots w_n$ of length $n$, $w \neq w^R$ if and only if $w_i \neq w_{n-(i-1)}$ for some $1 \leq i \leq \lfloor n/2 \rfloor$, where $\lfloor n/2 \rfloor$ denotes the maximum integer at most $n/2$.

(2) Give a context-free grammar that generates $\{w \in \{a, b\}^* \mid w \neq w^R\}$.

**4.5** Give context-free grammars that generate the following languages.

(1)   $\{w \in \{a, b, c, d\}^* \mid w = \varepsilon$ or $w = x_1 \cdots x_m y_1 \cdots y_m$ with $x_i \in \{a, b\}$, $y_i = \{c, d\}$ for $1 \leq i \leq m\}$

(2)   $\{w \in \{a, b, c\}^* \mid w = a^i b^j c^i$ for $i \geq 0, j \geq 0\}$

(3)   $\{w \in \{a, b, c, d\}^* \mid w = a^i b^j c^j d^i$ for $i \geq 0, j \geq 0\}$

(4)\*  $\{a^i b^j \mid 0 \leq i \leq j\}$

(5)\*  $\{a^i b^j \mid 0 \leq i \leq j \leq 2i\}$

(6)\*\* $\{a^i b^j c^k \mid i, j, k$ are nonnegative integers that satisfy $j = i + k\}$

**4.6**\*\* When string $w$ is expressed as a concatenation of two strings $w'$ and $w''$, $w'$ is called a *prefix* of $w$. Let $N_x(w)$ denote the number of $x$'s that appear in $w$. The language $L$ is defined as follows:

$$L = \big\{w \in \{(,)\}^* \mid N_{(}(w') - N_{)}(w') \geq 0 \text{ for any prefix } w' \text{ of } w,$$
$$\text{and } N_{(}(w) - N_{)}(w) = 0\big\}$$

Prove that the context-free grammar $G$ of Example 4.3 generates the language $L$.

**4.7**\* Answer the following questions for the context-free grammar $G$ given by the following rules:

$$S \to aSa \mid aSb \mid bSa \mid bSb \mid \varepsilon.$$

(1) Give informal English descriptions of the language $L(G)$ which the grammar generates.

(2) The language $L(G)$ turns out to be a regular language. Give a regular grammar that generates $L(G)$.

**4.8** Answer the following questions for the context-free grammar $G$ given by the following rules:

$$S \to aS \mid Sb \mid a \mid b \mid.$$

(1) Give informal English descriptions of the language $L(G)$ which the grammar generates.

(2) Give a regular grammar that generates $L(G)$.

**4.9**$^{**}$ Answer the following questions for the context-free grammar $G$ given by the following rules:

$$S \to aB \mid bA \mid \varepsilon,$$
$$A \to aS \mid bAA,$$
$$B \to bS \mid aBB.$$

(1) Give informal English descriptions of the language $L(G)$.
(2) Prove that the language described in (1) coincides with the language generated by the grammar $G$.

**4.10**$^{**}$ Answer the following.
(1) Give a context-free grammar $G$ that generates the language $L = \{w \in \{a, b\}^* \mid N_a(w) = N_b(w)\}$ and has the start symbol $S$ as the only nonterminal.
(2) Prove that $L = L(G)$.

**4.11**$^{**}$ Give a context-free grammar that generates

$$\{w \in \{a, b\}^* \mid N_a(w) = 2N_b(w)\},$$

where $N_x(w)$ is as in Problem 4.6$^{**}$.

**4.12**$^{*}$ Transform a context-free grammar given by the following rules to an equivalent one in Chomsky normal form

$$S \to aB \mid bA,$$
$$A \to a \mid aS \mid bbA,$$
$$B \to b \mid bS \mid aBB.$$

**4.13**$^{**}$ Prove that the following languages are not context-free languages where the alphabet is $\{a, b, c\}$.
(1) $\{a^i b^j c^k \mid i \le j \le k\}$
(2) $\{w \mid N_a(w) = N_b(w) = N_c(w)\}$

**4.14**$^{***}$ Answer the following questions for context-free languages.
(1) Show that the language $\{ww \mid w \in \{a, b\}^*\}$ is not a context-free language.
(2) Give a context-free grammar that generates the language given by $\{a, b\}^* - \{ww \mid w \in \{a, b\}^*\}$.
(3) Show that the class of context-free languages is not closed under the set operation of complement.

**4.15**[*]  Apply the algorithm *MEM* that solves the membership problem for a context-free language to the context-free grammar given by the following rules and construct a derivation table for substrings of *aaba*

$$S \to AA,$$
$$A \to AB \mid a,$$
$$B \to CA \mid b,$$
$$C \to AB \mid b.$$

# Pushdown Automata

5

In Chap. 3 we studied finite automata, which have a finite amount of memory. In this chapter we introduce a new type of computational model called a pushdown automaton, which can be obtained by giving a nondeterministic finite automaton a memory component called a *stack*. The stack memorizes an arbitrarily long string, but has the restriction that a symbol must be read and written at one end of the string. It will be shown that a pushdown automaton and a context-free grammar are equivalent in power to specify languages.

## 5.1    Pushdown Automaton

The finite automaton introduced in Chap. 3 can also be drawn as shown in Fig. 5.1. A symbol is written in a square of the input tape, and the input tape contains an input string. A state diagram is written in the control part which takes a state at each moment. A state diagram may be thought of as a program that describes the behavior of a finite automaton. The behavior is a sequence of moves, each being specified by a transition function that, given the present state and the symbol read by the input head, returns information on its next state.

On the other hand, as shown in Fig. 5.2, a pushdown automaton is essentially a nondeterministic finite automaton attached to a memory device, called a stack, which can contain an arbitrarily long string. The stack is divided into squares, each containing a symbol from a stack alphabet. The stack head can write symbols on the stack and read them back later, but only on the top of the stack. As for the finite automaton, when a transition function of a pushdown automaton is given the present state $q$, symbol $a$ read by the input head, and symbol $b$ read by the stack head, it returns information on its possible next action: a state $q'$ and string $u$. As is illustrated in Fig. 5.3 as a transition from (a) to (b), $q'$ is a state that the pushdown automaton can move into, and $u$ is a string that can be written on the top of the stack. To make such a transition possible, we have only to specify the transition function of a pushdown automaton $\delta$ as $(q', u) \in \delta(q, a, b)$. On the other hand, the transition function may be specified to make the transition

A. Maruoka, *Concise Guide to Computation Theory*,
DOI 10.1007/978-0-85729-535-4_5, © Springer-Verlag London Limited 2011

**Fig. 5.1**  Schematic of a
finite automaton

**Fig. 5.2**  Schematic of a
pushdown automaton

**Fig. 5.3**  One step of move of
pushdown automaton

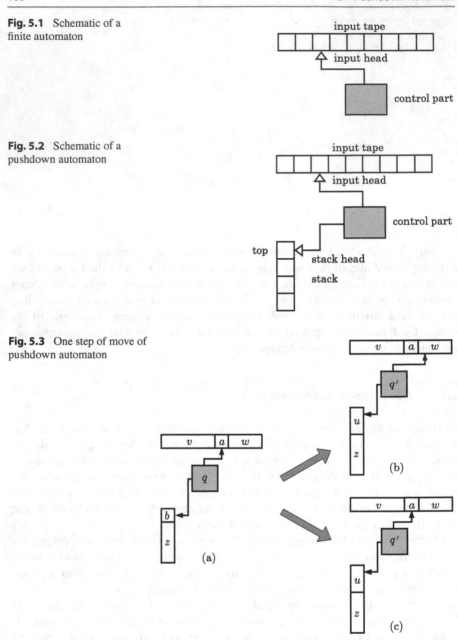

from (a) to (c) possible, which is a transition done without reading a symbol from
the input and thereby the input head stays at the same square, but otherwise the
same as the case from (a) to (b). To make the transition from (a) to (c) possible,
we only have to specify the transition function as $(q', u) \in \delta(q, \varepsilon, b)$. A symbol
read by the stack head may also be the empty symbol, which means that the stack

head does not read a symbol from the stack. As will be described later precisely, for each $q$, $a$, and $b$, $\delta(q, a, b)$ is a collection of items that specify possible actions as next moves provided that the present situation is given in terms of $q$, $a$, and $b$.

On the input tape the input head can only read a symbol from the tape, while on the stack the stack head can read and write on the top of the stack. To manipulate the contents of the stack the pushdown head *pushes* a symbol, which means to write the symbol on the stack, and *pops* a symbol, which means to remove the symbol from the stack. But the stack head can access information of a stack only on the top of the stack, so it can only access information stored last. Thus a stack is called a *"last in, first out"* storage device.

As for finite automata, we can think of deterministic and nondeterministic versions of pushdown automata, but in contrast to the case of finite automata deterministic and nondeterministic pushdown automata are not equivalent in power. In fact, there exist languages that can be accepted by nondeterministic pushdown automata, but cannot be accepted by any deterministic pushdown automata. In this book we study only nondeterministic pushdown automata, so when we refer to a pushdown automaton we mean a nondeterministic pushdown automaton unless stated otherwise.

Next we give the formal definition of a pushdown automaton together with a language accepted by a pushdown automaton.

**Definition 5.1** A *pushdown automaton* (PDA) is a 6-tuple $M = (Q, \Sigma, \Gamma, \delta, q_0, F)$, where
(1) $Q$ is the finite set of states.
(2) $\Sigma$ is the input alphabet.
(3) $\Gamma$ is the stack alphabet.
(4) $\delta : Q \times \Sigma_\varepsilon \times \Gamma_\varepsilon \to \mathcal{P}(Q \times \Gamma^*)$ is the transition function.
(5) $q_0 \in Q$ is the start state.
(6) $F \subseteq Q$ is the set of accept states.

In the definition above, (1), (2), (5), and (6) are the same as those items for a finite automaton, whereas (3) is the set of symbols that are written on the stack. Recall that $\Sigma_\varepsilon = \Sigma \cup \{\varepsilon\}$ and $\Gamma_\varepsilon = \Gamma \cup \{\varepsilon\}$. Since the transition function of (4) is expressed as $\delta : Q \times \Sigma_\varepsilon \times \Gamma_\varepsilon \to \mathcal{P}(Q \times \Gamma^*)$ for $q \in Q$, $a \in \Sigma_\varepsilon$, and $b \in \Gamma_\varepsilon$, in general, it is expressed as follows:

$$\delta(q, a, b) = \{(q_1, u_1), \ldots, (q_m, u_m)\},$$

where $m \geq 0$, $q_1, \ldots, q_m \in Q$, and $u_1, \ldots, u_m \in \Gamma^*$. In particular, if $m = 0$, the right-hand side of the above equation denotes the empty set, which means that no next move is defined.

Since $\delta(q, a, b)$ is expressed as the equation above and, hence, $m$ transitions to $(q_1, u_1), \ldots,$ and $(q_m, u_m)$ are possible, the PDA behaves nondeterministically.

**Fig. 5.4** Schematic of
$(q', u) \in \delta(q, a, b)$

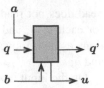

Figure 5.3 shows that if $a \in \Sigma$, then $(q_i, u_i) \in \delta(q, a, b)$ means the transition from
(a) to (b), whereas if $a = \varepsilon$, then $(q_i, u_i) \in \delta(q, \varepsilon, b)$ means the transition from (a)
to (c).

To summarize, we describe the two types of factors that cause nondeterministic
behavior as follows:

---

(1) If the present situation is given by $(q, a, b)$ such that $\delta(q, a, b) =$
$\{(q_1, u_1), \ldots, (q_m, u_m)\}$ for $a$ in $\Sigma_\varepsilon$, then any of the $m$ transitions $(q_1, u_1)$,
$\ldots, (q_m, u_m)$ is possible.
(2) For $a \in \Sigma$, if both $\delta(q, a, b)$ and $\delta(q, \varepsilon, b)$ are defined, that is, specified as
items different from $\emptyset$, then either of the moves with reading the symbol $a$
and without reading any symbol is possible. For $b \in \Gamma$, if both $\delta(q, a, b)$ and
$\delta(q, a, \varepsilon)$ are defined, then a similar nondeterministic behavior occurs.

---

Note that the nondeterministic behavior due to (1) and to (2) above may be allowed simultaneously. As mentioned, if $(q', u) \in \delta(q, a, b)$, each of $a$, $b$, and $u$ may
be the empty string, and the behaviors for these cases are summarized as follows:

---

(1) If $a = \varepsilon$, the input head does not move.
(2) If $b = \varepsilon$, the symbol at the top of the stack is not removed.
(3) If $u = \varepsilon$, nothing is added to the stack.

---

Furthermore, the move caused by $(q', u) \in \delta(q, a, b)$ turns out to be *pushing* if
$|u| \geq 1$, and to be *popping* if $|b| = 1$ and $u = \varepsilon$.

For ease of explanation, $(q', u) \in \delta(q, a, b)$ is expressed as in Fig. 5.4, where
$(q', u)$ is chosen from one of the possibilities given by $(q, a, b)$.

We are now ready to explain how a PDA accepts a string. A string $w$ placed in
the input tape is accepted by the PDA if there is a sequence of possible moves given
as follows: the sequence of moves begins at the start state $q_0$, with the empty stack
and with the input head placed on the leftmost square of the input tape; the sequence
of moves ends at one of the accept states, with the empty stack and with the input
head reading off the input tape. So we can say that the basic idea for acceptance
in the nondeterministic computational model of pushdown automata is the same as
that in finite automata. A PDA $M = (Q, \Sigma, \Gamma, \delta, q_0, F)$ accepts a string $w$, which
is precisely described as follows.

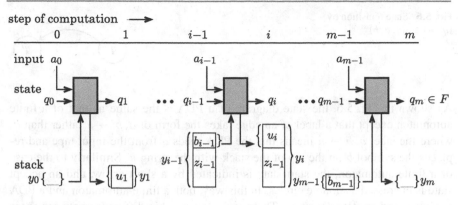

**Fig. 5.5** Moves to accept string $a_0 a_1 \cdots a_{m-1}$

---

$M$ accepts a string $w$

$\Leftrightarrow$ there exist $a_0, a_1, \ldots, a_{m-1} \in \Sigma_\varepsilon$, $q_0, q_1, \ldots, q_m \in Q$, and $y_0, y_1, \ldots, y_m \in \Gamma^*$ that satisfy the following four conditions:

(1) $w = a_0 a_1 \cdots a_{m-1}$

(2) $q_0$ is the start state of $M$ and $y_0 = \varepsilon$

(3) For $i = 1, \ldots, m$,

$$(q_i, u_i) \in \delta(q_{i-1}, a_{i-1}, b_{i-1}),$$

where $y_{i-1} = b_{i-1} z_{i-1}$, $y_i = u_i z_{i-1}$, $b_{i-1} \in \Gamma_\varepsilon$, and $z_{i-1} \in \Gamma^*$

(4) $q_m \in F$ and $y_m = \varepsilon$

---

Here, $w = a_0 a_1 \cdots a_{m-1}$ is a string obtained by inserting $m - n$ empty strings into the input string of length $n$ where $m \geq n$. So, $n$ transitions of the type from (a) to (b) and $m - n$ transitions of the type from (a) to (c) in Fig. 5.3 are executed before $w$ is accepted. Furthermore, (2) is the condition when $M$ starts transitions, whereas (4) is the condition when $M$ ends transitions, and finally (3) gives the condition imposed at each step in the course of the transitions. In the description of condition (3), the string $y_i$ expresses the contents of the stack just after the $i$th step where the leftmost symbol of $y_i$ corresponds to the top of the stack. Figure 5.5 illustrates these four conditions to accept string $a_0 a_1 \cdots a_{m-1}$.

So far we have defined a string that is accepted by a PDA. As in the case of a finite automaton, the language that is accepted by a PDA $M$ is defined to be the language that consists of the strings that $M$ accepts, which will be denoted by $L(M)$.

Next, we shall explain a state diagram for a PDA. First, a state transition is specified as $(q', u) \in \delta(q, a, b)$ or described as

$$q \xrightarrow{a, b \to u} q'.$$

**Fig. 5.6** State transition by
$(q', u) \in \delta(q, a, b)$

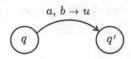

As shown in Fig. 5.6, the state diagram of a PDA is the same as that of a finite automaton except that a label of an edge takes the form of $a, b \rightarrow u$ rather than $a$, where the label $a, b \rightarrow u$ means that the PDA reads $a$ from the input tape and replaces the symbol $b$ on the top of the stack with the string $u$. Similarly to the case of a finite automaton, the start state is indicated by a short arrow, and an accept state is expressed as double circles. In this way, both a finite automaton and a PDA are expressed as state diagrams. There is, however, a big difference between these corresponding state diagrams in that a state diagram of a finite automaton has all information concerning a finite automaton but that of a PDA does not have information on the contents of the stack. It should be noted here that information on the input head positions will also be needed to completely specify situations of a finite state automaton and a PDA.

In the definition of the acceptance mentioned above, the two conditions are required for the acceptance when the input head reads out an entire input string: the machine is in an accept state; the stack is empty. But as mentioned below, either of these conditions can solely be the condition of acceptance. That is, it can be shown that even if the acceptance is defined based on any of the following three conditions, the class of the languages accepted by PDAs remains the same.

(1) The machine is in an accept state and the stack is empty.
(2) The machine is in an accept state.
(3) The stack is empty.

In the following, it is shown that conditions (1) and (2) are equivalent. Although the details are omitted here, condition (3) can also be shown to be equivalent to the others in a similar way.

In order to show that conditions (1) and (2) are equivalent, it suffices to construct an equivalent PDA based on one of the acceptance conditions from a PDA based on the other acceptance condition. First, we construct an equivalent PDA $M_2$ based on acceptance condition (2) from a PDA $M_1$ based on acceptance condition (1). The point of the construction is to modify $M_1$ to obtain $M_2$ by introducing a new start state $q_s$ and a new accept state $q_F$ so that $M_2$ moves to the accept state introduced only when the machine $M_1$ is in an accept state with the empty stack. Figure 5.7 illustrates how to modify $M_1$ based on this idea so as to obtain $M_2$. In this figure the hatched rectangle is the state diagram of $M_1$. As shown in the figure, $M_2$ begins at the introduced start state $q_s$, pushes the symbol \$ which means the bottom of the stack ($q_s \xrightarrow{\varepsilon, \varepsilon \rightarrow \$} q_0$), and simulates $M_1$ beginning at start state $q_0$ of $M_1$. $M_2$ obtained this way moves to the accept state $q_F$ newly introduced when $M_1$ not only is in one of its accept states, but also the stack is effectively empty ($q_{f_i} \xrightarrow{\varepsilon, \$ \rightarrow \varepsilon} q_F$), that is, the top of the stack is \$.

**Fig. 5.7** State diagram of
PDA $M_2$ that simulates
PDA $M_1$

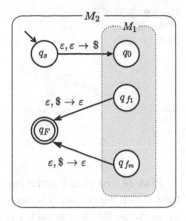

**Fig. 5.8** State diagram of
PDA $M_1$ that simulates
PDA $M_2$. For
$\Gamma = \{b_1, \ldots, b_k\}, \varepsilon, * \to \varepsilon$
means $\varepsilon, b_1 \to \varepsilon$,
$\varepsilon, b_2 \to \varepsilon, \ldots, \varepsilon, b_k \to \varepsilon$

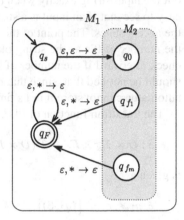

Conversely, the point in constructing an equivalent PDA $M_1$ from PDA $M_2$ is to modify $M_2$ to obtain $M_1$ by introducing the new start state $q_s$ and the new accept state $q_F$ so that $M_1$ at the accept state $q_F$ makes its stack empty even when $M_2$ at its accept state $q_{f_i}$ has its nonempty stack. Based on this idea, Fig. 5.8 illustrates how to modify $M_2$ to obtain $M_1$. As shown in this figure, $M_1$ moves from accept states $q_{f_i}$ of $M_2$ to the accept state $q_F$ ($q_{f_i} \xrightarrow{\varepsilon, * \to \varepsilon} q_F$) and repeats to pop ($q_F \xrightarrow{\varepsilon, * \to \varepsilon} q_F$) until the stack becomes empty. In this figure, if the stack alphabet is $\Gamma = \{b_1, \ldots, b_k\}$, the label $\varepsilon, * \to \varepsilon$ means

$$\varepsilon, b_1 \to \varepsilon, \ldots, \varepsilon, b_k \to \varepsilon.$$

Next, we discuss some examples of PDAs.

*Example 5.2* We consider the PDA $M$, given in Fig. 5.9, which accepts the language $\{0^n 1^n \mid n \geq 0\}$.

**Fig. 5.9** State diagram of
PDA that accepts
$\{0^n 1^n \mid n \geq 0\}$

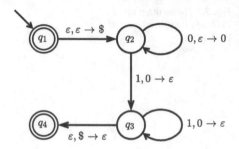

PDA $M$ first puts \$ at the bottom of the stack, then reads a string from the input tape. As each 0 is read, it is pushed on the stack. After the input changes from 0 to 1, $M$ works as follows: as each 1 is read, pop a 0 off the stack. If the input head reads out the input string exactly when the stack sees symbol \$, $M$ accepts the input. It is easy to see that if an input is a string other than $0^n 1^n$, then the PDA never moves to the accept states. The point of the behavior of $M$ is that it stores the information on the number of 0's as the string of 0's of the same number on the stack in order to check afterward if the number of 1's following is the same as the number stored. It should be noticed that, since that number becomes unboundedly large, a finite state automaton cannot store it in its finite memory.

The transition function of PDA $M$

$$\delta : Q \times \Sigma_\varepsilon \times \Gamma_\varepsilon \to \mathcal{P}(Q \times \Gamma^*)$$

is given as

$$\delta(q_1, \varepsilon, \varepsilon) = \{(q_2, \$)\},$$

$$\delta(q_2, 0, \varepsilon) = \{(q_2, 0)\}, \qquad \delta(q_2, 1, 0) = \{(q_3, \varepsilon)\},$$

$$\delta(q_3, 1, 0) = \{(q_3, \varepsilon)\}, \qquad \delta(q_3, \varepsilon, \$) = \{(q_4, \varepsilon)\}.$$

These specifications of the $\delta$'s are shown as the edges with the labels in Fig. 5.9. For

$$(q, a, b) \in Q \times \Sigma_\varepsilon \times \Gamma_\varepsilon$$

that does not appear in the above, let $\delta(q, a, b) = \emptyset$, where $Q = \{q_1, q_2, q_3, q_4\}$, $\Sigma = \{0, 1\}$, $\Gamma = \{0, \$\}$, and $F = \{q_1, q_4\}$.

*Example 5.3* We explain a PDA that accepts the language

$$\{w w^R \mid w \in \{0, 1\}^*\},$$

where $w^R = w_n w_{n-1} \cdots w_1$ for a string $w = w_1 w_2 \cdots w_n$ of length $n$. For example, if we have $w = 10011$ then $w w^R$ becomes 1001111001, which belongs to the language.

**Fig. 5.10** Checking the form of $ww^R$

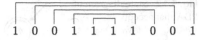

**Fig. 5.11** State diagram of PDA that accepts $\{ww^R \mid w \in \{0,1\}^*\}$

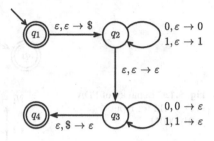

To decide whether or not a string is expressed as $ww^R$, we only have to check that the corresponding symbols in the pairs are identical, as shown in Fig. 5.10. To do so, assume that a string, say, $a_1 a_2 \cdots a_{10}$ is given as input. Store $a_1, \ldots, a_5$ in the first half part on the stack by pushing each one by one. Then as each of $a_6, a_7, \ldots, a_{10}$ is read, pop each of $a_5, a_4, \ldots, a_1$ off as long as the symbol read is the same as the symbol at the top. If the input head reads out the entire input string exactly when the stack becomes empty, accept the input. To test for an empty stack, the PDA pushes the new symbol $ at the beginning of the computation and checks if it sees the symbol $ at the end of the computation.

For the PDA to behave properly as mentioned above, the PDA seems to need to recognize where the center of an input string is, but it is unable to point out the center without seeing the entire input string. The trick in overcoming this difficulty is to make use of the nondeterministic behavior of a PDA. We make a PDA that behaves nondeterministically for every point in an input as if the point is the center, thereby accepting a string of the form $ww^R$. When the PDA guesses the right point between $w$ and $w^R$, the computation clearly ends up with an accept situation. On the other hand, if an input does not take the form $ww^R$, then any guess concerning the center does not lead to an accept situation no matter how the automaton guesses the center of the input because the automaton becomes stuck somewhere in the computation.

Figure 5.11 gives the state diagram of a PDA that behaves as described above. The first half of a string is pushed into the stack by the transitions $q_2 \xrightarrow{0,\varepsilon\to 0} q_2$ and $q_2 \xrightarrow{1,\varepsilon\to 1} q_2$. After that, the PDA performs the transition $q_2 \xrightarrow{\varepsilon,\varepsilon\to\varepsilon} q_3$. This transition corresponds to guessing the center, which may be everywhere in the input. During the second half of the input as the PDA reads a symbol from the input, it pops the symbol at the top off as long as these two symbols coincide. Furthermore, as in the previous example, in order to test for the empty stack, the PDA pushes $ by $q_1 \xrightarrow{\varepsilon,\varepsilon\to\$} q_2$ at the beginning of the computation and pops off the $ by $q_3 \xrightarrow{\varepsilon,\$\to\varepsilon} q_4$ at the end, leading to the situation that the stack is empty.

**Fig. 5.12** State diagram of
PDA that accepts
$\{w \in \{0, 1\}^* \mid N_0(w) = N_1(w)\}$

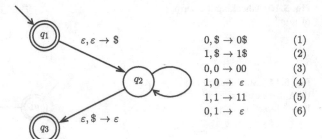

| | |
|---|---|
| $0, \$ \rightarrow 0\$$ | (1) |
| $1, \$ \rightarrow 1\$$ | (2) |
| $0, 0 \rightarrow 00$ | (3) |
| $1, 0 \rightarrow \varepsilon$ | (4) |
| $1, 1 \rightarrow 11$ | (5) |
| $0, 1 \rightarrow \varepsilon$ | (6) |

**Fig. 5.13** Behavior of PDA
in Fig. 5.12 for input
1101000011

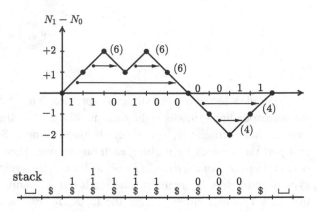

In a nondeterministic computational model, it is only required that, when an input is to be accepted, there exists at least one computation leading to an accept situation among all the possible computations. But, when explaining the behavior of a nondeterministic machine, it is often said that the machine by itself chooses or guesses a computation that leads to an accept situation, which is expressed as "choose nondeterministically" or "guess" the computation.

*Example 5.4* Consider the language consisting of strings that have equal numbers of 0's and 1's, which is expressed as

$$\{w \in \{0, 1\}^* \mid N_0(w) = N_1(w)\},$$

where $N_a(w)$ denotes the number of symbol $a$'s that appear in string $w$. Figure 5.12 shows a state diagram of a PDA that accepts the language.

Figure 5.13 illustrates the computation of the PDA when fed as input 1101000011 which belongs to the language. The horizontal axis represents the position in the string and the vertical axis represents the difference, given by $N_1(w) - N_0(w)$, between the numbers of 1's and 0's in the corresponding prefix $w$ of the input. Figure 5.13 also illustrates the contents of the stack at each step in the lower part.

The point of the behavior of the PDA in Fig. 5.12 is that the stack contains a certain number of the symbol that appears in prefix $w$ more often as compared to the other symbol, and the number is given by $|N_1(w) - N_0(w)|$. Next we shall explain the transitions $q_2 \rightarrow q_2$ with labels $(1), \ldots, (6)$ in Fig. 5.12.

If the top is $, then

(1) If a 0 is read, push it onto the stack.

(2) If a 1 is read, push it onto the stack.

If the top is 0, then

(3) If a 0 is read, push it onto the stack.

(4) If a 1 is read, pop a 0 off the stack.

If the top is 1, then

(5) If a 1 is read, push it onto the stack.

(6) If a 0 is read, pop a 1 off the stack.

Note that (4) and (6) above are the transitions where the height of the stack decreases by 1 because a symbol different from the symbol at the top is read. Such transitions are expressed as the arrows $\bullet\!\longrightarrow$ in Fig. 5.13, where the tail of the arrow corresponds to the symbol pushed while the head of the arrow corresponds to the symbol popped, thereby decreasing the height of the stack.

Observe that the state diagram in Fig. 5.12 can be modified by replacing the six transitions $q_2 \to q_2$ labeled by (1), ..., (6) with four transitions $q_2 \to q_2$ labeled by $0, \varepsilon \to 0$, $1, \varepsilon \to 1$, $0, 1 \to \varepsilon$, $1, 0 \to \varepsilon$ without changing the language accepted.

## 5.2 Equivalence of Pushdown Automaton and Context-Free Grammar

A pushdown automaton (PDA) and a context-free grammar (CFG) look completely different as computational models. But it turns out that these two models are equivalent in power in specifying languages, which we prove in this section. Once the equivalence is proved, a context-free language can be defined either to be a language generated by a CFG or to be a language accepted by a PDA. Consequently, when we need to prove that a language is a context-free language, we can prove either that the language is generated by a CFG or that it is accepted by a PDA. It would be helpful for us to have the two options because a certain language may be more easily described in terms of a CFG rather than a PDA, and *vice versa*.

The equivalence is proved in the next two subsections. We will show that, given either a CFG or a PDA, we can construct the other type behaving in substantially the same way as the given one, thereby concluding that these two specify the same language.

Before proceeding to prove these facts, we summarize what we want to prove in the next theorem.

**Theorem 5.5** *A language is generated by a context-free grammar if and only if the language is accepted by a pushdown automaton.*

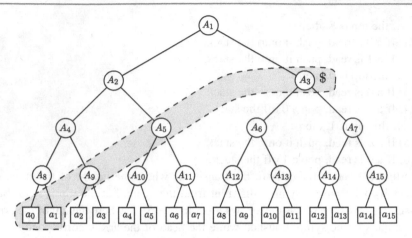

**Fig. 5.14** Parse tree for string $a_0 \cdots a_{15}$

## Simulating a Context-Free Grammar by a Pushdown Automaton

In this subsection, we shall construct a PDA $M$ that is equivalent to a given CFG $G$.

First, we intuitively explain how the PDA $M$ behaves. Let the parse tree given schematically in Fig. 5.14 be a typical one in which the CFG $G$ generates string $a_0 a_1 \cdots a_{15}$. In what follows, we shall explain how the PDA $M$ simulates the derivation based on CFG $G$. The transitions of $M$ are divided into two stages: the first consists of the transitions of first popping a nonterminal on the top of the stack and then pushing the right-hand side of a rule with that nonterminal on the left-hand side, which corresponds to growing the parse tree based on that rule; the second consists of the transitions of popping off a symbol when the symbol at the top coincides with the terminal read at the input head, which roughly corresponds to moving from the left to the right on the parse tree.

At a certain moment of the simulation the contents of the stack become $a_0 a_1 A_9 A_5 A_3 \$$, which is shown as surrounded by the dashed line in Fig. 5.14. Until that moment the $M$ uses only transitions of the first stage by following the leftmost derivation. Let us assume that the same string $a_0 a_1 \cdots a_{15}$ is given as input to $M$. Since the first two symbols of the input coincide with the first two symbols on the stack, these two symbols are popped off by performing the corresponding two transitions of the second stage. Then since $A_9$ is now on the top of the stack, $M$ is ready to pop $A_9$ and push $a_2 a_3$ to simulate the rule $A_9 \rightarrow a_2 a_3$. We now discuss more details of how $M$ works.

To begin with, $M$ pushes $A_1 \$$ on the empty stack. To do so, $\delta$ is specified as

$$(q_2, A_1 \$) \in \delta(q_1, \varepsilon, \varepsilon),$$

where $A_1$ is the start symbol of $G$, $q_1$ is the start state of $M$, and $q_2$ is the state that simulates derivations of $G$. Next, we show how to specify the transition function $\delta$ of $M$ that corresponds to the transitions of the first stage:

$$A_1 \Rightarrow A_2 A_3 \quad (q_2, A_2 A_3) \in \delta(q_2, \varepsilon, A_1)$$

$$\Rightarrow A_4 A_5 A_3 \quad (q_2, A_4 A_5) \in \delta(q_2, \varepsilon, A_2)$$

$$\Rightarrow A_8 A_9 A_5 A_3 \quad (q_2, A_8 A_9) \in \delta(q_2, \varepsilon, A_4)$$

$$\Rightarrow a_0 a_1 A_9 A_5 A_3 \quad (q_2, a_0 a_1) \in \delta(q_2, \varepsilon, A_8).$$

In general, a transition of the first stage is specified as

$$(q_2, u) \in \delta(q_2, \varepsilon, A)$$

for each rule $A \to u$ of $G$ so that the rules can be applied on the top of the stack without reading an input.

In order for $M$ to perform a transition of the first stage, a nonterminal to be replaced has to be on the top of the stack. So if any terminal is stored above this nonterminal in the stack, this terminal has to be removed.

For example, in the case of Fig. 5.14 $a_0$ and $a_1$ have to be removed for the nonterminal $A_9$ to be placed on the top so that transitions corresponding to the rule $A_9 \to a_2 a_3$ are performed. In general, a transition of the second stage is specified as

$$(q_2, \varepsilon) \in \delta(q_2, a, a)$$

for every terminal $a$ of $G$.

In this way $M$ repeats performing transitions chosen properly from the first and the second stages, depending on whether the symbol on the top is a nonterminal or a terminal, respectively. If the parse tree generates the string that is fed as input to $M$, then obviously $M$ can end up with $ left on the stack. If the transition function $\delta$ is specified as

$$(q_3, \varepsilon) \in \delta(q_2, \varepsilon, \$),$$

then $M$ moves to states $q_3$ with the empty stack, where $q_3$ is specified as the accept state of $M$. Note here that the PDA $M$ is not given the parse tree, so $M$ has to properly "guess" the parse tree and simulate the derivation by choosing transitions appropriately.

In the argument above, given a CFG $G$, we first constructed the PDA $M$ that simulates $G$. Then we assumed that the CFG $G$ generates string $a_0 \cdots a_{15}$ according to the parse tree based on $G$, which is given by Fig. 5.14, and derived that the PDA $M$ accepts that string. Conversely, assume that the PDA $M$ accepts string $a_0 \cdots a_{15}$. Then it is clear that there exists a parse tree based on the transitions of $M$, like the one given by Fig. 5.14, that generates the string. This is because, to make the stack empty at the end of computation, transitions of the first stage somehow simulate derivations of a certain parse tree, and an input string to $M$ coincides with the string

**Fig. 5.15** State diagram of
PDA equivalent to CFG

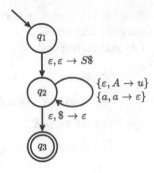

that the parse tree generates so as to pop off all the contents on the stack. It is easy
to see that the above arguments employing the parse tree and the string given by
Fig. 5.14 can be generalized to obtain the following theorem.

**Theorem 5.6** *If a language is generated by a context-free grammar, then there exists a pushdown automaton that accepts the language.*

*Proof* Let $L$ be generated by a context-free grammar $G = (\Gamma, \Sigma, P, S)$. Based
on $G$, a PDA $M = (Q, \Sigma, \Gamma', \delta, q_1, F)$ that accepts $L$ is defined as follows. Put

$$Q = \{q_1, q_2, q_3\}, \qquad F = \{q_3\}, \qquad \Gamma' = \Sigma \cup \Gamma \cup \{\$\},$$

and let the state transition function $\delta$ be given as shown in Fig. 5.15. Concerning the
transitions from $q_2$ to $q_2$, for every rule $A \rightarrow u \in P$ of $G$ it is specified as

$$q_2 \xrightarrow{\varepsilon, A \rightarrow u} q_2$$

and for every $a \in \Sigma$ it is specified as

$$q_2 \xrightarrow{a, a \rightarrow \varepsilon} q_2.$$

Then, clearly we have $L(M) = L(G)$, completing the proof of the theorem.    □

To obtain a PDA that accepts a given language, we may directly construct a PDA
that accepts the language. Alternatively, we can have a PDA by first constructing a
CFG that generates the language and then by transforming it to an equivalent PDA
in the way described in the proof of Theorem 5.6. In the next example we construct
two PDAs that accept the same language in these two ways.

*Example 5.7* Let $G = (\Sigma, \Gamma, P, S)$ be a CFG given by

$$\Sigma = \{(,)\}, \qquad \Gamma = \{(,), S, \$\},$$
$$P = \{S \rightarrow SS, S \rightarrow (S), S \rightarrow \varepsilon\}.$$

**Fig. 5.16** State diagram of
PDA that accepts strings of
well-nested parentheses,
being obtained by the
construction of the proof of
Theorem 5.6

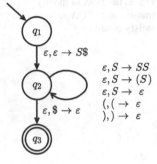

**Fig. 5.17** State diagram of
PDA that accepts strings of
well-nested parentheses,
being obtained by direct
construction

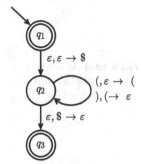

As mentioned in Example 4.3, the CFG generates strings of well-nested parentheses. According to the method given in the proof of Theorem 5.6, a PDA that simulates $G$ is given as shown in Fig. 5.16.

Next, we will directly construct a PDA that accepts strings of well-nested parentheses. In order to design such a PDA, suppose that we are given a string of parentheses of length, say, 100 and are asked whether or not the string is well-nested. Let us try to figure out how we decide it and construct a PDA that does the task by considering how we decide it. Since we cannot decide whether or not a string of that length is well-nested at a glance, we may decide it by deleting the corresponding left and right parentheses and see if there remains a parenthesis that does not have a counterpart. How can we find such corresponding parentheses? An easy way to do this is to repeat finding an innermost pair of left and right parentheses and cross them off. Based on that idea, we can obtain the PDA given in Fig. 5.17, which is a simplified version of the one given in Fig. 5.16.

The PDA given in Fig. 5.17 repeats the following depending on what symbol the input head reads: if a ( is read, push it; if a ) is read, pop off the (, which corresponds to the ) read. When the head reads out an input, *accept* if the stack is empty. As in the previous cases, $M$ can detect if the stack is empty by pushing $ initially and checking that the $ alone is on the top of the stack at the end of computation.

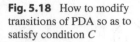

**Fig. 5.18** How to modify transitions of PDA so as to satisfy condition $C$

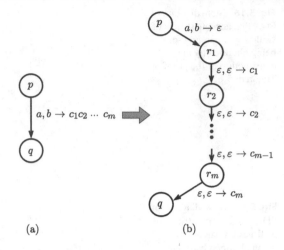

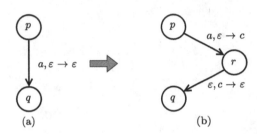

(a)                                           (b)

**Fig. 5.19** How to modify transition of PDA so as to satisfy condition $C$

(a)                                           (b)

## Simulating a Pushdown Automaton by a Context-Free Grammar

In this subsection, given a PDA $M = (Q, \Sigma, \Gamma, \delta, q_0, F)$, we shall construct a CFG that simulates the $M$.

First of all we can assume without loss of generality that a given PDA satisfies the following condition $C$ which makes it easy to construct an equivalent CFG.

**Condition $C$:**  A PDA either pushes or pops a symbol in one step. That is, any transition $p \xrightarrow{a,b \to c} q$ satisfies $|b| = 1$ and $c = \varepsilon$, or $b = \varepsilon$ and $|c| = 1$, where $a \in \Sigma_\varepsilon$.

This is because, as shown in Fig. 5.18, a general transition $p \xrightarrow{a,b \to c_1 c_2 \cdots c_m} q$ can be performed by $m + 1$ consecutive transitions that satisfy condition $C$, where $a \in \Sigma_\varepsilon$ and $b, c_1, c_2, \ldots, c_m \in \Gamma$. On the other hand, as shown in Fig. 5.19, a transition of the form $p \xrightarrow{a,\varepsilon \to \varepsilon} q$ can be performed by two transitions that satisfy condition $C$, where $c$ is any symbol in $\Gamma$. Furthermore, the intermediate states $r_1, \ldots, r_m, r$ introduced this way are set to be different from each other.

Next we explain an idea for constructing a CFG equivalent to a PDA that satisfies condition $C$. Figure 5.20 illustrates a typical computation of the PDA. The figure shows how the contents of the stack change as the computation proceeds step by

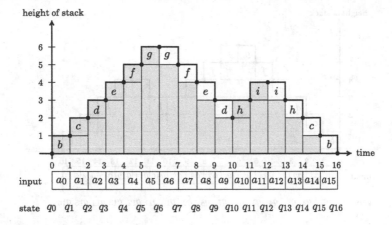

**Fig. 5.20** Moves of PDA to be simulated

step, where the horizontal axis is time and the vertical axis is the height of the stack. The gray-hatched vertical rectangles show the contents of the stack, whereas the symbols in the white cells denote the ones popped. Recall that since the PDA satisfies condition $C$, there is just one same symbol that is to be pushed and popped. For example, symbol $f$ is pushed at time 4 by $q_4 \xrightarrow{a_4, \varepsilon \to f} q_5$ and the same symbol $f$ is popped at time 7 by $q_7 \xrightarrow{a_7, f \to \varepsilon} q_8$. If state $q_{16}$ that finally appears is an accept state, then the string $a_0 a_1 \cdots a_{15}$ is accepted.

Next, we explain how to construct a CFG $G$ that simulates the PDA $M$. The point of the construction is that, introducing nonterminals $A_{pq}$ for every pair of states $p$ and $q$, we specify rules of $G$ so that $A_{pq}$ generates all the strings that cause $M$ to go from state $p$ with an empty stack to state $q$ with an empty stack. Notice that such strings also take $M$ from $p$ to $q$ no matter what contents the stack has at state $p$, and that the contents remain the same at state $q$; they are not changed during the computation. The followings are part of the derivations by $G$ constructed this way that correspond to the moves given in Fig. 5.20:

$$A_{q_0 q_{16}} \Rightarrow a_0 A_{q_1 a_{15}} a_{15}$$
$$\Rightarrow a_0 a_1 A_{q_2 q_{14}} a_{14} a_{15}$$
$$\Rightarrow a_0 a_1 A_{q_2 q_{10}} A_{q_{10} q_{14}} a_{14} a_{15}.$$

In these derivations, nonterminal $A_{q_0 q_{16}}$ generates $a_0$ and $a_{15}$ by itself and the remaining string $a_1 a_2 \cdots a_{14}$ is left for nonterminal $A_{q_1 q_{15}}$. The next derivation, where rule $A_{q_1 q_{15}} \to a_1 A_{q_2 q_{14}} a_{14}$ is applied, is similar to the previous derivation. In the next derivation, where $A_{q_2 q_{14}} \to A_{q_2 q_{10}} A_{q_{10} q_{14}}$ is applied, the task to derive $a_2 a_3 \cdots a_{13}$ is passed from $A_{q_2 q_{14}}$ to $A_{q_2 q_{10}}$ and $A_{q_{10} q_{14}}$, where $A_{q_2 q_{10}}$ is supposed to derive $a_2 a_3 \cdots a_9$ and $A_{q_{10} q_{14}}$ is supposed to derive $a_{10} a_{11} \cdots a_{13}$.

Figure 5.21 illustrates the derivations corresponding to Fig. 5.20. The symbols in the top cells in Fig. 5.20 are replaced by the corresponding symbols of the input. If you consider each gray-hatched rectangle as a nonterminal with suffixes $p$ and $q$,

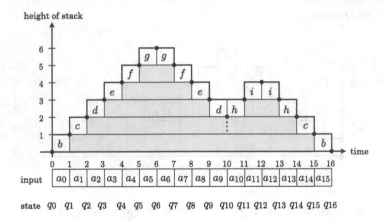

**Fig. 5.21** Moves of PDA to illustrate how it can be simulated by CFG

you can see that Fig. 5.21 illustrates the derivations. In particular, the rectangles at height 2 illustrate the derivation $A_{q_2q_{14}} \Rightarrow A_{q_2q_{10}} A_{q_{10}q_{14}}$, where the rectangle corresponding to $A_{q_2q_{14}}$ is divided into the two rectangles separated by the dashed line, each corresponding to $A_{q_2q_{10}}$ and $A_{q_{10}q_{14}}$. So we can say that the computation of PDA $M$ proceeds from the left to the right as shown in Fig. 5.20, whereas the computation of CFG $G$ proceeds from the bottom to the top as shown in Fig. 5.21. To make the simulation by CFG $G$ successful, we need to prepare four types of rules as follows.

The first type of rule means that a nonterminal generates two terminals on both sides and passes the task of generating a string between the two terminals to another nonterminal, provided that in the corresponding two transitions of $M$ the same stack symbol is pushed and then popped afterwards. More precisely, that type of rule is described as follows: if for each $b \in \Gamma$, $a, a' \in \Sigma$, and $p, q, r, s \in Q$ we have

$$p \xrightarrow{a,\varepsilon \to b} q, \qquad r \xrightarrow{a',b \to \varepsilon} s,$$

put
(1)  $A_{ps} \to a A_{qr} a'$
   As the second type of rule, for each $p, q, r \in Q$, put
(2)  $A_{pr} \to A_{pq} A_{qr}$
   Note that, as you can see in Figs. 5.22 and 5.23, we distinguish the case of rule (1) and that of rule (2) provided that the computation starts with empty stack: use rule (1) if the stack is empty only at the beginning of and the end of $M$'s moves: use rule (2) if the stack becomes empty at some point between the beginning and the end. Clearly in the former case, $M$'s first move must be a push, pushing a stack symbol, denoted by $b$ in Fig. 5.22, and the last move must be a pop, popping the same symbol that is pushed at the first move. On the other hand, in the latter case, there must be some point where the stack head touches the bottom of the empty stack. This is exactly the point where $M$ moves to state $q$ in Fig. 5.23.

**Fig. 5.22** Moves of PDA corresponding to $A_{ps} \rightarrow a A_{qr} a'$

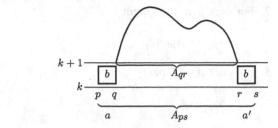

**Fig. 5.23** Moves of PDA corresponding to $A_{pr} \rightarrow A_{pq} A_{qr}$

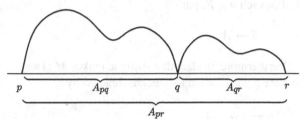

As shown in Fig. 5.21, applying the rules of types (1) and (2), we will end up with the positions that correspond to the peaks of the curve that traces the top position of the stack. Nonterminals $A_{q_6 q_6}$ and $A_{q_{12} q_{12}}$ are left in our case. For these nonterminals, since we do not expect to generate any more, we prepare the third type of rule as follows: for every $q \in Q$, put

(3) $A_{qq} \rightarrow \varepsilon$

Finally, as the fourth type of rule, for the start symbol $S$, the start state $q_0$, and each accept state $q$, put

(4) $S \rightarrow A_{q_0 q}$

Suppose that we construct CFG $G = (\Gamma', \Sigma, P, S)$ from PDA $M = (Q, \Sigma, \Gamma, \delta, q_0, F)$ as described so far. Intuitively, it is easy to see that if a string is accepted by $M$ as shown in Fig. 5.20, then the string is generated by $G$ as shown in Fig. 5.21. Conversely, we can see that, if a string is generated by $G$, then the string is accepted by $M$. The next Theorem 5.8 gives a proof of the equivalence between CFG $G$ and PDA $M$ described so far.

**Theorem 5.8** *If a language is accepted by a pushdown automaton, then it is generated by a context-free grammar.*

*Proof* Suppose that we are given a PDA $M = (Q, \Sigma, \Gamma, \delta, q_0, F)$. We define CFG $G = (\Gamma', \Sigma, P, S)$ based on $M$ as follows, which will be shown to generate the language $L(M)$, which denotes the language that $M$ accepts.

1. For each $p, q, r, s \in Q$, $a, a' \in \Sigma_\varepsilon$, and $c \in \Gamma$, if

$$p \xrightarrow{a, \varepsilon \rightarrow c} q, \qquad r \xrightarrow{a', c \rightarrow \varepsilon} s,$$

then put

$$A_{ps} \rightarrow a A_{qr} a'.$$

2. For each $p, q, r \in Q$, put

$$A_{pq} \to A_{pr} A_{rq}.$$

3. For each $p \in Q$, put

$$A_{pp} \to \varepsilon.$$

4. For each $q \in F$, put

$$S \to A_{q_0 q}.$$

Furthermore, the fact that string $w$ makes $M$ at state $p$ with the empty stack move to state $q$ with the empty stack is denoted by

$$p \xrightarrow[\text{emp}]{w} q.$$

In particular, let $p \xrightarrow[\text{emp}]{\varepsilon} p$ for any $p \in Q$. Observe that, if $p \xrightarrow[\text{emp}]{w} q$, then the string $w$ makes $M$ at state $p$ with any stack contents $u \in \Gamma^*$ move to state $q$ with the same stack contents $u$, where $M$ does not change the contents $u$ during the transitions.

The proof will be completed if we derive the following fact, because we have

$$w \in L(M) \quad \Leftrightarrow \quad \text{there exists } q \in F \text{ such that } q_0 \xrightarrow[\text{emp}]{w} q$$

$$\Leftrightarrow \quad \text{there exists } q \in F \text{ such that } S \Rightarrow A_{q_0 q} \text{ and } A_{q_0 q} \overset{*}{\Rightarrow} w$$

$$\Leftrightarrow \quad S \overset{*}{\Rightarrow} w$$

$$\Leftrightarrow \quad w \in L(G).$$

**Fact** For any $w \in \Sigma^*$ and any $p, q \in Q$, we have

$$A_{pq} \overset{*}{\Rightarrow} w \quad \Leftrightarrow \quad p \xrightarrow[\text{emp}]{w} q.$$

*Proof of $\Rightarrow$ of Fact* The proof is by induction on the number of steps of derivations. For $k = 1$, the only derivation of the form $A_{pq} \overset{*}{\Rightarrow} w$ is $A_{pp} \Rightarrow \varepsilon$. On the other hand, we have $p \xrightarrow[\text{emp}]{\varepsilon} p$ by definition.

Assume that $\Rightarrow$ of Fact holds for the number of steps of derivations less than or equal to $k$. We shall show that it holds for the number of steps $k + 1$. Suppose $A_{pq} \overset{(k+1)}{\Longrightarrow} w$, where $A_{pq} \overset{(k+1)}{\Longrightarrow} w$ means that $w \in \Sigma^*$ is derived from $A_{pq}$ by applying rules $k + 1$ times.

*Case 1* The case where $A_{pq} \Rightarrow a A_{rs} a' \overset{(k)}{\Rightarrow} w$ holds.

Let $w$ be expressed as $aya'$, thereby $A_{rs} \overset{(k)}{\Rightarrow} y$. By the inductive hypothesis we have

$$r \xrightarrow[\text{emp}]{y} s.$$

Since $A_{pq} \to a A_{rs} a'$ is a rule in this case, there exists a stack symbol $b \in \Gamma$ such that

$$p \xrightarrow{a, \varepsilon \to b} r,$$

$$s \xrightarrow{a', b \to \varepsilon} q.$$

Hence, we have

$$p \xrightarrow{a, \varepsilon \to b} r \xrightarrow[\text{emp}]{y} s \xrightarrow{a', b \to \varepsilon} q.$$

Thus, since $w = aya'$, we have

$$p \xrightarrow[\text{emp}]{w} q.$$

*Case 2* The case where $A_{pq} \Rightarrow A_{pr} A_{rq} \overset{(k)}{\Rightarrow} w$ holds.

Let $w$ be expressed as $xy$ so that $A_{pr} \overset{\le k}{\Rightarrow} x$ and $A_{rq} \overset{\le k}{\Rightarrow} y$, where $\overset{\le k}{\Rightarrow}$ means derivations of steps less than $k$. By the inductive hypothesis we have $p \xrightarrow[\text{emp}]{x} r$ and

$r \xrightarrow[\text{emp}]{y} q$. Thus, since $w = xy$, we have

$$p \xrightarrow[\text{emp}]{w} q.$$

*Proof of $\Leftarrow$ of Fact* The proof is by induction on the number $k$ of steps of transitions. For $k = 1$, the transition of the form $p \xrightarrow[\text{emp}]{w} q$ is only $p \overset{\varepsilon}{\to} p$. On the other hand, we have $A_{pp} \Rightarrow \varepsilon$ by definition, thereby completing the proof of $\Leftarrow$ for this case.

Next, assume that $\Leftarrow$ of Fact holds when the number of steps is less than or equal to $k$. Furthermore, suppose that $p \xrightarrow[\text{emp}]{w} q$ by transitions of $k + 1$ steps.

*Case 1* The case where the stack does not become empty during $k + 1$ transitions except for the beginning and the end.

Since in this case the first move is push and the last move is pop, there exist $q, r \in Q$, $a, a' \in \Sigma_\varepsilon$, and $b \in \Gamma$ such that

$$w = axa',$$

$$p \xrightarrow{a, \varepsilon \to b} r \xrightarrow[\text{emp}]{x} s \xrightarrow{a', b \to \varepsilon} q.$$

Then, by the inductive hypothesis we have

$$A_{rs} \overset{*}{\Rightarrow} x.$$

Furthermore, since $p \xrightarrow{a,\varepsilon \to b} r$ and $s \xrightarrow{a',b \to \varepsilon} q$, we have

$$A_{pq} \to a A_{rs} a',$$

by the definition of the rules of CFG $G$. Thus we have

$$A_{pq} \Rightarrow a A_{rs} a' \overset{*}{\Rightarrow} axa' = w.$$

*Case 2* The case where the stack becomes empty during the $k+1$ transitions except for the beginning and the end.
In this case there exist $r \in Q$ and $x, y \in \Sigma^*$ such that

$$p \xrightarrow[\text{emp}]{x} r, \qquad r \xrightarrow[\text{emp}]{y} q,$$

where $w = xy$. Then, since the numbers of steps of these transitions are less than or equal to $k$, we have

$$A_{pr} \overset{*}{\Rightarrow} x, \qquad A_{rq} \overset{*}{\Rightarrow} y,$$

by the inductive hypothesis. Thus we have

$$A_{pq} \Rightarrow A_{pr} A_{rq} \overset{*}{\Rightarrow} xy = w.$$

As stated in the paragraph before Fact, this completes the proof of the theorem. $\square$

## 5.3　Problems

**5.1** Answer the following questions about the pushdown automaton $M$ given by the state diagram below.

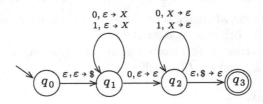

(1) Give concise descriptions of $L(M)$.

(2) Recall that Condition $C$ concerning PDA is such that any transition takes the form $p \xrightarrow{a,b \to c} q$ with $|b| = 1$ and $C = \varepsilon$, or $b = \varepsilon$ and $|c| = 1$, where $a$, $b$ and $c$ are in $\sum_{\varepsilon}$. Obviously, the $M$ does not satisfy Condition $C$. Give a PDA equivalent to the $M$ that satisfies Condition $C$.

(3) Give a context-free grammar equivalent to the pushdown automaton by using the method given in the proof of Theorem 5.8.

(4) Construct a context-free grammar directly that generates the language $L$ given in (1), and compare it with the context-free grammar given in (2).

**5.2\*** Let the pushdown automaton given below be denoted by $M$. Give concise descriptions of $L(M)$.

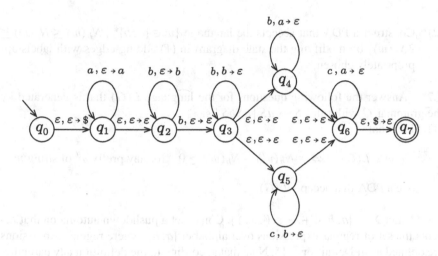

**5.3\*\*** Construct a pushdown automaton that accepts the language

$$\{a^i b^j c^k \mid i, j \text{ and } k \text{ are nonnegative integers such that } i = j \text{ or } j = k\}.$$

**5.4\*\*\*** Construct a pushdown automaton that accepts the language

$$\{a^i b^j c^k d^l \mid i, j, k \text{ and } l \text{ are nonnegative integers such that } i + j = k + l\}.$$

**5.5\*\*\*** Construct a pushdown automaton that accepts the language

$$\{a^i b^j \mid i \text{ and } j \text{ are nonnegative integers such that } i \neq j\}.$$

**5.6** Answer the following questions.

(1) Let the pushdown automaton given below be denoted by $M$. Give concise descriptions of $L(M)$.

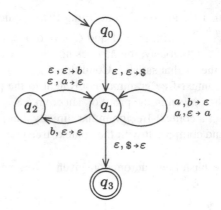

(2)* Construct a PDA that accepts the language $\{w \in \{a, b\}^* \mid N_b(w) \le N_a(w) \le 2N_b(w)\}$ by modifying the state diagram in (1) adding edges with labels appropriately chosen.

**5.7**** Answer the following questions for the language $L(G)$ that is generated by the grammar $G$ with rules $S \to aS \mid aSbS \mid \varepsilon$.
(1) Show that

$$w \in L(G) \quad \Leftrightarrow \quad N_a(w') - N_b(w') \ge 0 \quad \text{for any prefix } w' \text{ of string } w.$$

(2) Give a PDA that accepts $L(G)$.

**5.8***** Let $\Sigma = \{a, b, \cdot, +, *, \varepsilon, \emptyset, (, )\}$. Construct a pushdown automaton that accepts the set of regular expressions over alphabet $\{a, b\}$, where regular expressions are defined as in Definition 3.15. Note that according to the definition any parenthesis is not omitted.

# Part III

# Computability

# Turing Machine

6

The Turing machine is a model introduced by Alan Turing which performs simple
instructions for carrying out a computational task. Turing introduced that model to
clarify what a mechanical process computes. We can be sure that what a Turing machine can compute is considered to be computed mechanically. The point of Turing's
argument is its converse: what can be computed mechanically can be computed by
a Turing machine. Turing developed arguments to support this direction of thought.

## 6.1   Turing Machine

As shown in Fig. 6.1, the Turing machine is a computational model obtained from
a finite state automaton by replacing a tape of a finite length with a tape of infinite length. The tape head of the Turing machine can read and write symbols, and,
furthermore, can move both to the left and to the right on the tape. As in the case
of a finite automaton, a move of one step is specified by the transition function $\delta$.
Suppose that the machine is in state $q$ and the head is on the square containing 0, as
shown in Fig. 6.1. If $\delta$ is specified as $\delta(q, 0) = (q', 1, L)$, then the machine goes to
state $q'$, writes 1 replacing the 0, and moves the head to the left, which is shown in
Fig. 6.2.

Figure 6.1 shows the situation in which the state is $q$, the content of the tape
is string 110101, and the head is over the third square from the leftmost square. It
is assumed that all the squares except the squares containing the string 110101 are
placed with the blank symbols $\sqcup$. The computation of a Turing machine begins at
the following situation: consecutive leftmost squares of the tape contain an input
string followed by an infinite number of the blank symbols; the machine is in the
start state; the head is over the leftmost square of the tape. Once started, the computation proceeds: the machine repeats changing the tape contents, the state, and the
head position according to the rules specified by the transition function. For a finite automaton the computation halts when the automaton has read the entire input,
whereas for a Turing machine the computation halts when the machine goes to an
accept state or a reject state unless otherwise stated. As an outcome of computation,

A. Maruoka, *Concise Guide to Computation Theory*,
DOI 10.1007/978-0-85729-535-4_6, © Springer-Verlag London Limited 2011

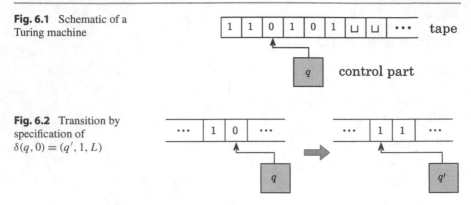

**Fig. 6.1**  Schematic of a Turing machine

**Fig. 6.2**  Transition by specification of $\delta(q, 0) = (q', 1, L)$

we have the three possibilities that the machine goes to an accept state, goes to a reject state, or never goes to an accept state or a reject state, moving on forever. Corresponding to these three possibilities, a Turing machine is said to accept, reject, or *loop*, respectively. We assume that there are three kinds of states, that is, accept states, reject states, and the remaining states which are neither accept states nor reject states. A string initially given on the tape is accepted if the machine eventually goes to an accept state and is rejected if it eventually goes to a reject state. A Turing machine fails to accept a string initially given on the tape if the machine eventually goes into a reject state or goes on forever, never going to an accept state or a reject state.

In this book we mostly think of a Turing machine as an accepting machine which accepts a language that consists of strings accepted by the Turing machine. However, we sometimes think of a Turing machine as a transformer which, given a string as input, yields a string as output when the machine eventually halts. Such a Turing machine can be thought of as computing a function $f : \Sigma^* \to \Sigma^*$ when the machine starts its computation with string $w$ on the tape, and eventually goes to a halting state with string $f(w)$ left on the tape, where the tape contains string $f(w)$ on its leftmost squares and blank symbols everywhere else.

A Turing machine can alternatively be defined as one that has two tapes, a finite input tape with a read-only head for an input string and an infinite work tape with a read-write head. When we think of a Turing machine this way, a pushdown automaton can be obtained from a Turing machine by imposing restrictions on how to access the work tape: the machine can read and write a symbol only on the rightmost square of the string on the work tape; the heads of the input tape can go only rightward. In this book we mostly deal with a Turing machine with a single tape that works as the input tape and the work tape as well.

As in the case of a pushdown automaton, a Turing machine has an input alphabet and a tape alphabet, where the input alphabet, denoted by $\Sigma$, is for input and the tape alphabet, denoted by $\Gamma$, is for read-write use on the tape. Let these alphabets be such that

$$\Sigma \subseteq \Gamma, \quad \sqcup \notin \Sigma, \sqcup \in \Gamma.$$

Putting together what we have described so far, we have the formal definition of a Turing machine as follows.

**Definition 6.1** A *Turing machine* (TM) is a 7-tuple $(Q, \Sigma, \Gamma, \delta, q_s, q_{\text{accept}}, q_{\text{reject}})$, where

(1) $Q$ is the finite set of states.
(2) $\Sigma$ is the input alphabet, where $\Sigma$ is a finite set such that $\sqcup \notin \Sigma$.
(3) $\Gamma$ is the tape alphabet, where $\Gamma$ is a finite set such that $\Sigma \subseteq \Gamma$ and $\sqcup \in \Gamma$.
(4) $\delta : Q \times \Gamma \to Q \times \Gamma \times \{L, R\}$ is the transition function.
(5) $q_s \in Q$ is the start state.
(6) $q_{\text{accept}} \in Q$ is the accept state.
(7) $q_{\text{reject}} \in Q$ is the reject state, where $q_{\text{accept}} \neq q_{\text{reject}}$.

The information that specifies the entire current situation of a Turing machine is called a *configuration*. Informally, a configuration contains all the information that is necessary for a Turing machine that is temporarily stopped to be restarted. More specifically, a configuration consists of the state, the current tape contents, and the current head location. We adopt a special way to represent a configuration. For example, in the case of Fig. 6.1, the configuration is represented as $11q0101$. Note that $q$ in the string indicates not only the current state, but also the current head position: the head is currently on the left 0. Since the transition is specified as

$$\delta(q, 0) = (q', 1, L),$$

the configuration changes as

$$11q0101 \Rightarrow 1q'11101.$$

In what follows we exclusively deal with the case where the entire tape contents are expressed as $w_{\sqcup\sqcup} \cdots$ for $w \in (\Gamma - \{\sqcup\})^*$ unless stated otherwise. In this case we simply say that the tape contents are $w$. So, we do not deal with a case where the configuration is expressed as, say, $01q_{\sqcup}10$. The tape contents of a configuration are assumed to be always supplied with the blank symbol from the right as needed at the right end of the contents.

We now explain generally how transition occurs in terms of configurations. Let $u, v \in \Gamma^*$ and $q, q' \in Q$. Then we have

$$\text{if } \delta(q, b) = (q', b', R), \quad \text{then let } uaqbv \Rightarrow uab'q'v.$$

And similarly,

$$\text{if } \delta(q, b) = (q', b', L), \quad \text{then let } uaqbv \Rightarrow uq'ab'v.$$

These do not include all cases of transitions. The transitions such that the head goes beyond the nonblank region remain unspecified. We specify such transitions below.

First we treat the case where the head is over the rightmost square of the nonblank region. The configuration in such a case is expressed as $uq$ for $u \in \Gamma^*$. Since $uq$ is equivalent to $uq_\sqcup$, we can apply the transition mentioned above. On the other hand, when the head is over the leftmost square and tries to move to the left off the tape we specify as follows:

$$\text{if } \delta(q, b) = (q', b', L), \quad \text{then let } qbv \Rightarrow q'b'v.$$

That is, in such a case a Turing machine is assumed to stay over the leftmost square and to change the current state and the current tape contents according to the transition function. But, as described below, it turns out that we can always design a Turing machine so as not to make the head move off the tape even if we do not adopt the convention described above about the head move going off the tape. Namely, by employing the technique for pushdown automata in Chap. 5, we can initially put a special symbol on the leftmost square showing the leftmost end of the tape and specify the transition function so that, when the head is over the special symbol afterward, the head never moves to the left off the tape. More specifically, if a Turing machine writes initially a symbol, say, $a$ on the leftmost square, we make the machine write $a_\vdash$ rather than $a$. Furthermore, if the Turing machine writes a symbol, say, $b$ replacing $a_\vdash$, then we make the machine write the symbol $b_\vdash$. If we specify the transition function this way, the symbols that appear on the leftmost square are always attached with $\vdash$ as a subscript so that the machine can prevent the head from moving off the tape by recognizing the leftmost square with subscript $\vdash$.

Next, we explain about a string and a language accepted by a Turing machine in terms of transition of configurations mentioned so far. Recall that a Turing machine eventually halts, moving to an accept state or a reject state, or never halts, going on forever. A Turing machine accepts a string if the machine starting its computation with the string on its tape eventually halts, moving to an accept state. A Turing machine never accepts a string if the machine continues transitions forever without moving to an accept state. More precisely, a Turing machine *accepts* string $w$ if there exist configurations $C_1, \ldots, C_m$ such that

$$C_1 \Rightarrow C_2 \Rightarrow \cdots \Rightarrow C_m,$$

and such that $C_1$ is expressed as $q_s w$ and $C_m$ is expressed as $uq_{\text{accept}}v$, where $q_s$ is the start state, $q_{\text{accept}}$ is an accept state, and $u$ and $v$ are strings in $\Gamma^*$. We call such configurations $C_1$ and $C_m$ a *start configuration* and an *accepting configuration*, respectively. A Turing machine $M$ *accepts* the language that consists of the strings accepted by $M$. The language accepted by Turing machine $M$ is denoted by $L(M)$.

Let us assume that a Turing machine $M$ eventually halts, moving either to an accept state or a reject state, but never continues to go on forever. If such a Turing machine $M$ accepts a language, then $M$ is said to *decide language $L(M)$*. That is, when $w$ is given as input, if $w \in L(M)$, then $M$ eventually goes to an accept state, whereas if $w \notin L(M)$, then $M$ eventually goes to a reject state. Then what is the difference between accepting a language and deciding a language? When a Turing machine $M$ decides a language, the machine $M$, when a string is given as input, decides whether the string is accepted or rejected because $M$ eventually halts, going to an accept state or a reject state. On the other hand, when a Turing machine $M$ accepts a language, the machine $M$, when a string is given as input, eventually goes to an accept state if the string belongs to $L(M)$, but if the string does not belong to $L(M)$, there is a possibility that $M$ never makes a decision, continuing transitions forever. In such a case, no matter how long we run the machine, it never tells whether the input is accepted or rejected. It will be shown in Chap. 7 that there exists a language that can be accepted by a Turing machine, but cannot be decided by *any* Turing machine. Such a language will be given in terms of a problem somehow defined.

**Definition 6.2** If a Turing machine accepts a language, the language is said to be *Turing-recognizable* or simply *recognizable*. If a Turing machine that eventually halts for all inputs accepts a language, the language is said to be *Turing-decidable* or simply *decidable*.

So far we have mostly thought of a Turing machine as a machine accepting or deciding a language. In what follows we occasionally think of a Turing machine as a machine that computes a function from $\Sigma^*$ to $\Sigma^*$ as described in the following definition.

**Definition 6.3** A Turing machine $M$ *computes* a function $f : \Sigma^* \rightarrow \Sigma^*$ if, given a string $w$ as input, $M$ starts with its start configuration and eventually halts with just $f(w)$ on its tape.

## Examples of Turing Machines

As is shown in Sect. 6.4, the Church–Turing thesis claims that what algorithms compute is exactly what Turing machines compute. Roughly speaking, an algorithm is a sequence of simple instructions for carrying out a certain computational task. To grasp the thesis intuitively, we study a couple of problems, and we present an algorithm in high-level description that solves each of these problems together with a Turing machine equivalent to the algorithm. Some evidence that supports the Church–Turing thesis is given in Sect. 6.4.

*Example 6.4* In Example 5.7, a pushdown automaton that accepts strings of well-nested parentheses was given. Here, we give a Turing machine that carries out the same task.

In order to decide whether a string of parentheses is well-nested or not, we only have to design an algorithm that works as follows: continue crossing off a pair of corresponding left and right parentheses until no further such pair remains; decide that the string is well-nested when all the parentheses in the input are crossed off. We present the algorithm *WELL-NESTED-PARENTHESES* based on the idea mentioned above. Note that in the algorithm replacing a parenthesis with $X$ means that

**Table 6.1**  Transition function of TM that accepts strings of well-nested parentheses

| State\Symbol | ) | ( | (⊢ | $X$ | $X_\vdash$ | ⊔ |
|---|---|---|---|---|---|---|
| $q_0$ | – | $(q_1, (_\vdash, R)$ | – | – | – | – |
| $q_1$ | $(q_2, X, L)$ | $(q_1, (, R)$ | – | $(q_1, X, R)$ | – | $(q_3, \llcorner, L)$ |
| $q_2$ | – | $(q_1, X, R)$ | $(q_1, X_\vdash, R)$ | $(q_2, X, L)$ | $(q_\text{reject}, -, -)$ | – |
| $q_3$ | – | $(q_\text{reject}, -, -)$ | $(q_\text{reject}, -, -)$ | $(q_3, X, L)$ | $(q_\text{accept}, -, -)$ | – |

the parenthesis is crossed off and that in stage 1 the indented portion "Replace the innermost right and left parentheses with two $X$'s" will be repeated as long as a pair of such right and left parentheses remains.

**Algorithm** *WELL-NESTED-PARENTHESES*
Input: a string in $\{(,)\}^*$.
**1**. Execute the following until no pair of the innermost right and left parentheses remains.
    Replace the innermost right and left parentheses with two $X$'s.
**2**. *Accept* if all the parentheses are replaced with $X$'s, and *reject* otherwise.

In Table 6.1 we give a precise description of a Turing machine that behaves as the algorithm *WELL-NESTED-PARENTHESES* does. Figure 6.3 gives a sequence of moves of the Turing machine that accepts $((())())$.

In Fig. 6.3 the vertical axis represents the progress of computation, whereas the horizontal axis represents position on the tape. The first row of the figure expresses the starting configuration with the start state $q_0$ and the head on the leftmost square. Then the head moves rightward searching for a right parenthesis. Through three transitions the machine gets to the configuration of the second row. Then, according to the specification of $\delta(q_1, )) = (q_2, X, L)$, the right parenthesis found is replaced with $X$ and the machine goes from $q_1$ to $q_2$. In state $q_2$, the head moves leftward, searching for a left parenthesis that corresponds to the right parenthesis just found, thus getting to the configuration on the third row. At this time the machine succeeds in finding the innermost right and left parentheses. According to $\delta(q_2, () = (q_1, X, R)$, the left parenthesis is replaced with $X$ and then the machine goes from $q_2$ to $q_1$. Then the machine continues searching the innermost right and left parentheses again. In searching such parentheses, $X$'s are skipped off and symbols $(_\vdash$ and $X_\vdash$ are regarded as the ones on the leftmost square on the tape.

In the case of the example, the input alphabet is $\Sigma = \{(,)\}$, the tape alphabet is $\Gamma = \{(, (_\vdash, ), X, X_\vdash, \llcorner\}$, and the set of states is $Q = \{q_0, q_1, q_2, q_3, q_\text{accept}, q_\text{reject}\}$. Furthermore, since the machine halts in the accept and reject states, we do not need to specify the transition function for these states. So the transition function takes the form $\delta : (Q - \{q_\text{accept}, q_\text{reject}\}) \times \Gamma \to Q \times \Gamma \times \{L, R\}$. In addition to accept and reject states, we also have some pairs $(q, a)$ that do not occur in any computation leading to an accepting configuration. For such

|   | 1 | 2 | 3 | 4 | 5 | 6 | 7 | 8 | 9 | 10 |
|---|---|---|---|---|---|---|---|---|---|----|
| 1 | ( [$q_0$] | ( | ( | ) | ) | ( | ) | ) | ⊔ | ⊔ |
| 2 | (⊢ | ( | ( | ) [$q_1$] | ) | ( | ) | ) | ⊔ | ⊔ |
| 3 | (⊢ | ( | ( [$q_2$] | X | ) | ( | ) | ) | ⊔ | ⊔ |
| 4 | (⊢ | ( | X | X | ) [$q_1$] | ( | ) | ) | ⊔ | ⊔ |
| 5 | (⊢ | ( [$q_2$] | X | X | X | ( | ) | ) | ⊔ | ⊔ |
| ⋮ | | | | | ⋮ | | | | | |
| 6 | (⊢ | X | X | X | X | X | X | ) [$q_1$] | ⊔ | ⊔ |
| 7 | (⊢ [$q_2$] | X | X | X | X | X | X | X | ⊔ | ⊔ |
| 8 | X⊢ | X | X | X | X | X | X | X | ⊔ [$q_1$] | ⊔ |
| 9 | X⊢ [$q_3$] | X | X | X | X | X | X | X | ⊔ | ⊔ |

**Fig. 6.3** Moves of Turing machine until ( ( ( ) ) ( ) ) is accepted

pairs we may specify the transition function vacuously so as to move to a reject state. But to keep things simple we leave the transition function unspecified for such pairs. In Table 6.1 "–" indicates such items being left unspecified.

It is not easy to see that the Turing machine expressed in Table 6.1 behaves like the algorithm *WELL-NESTED-PARENTHESES*. To make it easier to figure out how a Turing machine behaves, we introduce a state diagram for a Turing machine instead of a transition table.

The state diagram of a Turing machine is a graph such that each specification of $\delta(q, a) = (q', a', D)$ is expressed as a labeled edge between states like

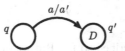

where $D \in \{L, R\}$. It is easy to see that in order for a transition function to be expressed by such a state diagram, the transition function must satisfy a certain condition: the direction of the head movement solely depends on the state that the

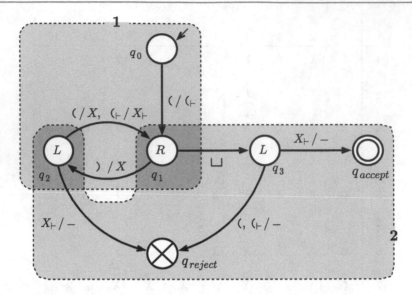

**Fig. 6.4** State diagram of TM that accepts strings of well-nested parentheses

machine goes to. That is, if

$$\delta(q_1, a) = (q, a', D'), \qquad \delta(q_2, b) = (q, b', D''),$$

then $D' = D''$ holds. In general, a Turing machine does not necessarily satisfy this condition, but it turns out that a Turing machine designed naturally to do a certain task mostly satisfies this condition. In fact, the state diagram in Fig. 6.4, which represents the Turing machine given by Table 6.1, satisfies the condition. In what follows we assume that a Turing machine satisfies this condition unless stated otherwise.

Here are some abbreviations for drawing state diagrams. If $\delta(q, a) = (q, a, D)$, then the transition is expressed as

according to the method described so far. But we omit this edge together with the label. So once a machine is in state $q$ with direction $D$, it continues moving toward $D$ until the head encounters a symbol on its label of an outgoing edge. Furthermore, if $\delta(q, a) = (q', a, D)$, then the label of the edge from $q$ to $q'$ is expressed as follows:

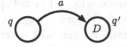

That is, in this case we write simply $a$ as the label instead of $a/a$ based on the general rule. The transition with label $a$ should be interpreted as saying that the machine goes from $q$ to $q'$ when the head encounters $a$, but without changing the symbol.

Finally the start state, the accept state, and the reject state are expressed as follows, respectively:

Figure 6.4 gives the state diagram which corresponds to the transition table in Table 6.1. In this diagram, when there are multiple edges between two states with different labels, these edges are put together into a single edge which has its own label representing all the labels of the original multiple edges. For example, the edges from $q_3$ and to $q_{reject}$ with labels $(/-$ and $(\vdash/-$ are put together to form label $(, (\vdash/-$.

The state diagram shown in Fig. 6.4 is the one that implements the algorithm *WELL-NESTED-PARENTHESES* in terms of a Turing machine. To begin with, note that portions 1 and 2 of the state diagram in Fig. 6.4 surrounded by the dashed lines correspond to stages **1** and **2** in the algorithm, respectively. The main parts of the state diagram are the transition from $q_1$ to $q_2$, crossing off a right parenthesis, and the transition from $q_2$ back to $q_1$, crossing off the corresponding left parenthesis. By repeating these transitions the machine continues crossing off the innermost right and left parentheses by writing $X$ to cross off these parentheses. In doing so the machine needs symbols that signify the right and left ends of the string. In our case $(\vdash$ and $X_\vdash$ signal the left end, whereas the blank $\sqcup$ signals the right end. Observe that while the machine is in state $q_1$ the head moves rightward searching for $)$, but skipping symbols except for $)$ and $\sqcup$. Similarly, while the machine is in state $q_2$ the head moves leftward searching for $($ and $(\vdash$, but skipping symbols except for $($, $(\vdash$, and $X_\vdash$. The string is accepted if all the parentheses are replaced with $X$ and $X_\vdash$ in this way, and it is rejected if either of the parentheses are left without being replaced with $X$. If all the right parentheses are replaced with $X$, then the machine goes from $q_1$ to $q_3$. Furthermore, if all the left parentheses are replaced as well, then the machine goes further to $q_{accept}$, thereby accepting the string. On the other hand, if, going leftward, the machine in state $q_3$ finds $($ or $(\vdash$ that is not replaced with $X$ or $X_\vdash$, then it goes to $q_{reject}$, thereby rejecting the string. Similarly, if, going leftward, the machine in state $q_2$ fails to find $($ or $(\vdash$, but finds $X_\vdash$, then it goes to $q_{reject}$. This is an outline of the behavior of the state diagram of Fig. 6.4. Finally, note that to simplify illustration the states lack some outgoing transitions in either

Table 6.1 or Fig. 6.4; for example, $\delta(q_0, )$ is not specified in both. To complete the specification it suffices to specify, e.g., $\delta(q_0, )) = (q_{\text{reject}}, ), R)$. To simplify things we sometimes omit transitions that will never lead to accepting configurations.

*Example 6.5* We explain a Turing machine $M$ that performs multiplication. Expressing a number as length of a string, when given $a^i b^j$ as input, $M$ produces $c^{i \times j}$ as output, where $i \geq 1$ and $j \geq 1$. To do this, when given $a^i b^j$ as input, $M$ repeats writing $c^j$ $i$ times. To explain how $M$ behaves, let us call the collection of consecutive squares containing symbol $a$ the $A$ region, and similarly for the $B$ region, $C$ region, and $\sqcup$ region. To count up to $i$, $M$ places a mark above an unmarked symbol in the $A$ region every time $M$ writes $c^j$. Likewise, to count up to $j$, $M$ places a mark above an unmarked symbol in the $B$ region every time $M$ places a single $c$. In other words, $M$ counts the number by repeatedly placing a mark above a symbol in a certain region. In the actual implementation, the machine uses two symbols, say, $b$ and $B$, where a $B$ means a marked $b$.

The idea above leads to the algorithm *MULT* below.

**Algorithm** *MULT*
Input: $a^i b^j$
Output: $c^{i \times j}$
1. If any unmarked $a$ remains, mark the leftmost of such $a$'s and go to **2**. Otherwise, go to **3**.
2. Repeat marking the leftmost unmarked $b$ and placing $c$ on the leftmost square of the $\sqcup$ region until no unmarked $b$ remains. Then, after unmarking all the marked $b$'s, go to **1**.
3. Move the string consisting of $c$'s leftward so that the string of $c$'s begins from the leftmost square.

Figure 6.5 gives a state diagram of a Turing machine that implements the algorithm *MULT*. Suppose that the Turing machine is given as input $a^3 b^4$, that is, $i = 3$ and $j = 4$. When *MULT* is executing stage 2 the second time yielding three $c$'s in the $\sqcup$ region, the Turing machine has

$$A \vdash Aa B B B b c c c c c c c_{\sqcup} \cdots$$

on its tape, because the first execution of stage 2 yields $c^4$ and the second yields $c^3$. So the total number of $c$'s written is $4 + 3 = 7$.

As shown in Fig. 6.5, the state diagram consists of three parts surrounded by dashed lines which correspond to stages **1**, **2**, and **3** of the algorithm *MULT*, respectively. We explain how these three parts behave as follows.

Behavior of **1**: Searching for an unmarked $a$, the head moves rightward from the leftmost square. If an unmarked $a$ remains, replace the $a$ with $A_\vdash$ or $A$ by $q_0 \xrightarrow{a/A_\vdash} q_2$ or $q_1 \xrightarrow{a/A} q_2$ and go to (2). Otherwise, replace $c$ with $b$ by $q_1 \xrightarrow{c/b} q_8$ and go to (3). The reason why $c$ is replaced with $b$ is explained in (3) below.

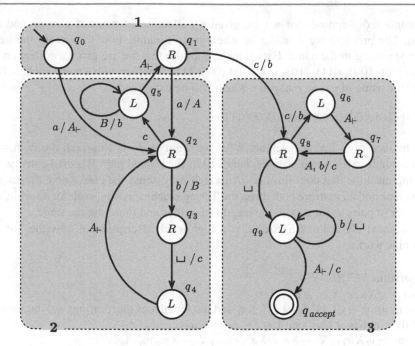

**Fig. 6.5** State diagram of TM that performs multiplication

**Behavior of (2)** This is the stage where, after marking an $a$ in stage (1), the machine copies in the $\sqcup$ region a string of $c$'s of the same length as that of the $B$ region. More specifically, repeat marking the leftmost $b$ in the $B$ region and then writing $c$ on the leftmost blank square through $q_2 \xrightarrow{b/B} q_3 \xrightarrow{\sqcup/c} q_4 \xrightarrow{A_\vdash} q_2$ until all $b$'s in the $B$ region are marked. Then, unmark all the marked $b$'s by $q_5 \xrightarrow{B/b} q_5$ and move to the leftmost $A_\vdash$. After preparing for the next computation this way, go to (1).

**Behavior of (3)** This is the stage where the machine moves the string of $c$'s in the $C$ region to the leftmost part of the tape and replaces all the remaining symbols with $\sqcup$. More specifically, repeat the following until no $c$ remains in the $C$ region: delete the leftmost $c$ by $q_8 \xrightarrow{c/b} q_6$; move the head leftward to the $A_\vdash$; then move the head rightward; and finally replace the first $A$ or $b$ with $c$. We must be careful to keep $A_\vdash$ as the leftmost mark throughout the computation. Usually deleting a $c$ and restoring it later is done by the pair of transitions $q_8 \xrightarrow{c/b} q_6$ and $q_7 \xrightarrow{A,b/c} q_8$, respectively. But in order that $A_\vdash$ on the leftmost square signals the leftmost end, we make the $A_\vdash$ survive until it is replaced with $c$ by $q_9 \xrightarrow{A_\vdash/c} q_{accept}$. For restoring the $c$, there corresponds the $c$ placed on the leftmost square of the $C$ region. For this case, deleting a $c$ and restoring it later is done by the pair of transitions $q_1 \xrightarrow{c/b} q_8$ and $q_9 \xrightarrow{A_\vdash/c} q_{accept}$. This is the reason why we replace the leftmost $c$ with $b$ by $q_1 \xrightarrow{c/b} q_8$ without restoring the corresponding $c$ in stage (1).

*Example 6.6* Suppose that we are given a collection of pairs of a name and an item. The problem we consider is, when given a name, how to pick up the item corresponding to the name. For example, suppose that we are given a collection of three pairs $(001, 011)$, $(101, 110)$, and $(000, 111)$ together with a name such as $101$ (i.e., the name of the second pair), which will be expressed as

$$\vdash 101X001011X101110X000111 \dashv$$

on the tape of a Turing machine. What we want to do is to search the item that corresponds to the name, namely, item $110$ of the second pair. We shall construct a Turing machine that does this task. Although we discuss the case where the length of a name and an item are both three, the Turing machine has to work for an arbitrary number of pairs as long as all the lengths of names and items are the same.

The next algorithm *RETRIEVE* is a high-level description of how the Turing machine works.

**Algorithm** *RETRIEVE*

Input: $\vdash uXu_1v_1X\cdots Xu_mv_m \dashv$
Here $u, u_i, v_i \in \{0, 1\}^*$ for $1 \le i \le m$ and the lengths of these strings are the same, and there exists at least one $1 \le j \le m$ such as $u = u_j$.
Output: $\vdash \hat{v}_iX\hat{u}_1\hat{v}_1X\cdots X\hat{u}_i\hat{v}_iXu_{i+1}v_{i+1}X\cdots Xu_mv_m \dashv$
Here $i$ is the least integer such as $u = u_i$.
1. Search for $u_i$ such as $u = u_i$ through $u_1, u_2, \ldots, u_m$.
2. Copy $\hat{v}_i$ in the place where $u$ originally was.

In this algorithm, $\hat{u}$ and $\hat{v}$ are the strings obtained by replacing the 0's and 1's in $u$ and $v$ with $A$'s and $B$'s, respectively. For example, if $v = 110$, then $\hat{v} = BBA$. $A$ and $B$ are interpreted as marked 0 and 1, respectively.

The Turing machine that implements the algorithm is shown in Fig. 6.6. When the machine starts in the configuration

$$\vdash 101X001011X101110X000111 \dashv$$
$$\Box$$

with $101$ between $\vdash$ and $X$ given as a name, it locates the second pair $(101, 110)$ and halts in the configuration

$$\vdash BBAXAABABBXBABBBAX000111 \dashv$$
$$\Box$$

placing $BBA$, which is marked $110$, between $\vdash$ and $X$.

The state diagram shown in Fig. 6.6 consists of the two parts corresponding to stages **1** and **2** in the algorithm, respectively. Figure 6.7 shows a sequence of configurations for the case mentioned above. In the figure (1) is the starting configuration. The head moves rightward until it reads $X$, and then moves leftward, replacing $101$

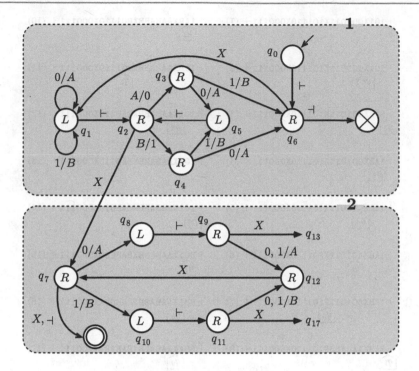

**Fig. 6.6** State diagram of TM that locates the item of a given name

with $BAB$. When $\vdash$ is encountered, the machine goes to $q_2$ and remembers a symbol encountered as a state. That is, when $A$ is encountered the machine goes to $q_3$, whereas when $B$ is encountered the machine goes to $q_4$. Then, going rightward the machine checks if the 0 or 1 first encountered coincides with what is remembered as a state. If they coincide, it goes to $q_5$. Otherwise, it goes to $q_6$. If the machine goes to $q_5$, it does the same check for the next bit. Once a certain bit is found not to be matched, then the machine skips off the current pair and proceeds to the next pair which will be checked in a similar way. Furthermore, strings of the skipped pair are converted to strings of $A$ and $B$, which shows that they have been checked already.

(11) shows the moment when the first pair (001,011) is skipped and, for the second pair (101,110), the first bit of 101 is checked for coincidence, and its second bit is about to be checked to see if it coincides with what is remembered, i.e., the 0 of $\vdash 101X$

$$\vdash \overbrace{10B}^{u} X \overbrace{AAB}^{\hat{u}_1} \overbrace{ABB}^{\hat{v}_1} X B\overbrace{01}^{u_2} \overbrace{110}^{v_2} X \overbrace{000}^{u_3} \overbrace{111}^{v_3} \dashv$$
$$\boxed{q_3}$$

It should be noted that marked symbols are expressed differently in $u$ and in $(u_1, v_1)$, $(u_2, v_2)$, and $(u_3, v_3)$. That is, for string $u$ marked symbols are denoted

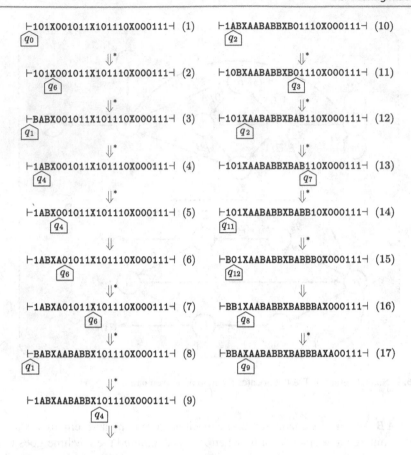

**Fig. 6.7** Transitions of configurations of TM in Fig. 6.6

by 0 and 1, whereas for strings of the pairs they are denoted by $A$ and $B$. Since the marked symbols are expressed this way, the machine in states $q_3$ or $q_4$ can skip through the unmarked portion of $u$ and the marked portion of the pairs, both portions being represented by $A$'s and $B$'s. Furthermore, when a name in the pairs is found to coincide with the name originally given, the machine proceeds to 2 in Fig. 6.6 by going from $q_2$ to $q_7$. In 2 the machine writes item $v_2$ found in the squares where the name 101 was originally placed.

Observe that the Turing machine constructed works for an arbitrary number of pairs as long as the lengths of names and items are the same. Finally note that the state diagram in Fig. 6.6 is modified and used as a part of the universal Turing machine which will be explained in Sect. 7.1. So we leave the state diagram as it is, although there remain states $q_{13}$ and $q_{17}$ which are not explained in this example.

## 6.2 Variants of Turing Machines

The definition of the Turing machine in Definition 6.1 is a typical and standard one, but it is by no means the only one. We can think of some variants of that definition by somehow generalizing the definition. In this and the next sections we generalize the definition in three ways: from a singly infinite tape to an infinite tape in both directions, from a single tape to multiple tapes, and from a deterministic Turing machine to a nondeterministic one. The point here is that Turing machines obtained by generalizing in this way are equivalent with each other in power. To show the equivalence we simply show that we can simulate one variant by the other.

### Simulating Bilaterally Infinite Tape TM by Unilaterally Infinite Tape TM

The Turing machine discussed so far has a tape infinite in the right direction, and is called a *unilaterally infinite tape* Turing machine. On the other hand, a Turing machine that has a tape infinite in both directions is called a *bilaterally infinite tape* Turing machine. We shall show that a unilaterally infinite tape Turing machine can simulate a bilaterally infinite tape Turing machine by folding the bilaterally infinite tape in two and treating it as a unilaterally infinite tape with two tracks.

Figure 6.8 illustrates a bilaterally infinite tape Turing machine $M$ which is simulated by a unilaterally infinite tape Turing machine $M'$. By folding the tape in two and overlapping the two as shown in Fig. 6.9, we make a unilaterally infinite tape Turing machine $M'$, as shown in Fig. 6.10. Machine $M'$ treats a pair of symbols $(a_{-i}, a_i)$ as a single symbol. For each state $q$ of $M$ there correspond two states $q_D$ and $q_U$ of $M'$. When the head is over $(a_{-i}, a_i)$, $q_D$ makes transitions relying on $a_i$, whereas $q_U$ does so relying on $a_{-i}$. That is, when $M'$ is in state $q_D$, it simulates $M$ on the lower track in Fig. 6.10, while when $M'$ is in state $q_U$, it simulates $M$ on the upper track. So, when $M'$ is in state $q_D$, its head moves in the same direction as the head of $M$, while when $M'$ is in state $q_U$, it moves in the opposite direction of the head of $M$.

Observe that in the tape of $M'$ shown in Fig. 6.10 the leftmost square contains a single symbol of $M$ rather than a pair of symbols so that $M'$ can recognize the leftmost square. To distinguish between the blank on the leftmost square and that on the other square, we write a symbol $a$ on the leftmost square within parentheses as $(a)$. In this way, $(\sqcup)$ means the blank on the leftmost square, whereas $\sqcup$ means the blank on the other square, hence being interpreted as $(\sqcup, \sqcup)$, which will be explained shortly. Note that $M'$ with its head on the leftmost square moves its head rightward whether $M$ moves its head rightward or leftward. What is described so far is the rough idea of how $M'$ simulates $M$. We now proceed to explain how to implement the idea in terms of a transition function.

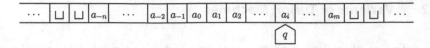

**Fig. 6.8**  Configuration of bilaterally infinite tape TM

**Fig. 6.9**  Bilaterally infinite
tape that is folded at $a_0$

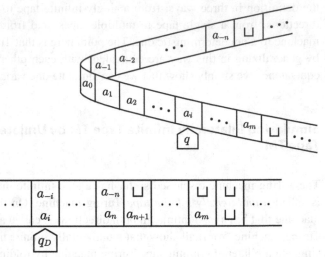

**Fig. 6.10**  Configuration of unilaterally infinite tape TM $M'$ that simulates bilaterally infinite tape
TM $M$

Let the transition function of Turing machine $M$ and that of Turing machine $M'$
be denoted by $\delta$ and $\delta'$, respectively. If $\delta$ is specified as $\delta(q, a) = (q', a', L)$, then $\delta'$
is defined as follows:

$$\delta'(q_D, (b, a)) = (q'_D, (b, a'), L) \quad \text{for each } b,$$
$$\delta'(q_U, (a, b)) = (q'_U, (a', b), R) \quad \text{for each } b.$$

Furthermore,

$$\delta'(q_D, (a)) = (q'_U, (a'), R) \quad \text{for each } b,$$
$$\delta'(q_U, (a)) = (q'_U, (a'), R) \quad \text{for each } b.$$

Since $(\sqcup, \sqcup)$ in Fig. 6.10 is initially written as $\sqcup$, we have to add the following in
defining $\delta'$: If $\delta(q, \sqcup) = (q', a', L)$, define

$$\delta'(q_D, \sqcup) = (q'_D, (\sqcup, a'), L),$$
$$\delta'(q_U, \sqcup) = (q'_D, (a', \sqcup), R).$$

It is similarly specified in the case of $\delta(q, a) = (q', a', R)$, and we omit it here.

For two Turing machines to become equivalent, the two must behave similarly when given essentially the same string initially. In our case, machine $M$ starts with tape $\cdots \sqcup\sqcup a_1 a_2 \cdots a_n \sqcup\sqcup \cdots$, whereas machine $M'$ starts with tape $a_1 a_2 \cdots a_n \sqcup\sqcup \cdots$. So $M'$ has to convert initially the tape to $(a_1)(\sqcup, a_2) \cdots (\sqcup, a_n)\sqcup\sqcup \cdots$. We omit showing how to specify $\delta'$ in order to do it.

As described above, when given a Turing machine with a infinite tape in both directions, we can construct an equivalent Turing machine with a unilaterally infinite tape. This is described as the next theorem, whose proof is omitted.

**Theorem 6.7** *For any bilaterally infinite tape Turing machine, there exists an equivalent unilaterally infinite tape Turing machine.*

## Simulating Multitape TM by Single-Tape TM

A multitape Turing machine is like an ordinary Turing machine except that it has several tapes, each having its own head for reading and writing. A multitape Turing machine starts its computation with an input being placed on the first tape, and the blank symbol being placed on every square of the remaining tapes. The transition function $\delta_k$ of a Turing machine with $k$ tapes is specified as follows:

$$\delta_k(q, a_1, \ldots, a_k) = (q', a_1', \ldots, a_k', D_1, \ldots, D_k)$$

means that if the machine is in state $q$, and $k$ heads are reading symbols $a_1$ through $a_k$, respectively, then the machine goes to state $q'$, the $k$ heads write $a_1'$ through $a_k'$, respectively, and move to the left or to the right as specified by $D_1, \ldots, D_k$, respectively.

Since a multitape Turing machine has several tapes, it would seem to be more powerful than a single-tape Turing machine. It turns out, however, that they are equivalent in power to accept languages, which will be proved in this subsection. To prove this, it suffices to show that, given a $k$-tape Turing machine $M_k$, we can construct an equivalent single-tape Turing machine $M_1$. The point of the simulation is that we think of a single tape with $k$ tracks, each corresponding to a tape of $M_k$ and each equipped with a virtual head by marking a symbol with " ^ ". Then one step of $k$-tape machine $M_k$ can be simulated by single-tape machine $M_1$ by a sequence of transitions: scan $k$ tracks all the way; store information on $k$ symbols under the $k$ virtual heads on $k$ tracks that simulate $k$ heads of $M_k$; perform transitions based on information stored by updating the state of $M_1$, the contents of $k$ tracks, and the positions of the $k$ virtual heads. The details of the simulation are explained in the proof of the next theorem.

**Theorem 6.8** *For any multitape Turing machine, there exists an equivalent single-tape Turing machine.*

*Proof* Let us denote a $k$-tape Turing machine by $M_k$. We shall show how to construct $M_1$ that simulates arbitrarily given $M_3$ and then generalize the construction to arbitrary $k$.

**Fig. 6.11** One step of
three-tape TM $M_3$

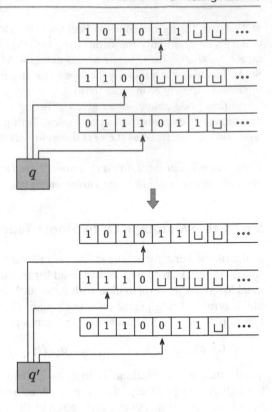

Figure 6.12 shows how $M_1$ simulates one step of $M_3$, which is given in Fig. 6.11.
As mentioned before, a symbol marked with " ˆ " in Fig. 6.12 indicates the square
where a virtual head is currently located.

Let $Q$ denote the set of states of $M_3$ and let $\Gamma$ denote the tape alphabet of $M_3$.
The state of Turing machine $M_1$ is defined to be $(q, a_1, a_2, a_3, D_1, D_2, D_3)$, where
$q \in Q$, $a_i \in \Gamma \cup \{\natural\}$, and $D_i \in \{L, R, \natural\}$ for $1 \leq i \leq 3$. Symbol $\natural$ means that the
update is not yet done after the previous step of the simulation. $M_1$ starts the sim-
ulation with the state denoted by $(q, \natural, \natural, \natural, \natural, \natural, \natural)$. Starting in this state, $M_1$ scans
through the whole nonblank portion of its tape to store the three symbols under the
virtual heads. These symbols are indicated by "ˆ". Then state $(q, \natural, \natural, \natural, \natural, \natural, \natural)$ will
be updated to $(q, 1, 0, 1, \natural, \natural, \natural)$ when $M_1$ finishes storing information on the symbols under the virtual heads shown at the upper part of Fig. 6.12. To update the state
this way, $M_1$ needs to scan the whole nonblank portion of the tape. Suppose the
transition function of $M_3$, denoted $\delta_3$, is specified as

$$\delta_3(q, 1, 0, 1) = (q', 1, 1, 0, L, L, R).$$

Then $M_1$ goes from $(q, 1, 0, 1, \natural, \natural, \natural)$ to $(q, 1, 1, 0, L, L, R)$ in one step. Then $M_1$
needs to update the contents and virtual head positions on the three tracks according
to $(q', 1, 1, 0, L, L, R)$. When the simulation for one step of $M_3$ is finished in this

**Fig. 6.12** One step of
one-tap TM $M_1$

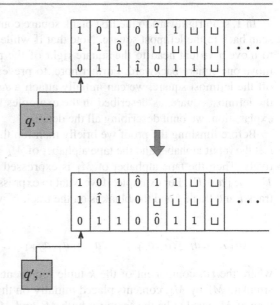

way, $M_1$ goes to state $(q', \sharp, \sharp, \sharp, \sharp, \sharp, \sharp)$, thereby preparing for the next cycle of the
simulation, where $\sharp$ indicates that the update is completed.

It is easy to generalize the argument so far to the case of a $k$-tape Turing machine for arbitrary number $k$. We summarize how to simulate $k$-tape machine $M_k$
by single-tape machine $M_1$ as follows. In the general case a state of $M_1$ is represented by $(2k + 1)$-tuple $(q, a_1, \ldots, a_k, D_1, \ldots, D_k)$ and the start state of $M_1$ is
$(q_s, \sharp, \ldots, \sharp, \sharp, \ldots, \sharp)$, where $q_s$ is the start state of $M_k$. The transition function of
$M_k$ is denoted by $\delta_k$.

The behavior of machine $M_1$ that simulates one step of machine $M_k$ is detailed
as follows:

1. Set the state of $M_1$ to $(q, \sharp, \ldots, \sharp, \sharp, \ldots, \sharp)$. Scan the tape rightward from the
   leftmost square to the first blank square and store the $k$ symbols read by the $k$
   virtual heads (i.e., marked with " $\hat{\phantom{a}}$ "). Denoting the symbol stored on the track $i$
   by $a_i$, go to state $(q, a_1, \ldots, a_k, \sharp, \ldots, \sharp)$.
2. Change the state to the new state determined by the current state of $M_k$ and
   $k$ symbols where the virtual heads are located, according to $\delta_k$. More specifically, if $\delta_k(q, a_1, \ldots, a_k) = (q', a'_1, \ldots, a'_k, D_1, \ldots, D_k)$, then change the state
   from $(q, a_1, \ldots, a_k, \sharp, \ldots, \sharp)$ to $(q', a'_1, \ldots, a'_k, D_1, \ldots, D_k)$. Furthermore, if $M_k$
   halts, then halt.
3. Moving leftward until the leftmost square is encountered, do the following for
   each track $i$: replace the symbol $\hat{a}_i$ with $a'_i$ and replace the symbol, say, $b_i$ on
   the new virtual head position with $\hat{b}_i$, where the new virtual head position is the
   square one unit to the direction indicated by $D_i$ from the previous virtual head
   position. When all of the update is done, go to state $(q', \sharp, \ldots, \sharp, \sharp, \ldots, \sharp)$. Go
   to **1**.

In **1**, scan rightward to the first blank square containing ⊔ or (⊔, ..., ⊔), and in **3**, scan back to the leftmost square. Note that if while scanning leftward it is required to move a virtual head to the square right of the previous one, the head needs to move one square backward. Furthermore, to prevent the head from going leftward off the leftmost square, we can initially attach a subscript like ⊢ to the symbol on the leftmost square, as described in the examples in Sect. 6.1. But to simplify the explanation, we omit describing all the details.

Before finishing the proof we briefly mention the tape alphabets of $M_1$ and $M_k$. Let the input alphabet and the tape alphabet of $M_k$ be denoted by $\Sigma$ and $\Gamma$, respectively. Then the tape alphabet of $M_1$ is expressed as $(\Gamma \cup \hat{\Gamma})^k \cup \Sigma \cup \{⊔\}$, where $\hat{\Gamma} = \{\hat{a} \mid a \in \Gamma\}$. For example, if we want to express that the virtual heads are on the track $i$ and track $j$ and the blank is on the track $k$, we can use the symbol expressed as

$$(a_1, \ldots, a_{i-1}, \hat{a}_i, a_{i+1}, \ldots, a_{j-1}, \hat{a}_j, a_{j+1}, \ldots, a_{k-1}, ⊔),$$

where the $i$th component of the $k$-tuple represents the symbol on the track $i$. To simulate $M_k$ by $M_1$, contents placed initially on the first tape of $M_k$ and the single tape of $M_1$ need to be the same for both $M_k$ and $M_1$. This is why $\Sigma \cup \{⊔\}$ is added to the tape alphabet of $M_1$. Let the input placed on the first tape of $M_k$ be

$$a_1 a_2 \cdots a_n ⊔ \cdots,$$

initially. First $M_1$ starts with the same input placed on the single tape and converts the content on the tape to

$$(\hat{a}_1, \hat{⊔}, \ldots, \hat{⊔})(a_2, ⊔, \ldots, ⊔) \cdots (a_n, ⊔, \ldots, ⊔)(⊔, \ldots, ⊔) \cdots,$$

then proceeds to the simulation described above. The blank ⊔ in the string first will be converted to the $k$-tuple $(⊔, \ldots, ⊔)$.                                  □

## 6.3    Nondeterministic Turing Machines

So far we have described a deterministic Turing machine. As in the case of a finite automaton or a pushdown automaton, we can think of a nondeterministic Turing machine which is allowed to choose the next move at any moment from among several possibilities. We study nondeterministic Turing machines and show that they are equivalent in computation power to deterministic Turing machines. Nondeterministic Turing machines play a very important role in Part IV.

A transition function of a deterministic Turing machine takes the form $\delta :$ $Q \times \Gamma \to Q \times \Gamma \times \{L, R\}$, whereas that of a nondeterministic one takes the form $\delta : Q \times \Gamma \to \mathcal{P}(Q \times \Gamma \times \{L, R\})$. A typical specification of $\delta$ is given by $\delta(q, a) = \{(q_1, a_1, D_1), \ldots, (q_m, a_m, D_m)\}$, which means that there are $m$ possibilities of the next move when the machine in state $q$ reads symbol $a$. In this chapter we first defined a deterministic Turing machine. Alternatively, we can first define a

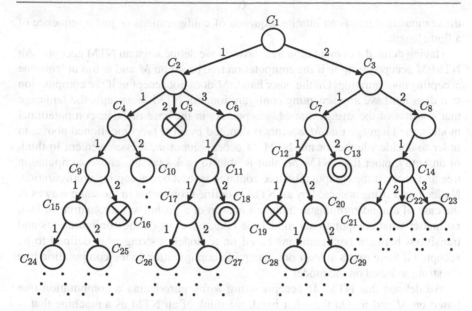

**Fig. 6.13** Example of computation tree

nondeterministic Turing machine, and then define a deterministic one as a special nondeterministic machine for which the number of possibilities $m$ is restricted to be 0 or 1 for every pair of $q$ and $a$. Recall that $m = 0$ is typically the case where the current state is the accept state or the reject state, and hence the next state is not defined. Following the alternative argument, it is easy to see that a deterministic machine is a special case of a nondeterministic machine. In our discussion a nondeterministic Turing machine will be abbreviated NTM, while a deterministic Turing machine will be abbreviated DTM.

For an NTM each configuration has, in general, several subsequent configurations to which the former configuration moves. So, when given an NTM and its input, we can draw a tree which shows, beginning with the start configuration, how each configuration proceeds to subsequent configurations, splitting according to several possible transitions. Such a tree is called a *computation tree*. Figure 6.13 illustrates a typical example of such a tree. In this figure $C_1$ is the start configuration: if the start state and the input are denoted by $q_s$ and $a_1 \cdots a_n$, respectively, then $C_1$ is written $q_s c_1 \cdots c_n$. In this case, since $\delta(q_s, a_1)$ is assumed to have two specifications, $C_1$ splits into two configurations $C_2$ and $C_3$. As in the case of state diagrams, $C_{13}$ and $C_{18}$ denote *accepting configurations*, whereas $C_5$, $C_{16}$, and $C_{19}$ denote *rejecting configurations*.

Whether it's a finite automaton, a Turing machine, or something else, or whether it's deterministic or nondeterministic, we can introduce a corresponding computation tree by defining a configuration and a transition between configurations as well. When the computational model is deterministic, the corresponding computation tree turns out to be a straight sequence of configurations without splitting. In this case the computation tree is an infinite sequence of configurations or just a sequence of a finite length.

Having defined a computation tree, how do we define what an NTM accepts? An NTM $M$ accepts string $w$ if the computation tree based on $M$ and $w$ has at least one accepting configuration. On the other hand, $M$ does not accept $w$ if the computation tree does not have any accepting configuration. An NTM $M$ accepts the language that consists of the strings that $M$ accepts. As in the case of other computational models, the language that $M$ accepts is denoted by $L(M)$. As mentioned above, in order to decide whether or not NTM $M$ accepts input $w$, it is convenient to think of an entity, other than NTM $M$, that is able to look over the entire computation tree and check if the computation tree contains at least one accepting configuration. Notice that a string accepted by an NTM is defined somewhat in the same way as in the case of a nondeterministic finite automaton or a pushdown automaton. In fact, no matter what computational model we take, once we define configurations and transitions between configurations based on a model, a string $w$ is defined to be accepted if there exists at least one accepting configuration in the computation tree for string $w$ based on the model.

We defined that NTM $M$ accepts string $w$ by introducing a computation tree based on $M$ and $w$. On the other hand, we think of an NTM as a machine that is drawn schematically as in Fig. 6.1. How does the machine depicted in Fig. 6.1 work in terms of the computation tree? Given a computation tree, we can interpret the behavior of the machine as follows: the machine with the start configuration starts its computation at the root of the tree; starting at the root this way, the machine repeats branching into a couple of its copies at a succeeding node, each corresponding to one of the possible transitions as long as such transitions exist. The copies made this way can be thought of as offsprings of the parent machine, where each offspring is associated with a configuration corresponding to one of the possible transitions. In our case for the computation tree in Fig. 6.13, starting at the root, $C_1$ branches into $C_2$ and $C_3$. Likewise, $C_2$ branches into $C_4$, $C_5$, and $C_6$, and $C_3$ branches into $C_7$ and $C_8$, and so on. In the course of the process the newly replicated copies behave independently in parallel. In Chaps. 9 and 10, we shall give yet another computational model, namely, Boolean circuits, for an NTM.

In what follows we shall construct a DTM $D$ that simulates an arbitrarily given NTM $N$. The idea of how $D$ works is that $D$ goes around the entire computation tree defined based on NTM $N$ and input $w$ searching for an accepting configuration on the tree. If the computation tree has at least one accepting configuration, then $D$ halts, accepting the input. On the other hand, if the tree does not have any accepting configuration, then $D$ does not accept the input. If there exists no accepting configuration, there are two cases depending on whether the number of nodes in the computation tree is finite or infinite. In the finite case, $D$ halts, rejecting the input. On the other hand, in the infinite case, $D$ continues searching for an accepting configuration, but fails to find it, never halting to reject the input. The crucial point in this argument is that we have to search the computation tree in such a way that if there exists an accepting configuration at all in the tree, $D$ must eventually visit that configuration. If the condition described above is satisfied, DTM $D$ is equivalent to NTM $N$ in the sense that $D$ accepts input $w$ if and only if $N$ accepts input $w$.

Now we proceed to explain in what order $D$ visits the computation tree so that the condition described above is satisfied. In order to represent a path in a computation tree by a sequence of integers, we assign integers to the possible transitions for each node in the tree. More precisely, let $m$ denote the maximum of the numbers of possible transitions for the pairs expressed as $(q, a)$'s for state $q$ and symbol $a$. Namely, let

$$m = \max_{q \in Q, a \in \Gamma} |\delta(q, a)|.$$

Furthermore, let us assume that for each pair $(q, a)$ we assign integers 1 through $j$ to the 3-tuples in $\delta(q, a)$, where $j = |\delta(q, a)| \leq m$. Then, obviously, a string $u$ in $\{1, \ldots, m\}^*$ specifies the path from the root to a node in the tree. So we can say such a string $u$ indicates the node at the end point of the path. Of course, there might be a string in $\{1, \ldots, m\}^*$ that does not correspond to any path in the tree, and hence does not correspond to any node because some edges are lacking in the computation tree. While enumerating strings $u$ in $\{1, \ldots, m\}^*$ in *lexicographical order* DTM $D$ visits the corresponding nodes in that order. In the course of the enumeration, if there exists no path that corresponds to string $u$ or there exists a rejecting configuration on the path $u$, skip to the string that is lexicographically next to $u$.

For example, in the case of $m = 3$, the lexicographical order of strings in $\{1, 2, 3\}^*$ is given by

$$\varepsilon, 1, 2, 3, 11, 12, 13, 21, 22, 23, 31, 32, 33, 111, \ldots.$$

In the lexicographical order of strings shorter strings precede longer strings. Among strings of the same length the order is the same as the familiar dictionary ordering. For example, for Fig. 6.13, 131 indicates $C_{11}$, and 1312 indicates $C_{18}$. 132 and 1313 correspond to no node. Given a string $u$ in $\{1, 2, 3\}^*$, DTM $D$ can simulate NTM $N$ along the path associated with the string $u$ on the computation tree. In simulating $N$ this way, each digit of $u$ tells us which choice to make next in the computation tree.

We summarize what is described so far in the next theorem.

**Theorem 6.9** *For any nondeterministic Turing machine, there exists an equivalent deterministic Turing machine.*

*Proof* We shall construct DTM $D$ that is equivalent to given NTM $N$, as described before the theorem. Let

$$m = \max_{q \in Q, a \in \Gamma} |\delta(q, a)|.$$

Until an accepting configuration is encountered, DTM $D$ repeats generating string $u$ in $\{1, \ldots, m\}^*$ in lexicographical order to see if there exists an accepting configuration on the path $u$ from the root in the computation tree by running NTM $N$ choosing nondeterministic transitions according to $u$. To do so, $D$ uses three tapes:

tape 1 is used to store input $w$ throughout the computation; tape 2 is used as $N$'s tape in simulating $N$ along path $u$; tape 3 is used to generate string $u$ in $\{1, \ldots, m\}^*$ in the lexicographical order.

What is described informally so far is summarized as follows:

Three-tape DTM $D$ that simulates NTM $N$:

1. Place the input $w$ in tape 1, make tape 2 empty, and place the empty string in tape 3.
2. Copy the input $w$ to tape 2.
3. Let $u$ be the string in tape 3. Simulate $N$ with input $w$ by choosing the nondeterministic transitions from the starting configuration successively according to $u$. If there exists no path that corresponds to $u$ or a rejecting configuration is encountered in the course of the simulation, go to **4**. If an accepting configuration is encountered, *accept* the input.
4. Replace the string on tape 3 with the string lexicographically next to the present string $u$ and go to **2**.

Furthermore, by Theorem 6.8 there exists a single-tape deterministic Turing machine that is equivalent to the DTM $D$ with three tapes, completing the proof.   □

## 6.4   Church–Turing Thesis

We have an intuitive notion of an *algorithm* which says that it is a sequence of simple instructions for carrying out a computational task. The intuitive notion tells us that an algorithm must be described unambiguously and hence be able to be executed mechanically to yield an output. So the intuitive notion is adequate when we are required to design an algorithm to compute a certain problem. We are always sure about whether a procedure designed to compute a given problem could be an algorithm or not. In contrast, when we are required to prove that there exists no algorithm to compute a certain problem, the intuitive notion of algorithms is not adequate. In such a situation we need to have a precise definition of an algorithm. Informally speaking, this is because, if we are able to prove that a certain problem cannot be computed by *any* algorithm, that argument necessarily somehow refers to *all* of what is called an algorithm, which is impossible unless we have a precise definition of what an algorithm is. In Chap. 7 we consider what is called the halting problem to decide whether a Turing machine eventually halts or runs forever, and prove that there exists no algorithm that computes the halting problem. To prove that fact, we need to give a precise definition of an algorithm.

The precise definition of an algorithm is given based on the *Church–Turing thesis*, which gives the connection between the informal notion of an algorithm and its mathematical counterpart of the precise definition. That is, the Church–Turing thesis claims that what algorithms compute is exactly what Turing machines compute, which is rewritten as follows:

Being computed by an algorithm $\Leftrightarrow$ Being computed by a Turing machine

In the next chapter it turns out that the notion of Turing machines introduced by Turing plays an important role of exploring the limits of algorithmic solvability in terms of the Church–Turing thesis.

We can intuitively accept the direction $\Leftarrow$ of the Church–Turing thesis because no matter what a Turing machine is given we can execute it mechanically to obtain an output. On the other hand, the direction $\Rightarrow$ asserts that no matter what an algorithm is given we can describe the algorithm in terms of a Turing machine. But it is impossible to prove this direction because we do not have any precise definition of algorithms. In other words, we cannot provide convincing arguments to deny the possibility that there exists a procedure that we could call an algorithm, but cannot be described in terms of any Turing machine. The Church–Turing thesis proposes to take the direction $\Rightarrow$ for granted.

As mentioned above, the Church–Turing thesis is not what we can prove, but is what we admit. But there are a couple of evidences that support the thesis. First, there has been no algorithm that cannot be described in principle in terms of any Turing machine. Furthermore, after we have had enough experience to write Turing machines to solve many concrete problems, we come to have a feeling that any algorithm can be described in terms of a Turing machine. Second, several mathematical models have been proposed independently to define algorithms, such as the $\lambda$-calculus and recursive functions, which have been proved to be equivalent with each other in power to the Turing machines. The fact that several mathematical models proposed independently are equivalent with each other reveals that the models are natural, and *robust* in the sense that variance in the model keeps invariance in power.

The readers are expected to solve and study the details of the examples and problems in this chapter and in the next chapter on the universal Turing machine in order to develop a feeling for accepting the Church–Turing thesis. Readers who do not wish to take the time to do this can skip the explanation of the details of behaviors of the universal Turing machine in the next chapter, admitting the claim that whatever can be expressed as an algorithm can also be described in terms of a Turing machine.

## 6.5 Problems

**6.1** Give concise descriptions for the language that the nondeterministic Turing machine (NTM) given in the figure below accepts. Recall that for an NTM all the transitions specified by a transition function are expressed as the edges in the state diagram.

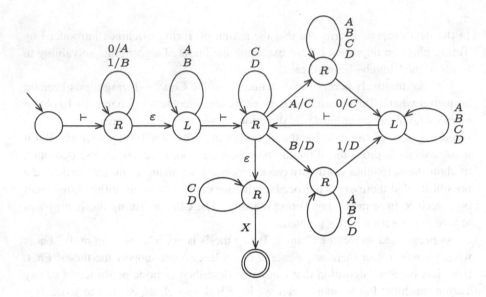

**6.2\*** Let us allow the head of a Turing machine not only to move to the left or to the right, but also to stay put. This model is called a three-way model, whereas the model given in Definition 6.1 is called a two-way model. Given a three-way model, construct a two-way model equivalent to the three-way model. (Hint: A two-way model can simulate a three-way model by spending two steps for each step of staying put.)

**6.3\*** Give state diagrams of Turing machines that accept the following languages. For notation $w^R$, refer to Problem 4.3.
(1) $\{ww^R \mid w \in \{a, b\}^* - \{\varepsilon\}\}$
(2) $\{a^n b^n c^n \mid n \geq 1\}$

**6.4** Give a state diagram of a nondeterministic Turing machine that accepts the language $\{\vdash wXyY \mid w, y \in \{0, 1\}^*$ and $w = uyv$ holds for some $u, v \in \{0, 1\}^*\}$. Notice the comment in Problem 6.1 for state diagrams of NTMs.

**6.5\*\*** Give a state diagram of a Turing machine that accepts the language $\{\vdash 1^{2^n} X \mid n \geq 0\}$.

**6.6** Describe an outline of the behavior of a Turing machine that computes the reachability problem in Sect. 2.7.

**6.7** Show the computation tree of the NTM given in the following state diagram provided that the starting configuration is $q_0 001$.

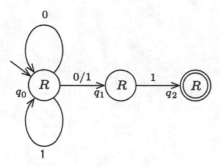

# Universality of Turing Machine and Its Limitations

# 7

The universal Turing machine is the one that, when given a description of Turing machine $M$ and input $w$, can simulate the behavior of $M$ with input $w$. In this chapter we construct the universal Turing machine. This means that the universal Turing machine can behave like any Turing machine with any input if the description of the Turing machine to be simulated is given together with its input. On the other hand, there exists a limit to the computational power of Turing machines. This limitation is shown by giving the halting problem, which any Turing machine can never solve. The halting problem asks whether, when a Turing machine $M$ and an input $w$ are given, $M$ with input $w$ will eventually halt or continue to run forever. As another example of a problem that any Turing machine cannot solve, we give the Post correspondence problem. The Post correspondence problem asks whether or not, when a collection of pairs of strings is given, there exists a sequence of the pairs (repetitions permitted) that has certain properties of a match.

## 7.1 Universal Turing Machine

When given a description of a Turing machine (TM) and its input, it can be mechanically performed to simulate its moves in steps. Therefore, according to the Church–Turing thesis, there *exists* a Turing machine that performs so. But how can we construct such a Turing machine? That's another story. The Turing machine that does this task is called the *universal Turing machine*. As will be shown, the universal Turing machine behaves essentially the same way as we trace a program of a Turing machine written in terms of 5-tuples. Let $U$ denote the universal Turing machine and let $M$ denote any Turing machine with input $w$. When $U$ is given as input $M$, or more precisely a description of $M$, and $w$, $U$ simulates the behavior of $M$ with input $w$, by examining descriptions of $M$ and its input $w$. Shortly we explain how to describe a Turing machine as a string.

The notion of the universal Turing machine reminds us of the fact that a modern computer has its hardware fixed and what we want to perform is stored as a program in its memory; this is called a stored-program computer. You can easily see that

A. Maruoka, *Concise Guide to Computation Theory*,
DOI 10.1007/978-0-85729-535-4_7, © Springer-Verlag London Limited 2011

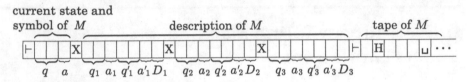

**Fig. 7.1** Tape of universal TM $U$

there exists a correspondence between the stored-program computer and the universal Turing machine: the hardware corresponds to the universal Turing machine, while the stored program itself corresponds to a description of a Turing machine to be simulated which is contained in the tape of the universal Turing machine initially. Von Neumann wrote a report on the EDVAC (Electronic Discrete Variable Automatic Computer)—one of the earliest electronic computers—in which he introduced the idea of a stored-program computer, which owes to the concept of the universal Turing machine due to Alan M. Turing. In fact, following the behavior of the universal Turing machine described later, the readers are expected to grasp intuitively the core idea of the stored-program computers.

Let us proceed to the details of the universal Turing machine. Throughout this chapter we assume that a Turing machine $M$ that we want to simulate is deterministic unless otherwise stated. Initially, the universal Turing machine has on its tape the description of a Turing machine $M$, denoted by $\langle M \rangle$, and an input to $M$, denoted by $w$, where $\langle M \rangle$ is the string that somehow encodes a Turing machine $M$. More specifically, as illustrated in Fig. 7.1, the contents of the tape of the universal Turing machine is stored in three regions, each containing the following information: the current state and the symbol under the head of $M$; the description of $M$; the tape contents of $M$. As illustrated in Fig. 7.1, in the machine description region the sequence of 5-tuples separated by $X$'s is placed, where each 5-tuple is expressed as the binary sequence representing $(q, a, q', a', D)$ such that $\delta(q, a) = (q', a', D)$, where $\delta$ denotes $M$'s transition function. To the right of the machine description region there exists $M$'s tape region, where the current content of $M$'s tape is placed except for the square on which the head is located. On that square the marker $H$ is placed instead of the symbol under the head. Finally, to the left of the machine description region, there exists the machine condition region, where we find $M$'s current state $q$ and the current symbol $a$ under the head of $M$, namely, the symbol that is supposed to be placed on the square of $H$. So the head position and the symbol under the head are stored separately. It will be explained shortly that the current symbol square in the machine condition region contains the symbol that $M$ has just read or is just about to write, depending on the phase of the simulation.

It is easy to figure out how the universal Turing machine $U$ behaves by consulting the contents of the tape of the universal Turing machine just described. That is, the Turing machine $U$ continues simulating the behavior of $M$ one step at a time: by remembering $M$'s current state and the current symbol in the machine condition region and by searching for the 5-tuple that is compatible with the current state and the current symbol, $U$ accordingly finds out what state to go into, what symbol to write, which way to move the head, and how to update the tape content. More specifically, $U$ takes up and memorizes $(q, a)$ in the machine condition region, finds out the 5-tuple $(q, a, q', a', D)$ whose first and second components coincide with $(q, a)$, and updates the machine condition region and $M$'s tape region according to $(q', a', D)$.

Before proceeding to more details of the simulation, we should explain about the tape alphabet of the universal Turing machine $U$. First of all, $U$ is just one fixed machine; hence, it has a fixed tape alphabet. On the other hand, the machines that are simulated by $U$ are arbitrary and hence may have arbitrarily large numbers of states, although each of them is finite. So $U$ cannot use a different symbol to represent each state of the machine to be simulated. How can we designate an arbitrarily large number of states by a fixed tape alphabet? The trick is as follows. When the machine to be simulated has $m$ states and $l$ is the least integer such that $m \le 2^l$, we use binary sequences of length $l$ to represent the states. In Fig. 7.1 this $l$ is assumed to be 2. As for a tape alphabet of a machine to be simulated we can argue similarly. Namely, we can use binary sequences to represent an arbitrary large number of the tape symbols of the machines to be simulated. But to make matters simple we assume instead that the tape alphabet of any machine to be simulated is fixed as binary symbols. Furthermore, the directions $L$ and $R$ are represented by 0 and 1, respectively. Then, by adding the symbols $A, B, X, H, S, \vdash$ to the binary symbols 0 and 1, we set the tape alphabet of the $U$ as follows:

$$\Gamma = \{0, 1, A, B, X, H, S, \vdash, \sqcup\}.$$

Before proceeding to explain the state diagram of the universal Turing machine $U$ in Fig. 7.3, it would be helpful to trace a higher level description of $U$, which is given as follows.

The behavior of the universal Turing machine $U$:
1. Search for a 5-tuple $(q, a, q', a', D)$ in the machine description region such that its first and second elements coincide with the current state and the current symbol in the machine condition region.
2. Write $(q', a')$ of the 5-tuple $(q, a, q', a', D)$ found in **1** into the machine condition region. Furthermore, memorize as a state the direction $D$ of that 5-tuple.
3. Memorize as a state the $a'$ written on the current symbol square in **2** and write symbol $S$ on that square. Write the $a'$ memorized on the square on which $H$ is placed, and move $U$'s head one square to the direction indicated by $D$ that is memorized in **2**.
4. Memorize as a state the symbol $a''$ that the head reads, and write symbol $H$ on the square under the head. Then search for the square with symbol $S$ and write on that square the $a''$ memorized. Go to **1**.

Figure 7.2 illustrates how the contents of the tape will change when $U$ simulates $M$. The four tapes in the figure represent some parts of the contents just at the end of the four stages of the behavior of $U$. Among the three 5-tuples $(q_1, a_1, q_1', a_1', D_1), \ldots, (q_3, a_3, q_3', a_3', D_3)$ in the machine description region, only

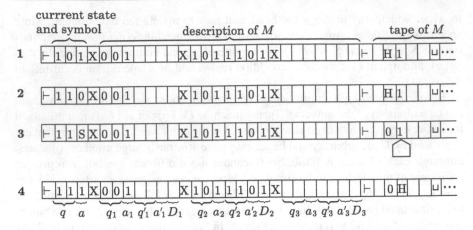

**Fig. 7.2** Transitions of the contents of the tape of the universal TM $U$

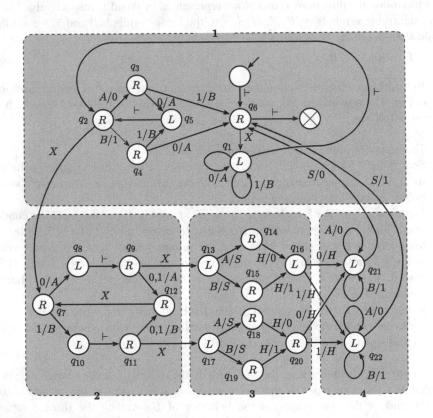

**Fig. 7.3** State diagram of the universal TM $U$

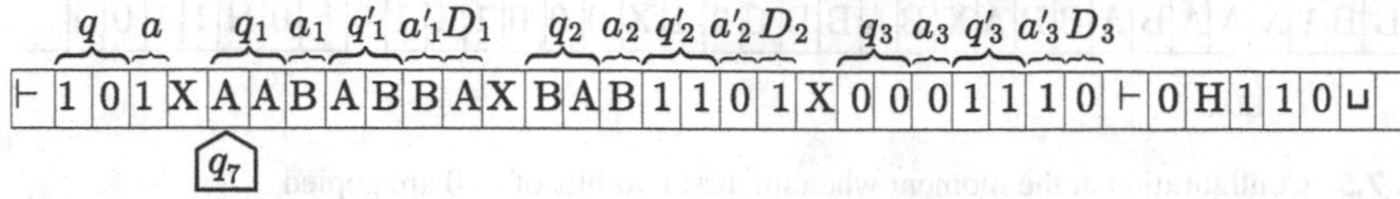

$$\overbrace{q\ \ a}\quad \overbrace{q_1\ a_1\ q'_1\ a'_1 D_1}\quad \overbrace{q_2\ a_2\ q'_2\ a'_2 D_2}\quad \overbrace{q_3\ a_3\ q'_3\ a'_3 D_3}$$

⊢ 1 0 1 X A A B A B B A X B A B 1 1 0 1 X 0 0 0 1 1 1 0 ⊢ 0 H 1 1 0 ⊔

$q_7$

**Fig. 7.4** Configuration at the moment when a transition from $q_2$ to $q_7$ occurs

the second is shown in the figure. The states are represented by strings in $\{0, 1\}^2$ and both the directions and the tape symbols are represented by binary symbols in $\{0, 1\}$. As mentioned above, in **1**, the 5-tuple $(q_2, a_2, q'_2, a'_2, D_2)$ is searched for because the first and the second components coincide with the pair of the current state and symbol $(10, 1)$. In **2**, the $q'_2 = 11$ and $a'_2 = 0$ of that 5-tuple are written in the machine condition region. The direction $D_2$ (in this case, 1 meaning $R$) is memorized as a state. In **3**, $a'_2 = 0$ is memorized and $S$ is written in that square on which the $a'_2$ was temporarily written. The memorized symbol $a'_2$ is written on the square on which symbol $H$ is placed. Then the head of $U$ is moved one square to the direction of $D_2$. In **4**, the symbol under the current head is memorized as a state and $H$ is written on that square which is the new head position. The memorized symbol is written on the square on which symbol $S$ is placed as the new current symbol.

Figure 7.3 gives the state diagram of the universal Turing machine $U$. The diagram is divided into four parts attached with **1** through **4**, each corresponding to one of the four stages of the behavior of the $U$. The readers who have grasped intuitively how the universal Turing machine works so far may skip the rest of this section. The readers who are not yet convinced may figure it out by tracing the more detailed behavior of $U$, which will be given in what follows.

We begin with the behavior of stages **1** and **2**. These parts of the diagram of Fig. 7.3 coincide with the diagram of Fig. 6.6, which stores and retrieves information. In the case of $U$, the three bits of $(q_i, a_i)$ correspond to a name, whereas the four bits of $(q'_i, a'_i, D_i)$ correspond to an item. So, as opposed to the former case of the diagram of Fig. 6.6, the length of a name is not equal to that of an item. The point is that $U$ makes use of the difference of the lengths as follows. Figure 7.4 shows the moment when $U$ has found out $(q_2, a_2)$ of the second 5-tuple that coincides with the current state and symbol in the machine condition region. Notice that, to keep the symbols used simple, the same symbol $q_i$ is used for the states of both $M$ and $U$. The machine in the configuration shown in Fig. 7.4 is about to copy the part $(q'_2, a'_2, D_2)$ into the machine condition region. In doing so the squares containing $(q_1, a_1, q'_1, a'_1, D_1)$ and $(q_2, a_2)$ will be skipped because the symbols 0 and 1 on these squares are changed to the symbols $A$ and $B$, respectively. In general, the symbols 0 and 1 that are already scanned will be changed to $A$ and $B$ so that they will be skipped in later searching and storing. In this figure note that 0 and 1 correspond to $A$ and $B$ as well as to $L$ and $R$, respectively.

Let's see more precisely how to copy the first three bits of $q'_2 a'_2 D_2 (= 1101)$, namely 110, into the machine condition region. Starting on the leftmost $X$ with state $q_7$, the machine copies 110 one symbol at a time repeatedly as follows: shift to the right until $U$ reads for the first time some 0 or 1; memorize the symbol just

⊢ B B 1 X A A B A B B A X B A B B B 0 1 X 0 0 0 1 1 1 0 ⊢ 0 H 1 1 0 ⊔
$q_7$

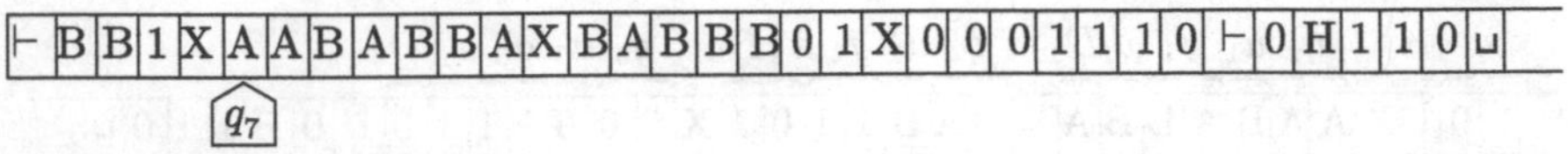

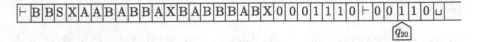

**Fig. 7.5** Configuration at the moment when the first two bits of 110 are copied

⊢ B B S X A A B A B B A X B A B B B A B X 0 0 0 1 1 1 0 ⊢ 0 0 1 1 0 ⊔
$q_{20}$

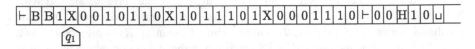

**Fig. 7.6** Configuration at the moment when $U$ is in $q_{20}$

⊢ B B 1 X 0 0 1 0 1 1 0 X 1 0 1 1 1 0 1 X 0 0 0 1 1 1 0 ⊢ 0 0 H 1 0 ⊔
$q_1$

**Fig. 7.7** Configuration at the moment when $U$ is in $q_1$

read and replace the symbol with $A$ or $B$ ($q_7 \xrightarrow{0/A} q_8$ or $q_7 \xrightarrow{1/B} q_{10}$); shift to the left until $U$ reads the leftmost ⊢ ($q_8 \xrightarrow{\vdash} q_9$ or $q_{10} \xrightarrow{\vdash} q_{11}$); shift to the right until $U$ reads for the first time some 0 or 1 and write on the square the symbol memorized with $A$ or $B$ ($q_9 \xrightarrow{0,1/A} q_{12}$ or $q_{11} \xrightarrow{0,1/B} q_{12}$); shift to the right until $U$ reads the first $X$ to prepare for the next cycle ($q_{12} \xrightarrow{X} q_7$). Figure 7.5 shows the tape contents at the time when the first two bits of $q'_2a'_2 = 110$ have been copied as $BB$ on the leftmost squares between ⊢ and $X$. When starting at the configuration of Fig. 7.4 with $q'_2a'_2D_2 = 1101$, the machine repeats the cycle three times, and the content on the leftmost five squares becomes $\vdash BBAX$. Then the machine remembers the fourth bit 1 of $q'_2a'_2D_2 = 1101$ by going from $q_7$ to $q_{10}$ and moves to the leftmost ⊢ and then searches for 0 or 1 encountered for the first time going rightward. But this time before the machine encounters 0 or 1 it reads $X$ and does transition $q_{11} \xrightarrow{X} q_{10}$, thereby proceeding to **3**.

In stage **3** the machine writes the symbol $a'_i$ written on the current symbol square onto the square on which symbol $H$ is placed and moves the head one square to the direction of $D_i$ (in our case to the right) that is memorized as a state in **2**. That head position is the new head position for the next step, whose configuration is shown in Fig. 7.6. While doing what has been described so far, the machine writes $S$ on the current symbol square. Then the machine goes to **4**.

Finally in **4**, the machine memorizes the symbol on the new head position as a state (state $q_{21}$ or $q_{22}$) and moves its head to the left, restoring 0 and 1 until it encounters $S$, which will be replaced with the new memorized symbol $a'_i$ (in our case $a'_i = 1$). Figure 7.7 shows the configuration at this moment of the simulation. Then the machine goes to **1**. We omit the accept and reject states in Fig. 7.3. To complete the diagram, we must modify it by adding, for example, the transitions to the accept state of $U$ when in the course of the simulation an accept state of $M$ is encountered. But the details of the modification are omitted.

## 7.2    Undecidability of the Halting Problem

In general, if there exists no Turing machine that eventually halts and decides YES/NO for a given problem, the problem is said to be *undecidable*. The problem, called the *halting problem*, to tell whether any Turing machine will eventually halt or run on forever is such a problem. In this section we prove that the halting problem is undecidable. At first glance, the halting problem seems to be decidable. It seems like we can let the universal Turing machine $U$ solve the halting problem by running $U$ for an arbitrarily given Turing machine $M$ together with its input $w$ and yielding YES when $U$ halts in the course of the simulation. In fact, the output YES is correct as long as the machine $M$ with input $w$ eventually halts. But, if the machine $M$ with input $w$ runs forever, then the universal Turing machine $U$ will also run forever, and hence can never output NO forever. So we cannot solve the halting problem by simply running the universal Turing machine. But this is by no means a proof for the fact that the halting problem cannot be decided by *any* Turing machine.

We shall prove that the halting problem is undecidable. The proof is by contradiction. The argument for the proof is delicate and hard to understand. So, to begin with, we explain informally the point of the argument of the proof, which is referred to as the *barber paradox*. The barber in a small town declares that he shaves everyone in the town who does not shave himself, and that he does not shave anyone who shaves himself. It will be shown shortly that the barber's declaration leads to a contradiction, and this contradiction is at the heart of the argument to prove by contradiction that the halting problem is undecidable.

Let us examine how the barber's declaration leads to a contradiction. Let $a$ denote how the barber behaves, $a = 1$ meaning to shave a resident in the town and $a = 0$ meaning not to shave. Similarly, let $b$ denote how a resident in the town behaves, $b = 1$ meaning to shave himself and $b = 0$ meaning not to shave. Clearly, what the barber declared is that for any resident including himself $(a, b)$ must be $(1, 0)$ or $(0, 1)$. On the other hand, the barber shaves not only as the barber but also as one of the residents. Therefore, if the barber shaves himself, $(a, b)$ becomes $(1, 1)$, whereas if the barber does not shave himself, $(a, b)$ becomes $(0, 0)$. In either case, this contradicts what the barber declared. If the barber is not thought of as being a resident as well, the contradiction does not occur. In this way, the barber's declaration leads to a contradiction because of the feature of self-reference in the declaration. In other words, the barber referred to himself as one of the residents. Observe that the point of the argument leading to a contradiction is that by self-reference an entity must satisfy two requirements that are not consistent with each other. Observe that in order to lead to a contradiction we need not only self-reference but also a situation that invokes requirements that are inconsistent. For example, if the barber's declaration was, "I shave the beard of anyone who shaves himself," then it has self-reference, but in this case the requirements imposed on the barber are consistent and hence are unable to lead to a contradiction.

**Fig. 7.8** State diagram of
TM $H$ with all the states
omitted except for the start
state, the accept state, and the
reject state

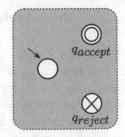

**Fig. 7.9** State diagram of
TM $N$ obtained by modifying
TM $H$ as indicated

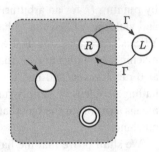

The argument described so far in terms of the barber's declaration underlies the argument to prove that the halting problem is undecidable. So before proceeding to the proof of the undecidability, the readers are expected to understand thoroughly the argument above leading to a contradiction from the barber's declaration.

We now explain an idea for proving that the halting problem is undecidable. The proof is by contradiction. Assume that there exists a Turing machine, denoted $H$, that decides the halting problem. We shall show that the assumption leads to a contradiction. In general, when fed $w$ as input, TM $M$ will either eventually halt or go on forever, which will be denoted by $M(w)$. That is, let $M(w) = halt$ if TM $M$ with input $w$ eventually halts, and $M(w) = loop$ otherwise. An input to the TM $H$ takes the form of $\langle M, w \rangle$ for TM $M$ and an input $w$. By the assumption, when fed $\langle M, w \rangle$ as input, TM $H$ computes $M(w)$ by going to an accept state if $M(w) = halt$ and going to a reject state if $M(w) = loop$. In either case, $H$ itself will eventually halt. The accept state and the reject state will be denoted by $q_{accept}$ and $q_{reject}$, respectively. Figure 7.8 depicts the state diagram of $H$ schematically, omitting all the states except for the start state, the accept state, and the reject state.

To make incompatible situations, first, we modify TM $H$ to obtain TM $N$, which is shown in Fig. 7.9. Roughly speaking, TM $N$ is a machine that does the opposite of what TM $M$ does concerning whether the machines eventually halt or go on forever. To make this TM $N$, as shown in Fig. 7.9, we introduce a new state so that, once the machine goes to $q_{accept}$, transitions between $q_{accept}$ and the new state are repeated forever. In the state diagram the edges with label $\Gamma$ indicate that the transitions take place unconditionally, that is, no matter what symbol the head reads. By the construction described so far, we can claim that concerning whether a machine halts or goes on forever, the situation when $N$ is fed as input $\langle M, w \rangle$ is opposite to the

**Fig. 7.10** TM $E$ transforming $\langle M \rangle$ to $\langle M, \langle M \rangle \rangle$, then feeding $\langle M, \langle M \rangle \rangle$ to TM $N$

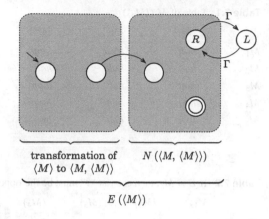

$$\underbrace{\underbrace{\text{transformation of}}_{\langle M \rangle \text{ to } \langle M, \langle M \rangle \rangle} \qquad N\left(\langle M, \langle M \rangle \rangle\right)}_{E\left(\langle M \rangle\right)}$$

situation when $M$ is fed as input $w$. That is,

$$N\left(\langle M, w \rangle\right) = \overline{M(w)},$$

where $\overline{halt} = loop$ and $\overline{loop} = halt$. So what remains to show in order to obtain a contradiction is to arrange to make the situation of TM $N$ with input $\langle M, w \rangle$ coincide with that of TM $M$ with input $w$. To do so we set $M$ and $w$ as follows. As illustrated in Fig. 7.10, we introduce TM $E$ which works as follows: when fed $\langle M \rangle$ as input, make $\langle M, \langle M \rangle \rangle$; feed $\langle M, \langle M \rangle \rangle$ as input and run TM $N$. Replacing TM $N$ with TM $E$ defined this way and replacing $\langle M, w \rangle$ with $\langle E, \langle E \rangle \rangle$ in the argument above, we can have situations which conflict with each other.

**Theorem 7.1** *The halting problem is undecidable.*

*Proof* The proof is by contradiction. Assume that TM $H$ decides the halting problem. TM $E$ is defined in terms of TM $N$ in Fig. 7.9 as follows:

> Turing machine $E$:
> Input: $\langle M \rangle$
> **1.** Transform input $\langle M \rangle$ to $\langle M, \langle M \rangle \rangle$.
> **2.** Run TM $N$ on input $\langle M, \langle M \rangle \rangle$.

Having defined TM $E$ this way, let us see what happens if we feed its own description $\langle E \rangle$ as input to TM $E$ and then run TM $E$. Then we have

$$E\left(\langle E \rangle\right) = N\left(\langle E, \langle E \rangle \rangle\right) \quad \text{(by the definition of TM } E\text{)}$$

$$= \overline{E\left(\langle E \rangle\right)} \quad \text{(by the definition of TM } N\text{)}.$$

Thus we obtain a contradiction concerning whether TM $E$ with input $\langle E \rangle$ eventually halts or runs on forever, completing the proof. $\qquad \square$

The proof by contradiction described so far can be replaced with a proof based on a proof technique called *diagonalization*. To see how another proof based on diagonalization goes, see Table 7.1, constructed with all TM's $M_1$, $M_2$, ... listed

**Table 7.1**  Table of $M_i(\langle M_j \rangle)$

|        | $\langle M_1 \rangle$ | $\langle M_2 \rangle$ | $\langle M_3 \rangle$ | $\langle M_4 \rangle$ | $\langle M_5 \rangle$ | $\ldots$ |
|--------|------|------|------|------|------|-----|
| $M_1$  | **halt** | loop | loop | halt | halt | $\ldots$ |
| $M_2$  | halt | **halt** | halt | halt | loop | $\ldots$ |
| $M_3$  | halt | halt | **loop** | loop | loop | $\ldots$ |
| $M_4$  | halt | loop | loop | **halt** | halt | $\ldots$ |
| $M_5$  | loop | halt | halt | halt | **loop** | $\ldots$ |
| $\vdots$ | $\vdots$ | $\vdots$ | $\vdots$ | $\vdots$ | $\vdots$ | $\ddots$ |

**Table 7.2**  If $E = M_k$, then entry at "?" must be the opposite of itself

|        | $\langle M_1 \rangle$ | $\langle M_2 \rangle$ | $\langle M_3 \rangle$ | $\langle M_4 \rangle$ | $\langle M_5 \rangle$ | $\ldots$ | $\langle M_k \rangle$ | $\ldots$ |
|--------|------|------|------|------|------|-----|------|-----|
| $M_1$  | **halt** | loop | loop | halt | halt | $\ldots$ | | |
| $M_2$  | halt | **halt** | halt | halt | loop | $\ldots$ | | |
| $M_3$  | halt | halt | **loop** | loop | loop | $\ldots$ | | |
| $M_4$  | halt | loop | loop | **halt** | halt | $\ldots$ | | |
| $M_5$  | loop | halt | halt | halt | **loop** | $\ldots$ | | |
| $\vdots$ | $\vdots$ | $\vdots$ | $\vdots$ | $\vdots$ | $\vdots$ | $\ddots$ | | |
| $M_k$  | loop | loop | halt | loop | halt | | **?** | |
| $\vdots$ | $\vdots$ | $\vdots$ | $\vdots$ | $\vdots$ | $\vdots$ | | | $\ddots$ |

down the rows and all their descriptions $\langle M_1 \rangle$, $\langle M_2 \rangle$, $\ldots$ across the columns. The entry in the $i$th row and the $j$th column gives $M_i(\langle M_j \rangle)$. In other words, the entry tells whether the corresponding machine with the input eventually halts or runs on forever.

We are now ready to proceed with the proof based on the diagonalization. By the construction of $E$ described in the proof of Theorem 7.1, we have $E(\langle M_i \rangle) = \overline{M_i(\langle M_i \rangle)}$, which means that TM $E$ with input $\langle M_i \rangle$ does the opposite of what TM $M_i$ with input $\langle M_i \rangle$ does concerning whether the machines eventually halt or go on forever. On the other hand, since the table lists all the Turing machines, TM $E$ itself appears somewhere in the list. So let TM $E$ be the $k$th Turing machine in the list. Then we add the entries in the $k$th row as shown in Table 7.2. As shown in the table, since $M_k$ is TM $E$ and $M_k(M_i)$ turns out to be the opposite of $M_i(\langle M_i \rangle)$, the entries in the $k$th row become the opposites of the corresponding entries on the diagonal. But we cannot fill out the entry on the $k$th row and the $k$th column, because the entry must be the opposite of itself. We can state the argument above in a more formal way as follows:

$$M_k(\langle M_k \rangle) = E(\langle M_k \rangle) \quad \text{(by } E = M_k)$$

$$= \overline{M_k(\langle M_k \rangle)} \quad \text{(by the construction of } E),$$

which leads to a contradiction.

Observe that the proof based on diagonalization requires the fact that all the Turing machines can be enumerated like $M_1, M_2, \ldots$. We can see that Turing machines can be enumerated by listing all the strings in $\{0, 1, X\}^*$ in lexicographical order so that the valid string among such strings can be interpreted as Turing machines as in the case of the strings in the machine description region of the universal Turing machine. We prefer the proof of Theorem 7.1 because it does not require such enumeration.

## 7.3 Reduction

Let us suppose that solving problem $A$ cannot be harder than solving problem $B$ which is denoted by $A \leq B$. It is reasonable to say that the fact that $A \leq B$ and $A$ is undecidable implies that $B$ is undecidable. Similarly, the fact that $A \leq B$ and $B$ is decidable naturally implies that $A$ is decidable. In this way, from a hard-to-solve relation between two problems and either a hardness or easiness condition to solve on one of the two problems, we may naturally induce a new hardness or easiness condition on the remaining problem. In this section we formalize a relation, called reduction, among problems in terms of computational difficulty and establish a couple of theorems, each stating positive or negative results concerning the difficulty of solving problems.

To make such reduction valid, we must formalize the hard-to-solve relation denoted by $\leq$. Roughly speaking, the relation $A \leq B$ means that problem $A$ can be solved by using a solution to problem $B$. More precisely, $A \leq B$ means that, in order to get a solution of $A$ for an instance $w$, all we have to do is to modify $w$ somehow to obtain $w'$ and obtain a solution of $B$ for the instance $w'$ so that the solution of $B$ for $w'$ turns out to be the solution of $A$ for $w$ as well. In this case we say that problem $A$ is *reducible* to problem $B$. Furthermore, the correspondence between input $w$ of $A$ and input $w'$ of $B$ can be thought of as a function $f$ which is called the reduction, where $w' = f(w)$.

Having defined the notion of reducibility this way, we can say that $A \leq B$ means that $A$ is not harder than $B$ to solve. Then from $A \leq B$ and a certain negative statement on solving $A$ it follows that the negative statement on solving $B$ is derived (Theorem 7.5). Similarly, from $A \leq B$ and a certain positive statement on solving $B$ it follows that the positive statement on solving $A$ is derived (Theorem 7.4).

**Definition 7.2** Let $A$ and $B$ be *YES/NO* problems. Problem $A$ is reducible to problem $B$, denoted $A \leq B$, if there exists a function $f : \Sigma^* \to \Sigma^*$ that is computable in a sense specified somehow such that

$$A(w) = YES \quad \Leftrightarrow \quad B\big(f(w)\big) = YES,$$

for all $w$ in $\Sigma^*$. The function $f$ is called the *reduction* of $A$ to $B$. In particular, when $A$ and $B$ are the membership problem of languages, language $A$ is said to be reducible to language $B$, where the languages are thought of as the membership

**Fig. 7.11** Membership
problem for language $A$
reduces to language $B$

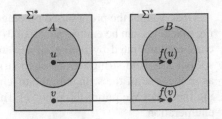

problems. Whether $A$ and $B$ are problems or languages, if the reduction is computable by a Turing machine, $A$ is said to be *mapping reducible* to $B$, which is denoted by $A \leq_{\text{mapping}} B$. Similarly, if the reduction is computable by a Turing machine in polynomial time, then $A$ is said to be *polynomial time reducible* to $B$, which is denoted by $A \leq_{\text{poly}} B$. If it is clear from the context in what sense the reduction is computable, we refer to mapping reducibility and polynomial time reducibility as just reducibility, and denote it by $A \leq B$ rather than $A \leq_{\text{mapping}} B$ or $A \leq_{\text{poly}} B$. In the definition above, a Turing machine computable reduction $f : \Sigma^* \to \Sigma^*$ is computed in polynomial time if there exist integers $a$ and $k$ such that for any $w$ in $\Sigma^*$ $f(w)$ is computed in at most $a|w|^k$ steps, where $|w|$ denotes the length of string $w$. (The notion of polynomial time will be explained in detail in Chap. 8.)

There is yet another notion of reducibility called *Turing reducibility*, which is obtained by generalizing mapping reducibility. Turing reducibility is defined in terms of an oracle. An *oracle* is a device that is equipped to a Turing machine. When the Turing machine asks an oracle whether any string $w$ is a member of a certain language, say, $B$ the oracle is supposed to answer the question by reporting *YES/NO* to the Turing machine depending on $w \in B$ or $w \notin B$. Such a Turing machine that has the additional capability of querying an oracle about the membership of language $B$ is called an *oracle Turing machine* and is denoted by $M^B$. More specifically, an oracle Turing machine queries an oracle by placing a string on a special oracle tape and then obtains a *YES/NO* answer from the oracle in a single computation step. Language $A$ is *Turing reducible* to language $B$ if there exists an oracle Turing machine with an oracle of $B$ that decides $A$, which is denoted by $A \leq_{\text{Turing}} B$. If we restrict Turing reducibility by imposing the conditions that an oracle Turing machine queries an oracle only once at the end of computation and yields as output the answer from the oracle, then the Turing reducibility turns out to be the mapping reducibility. Thus Turing reducibility can be thought of as a generalization of mapping reducibility.

If $A \leq B$ holds for languages $A$ and $B$, we have

$$w \in A \quad \Leftrightarrow \quad f(w) \in B,$$

where $f$ denotes the reduction. Figure 7.11 illustrates schematically the equivalence of the membership problems.

Furthermore, Fig. 7.12 illustrates how to construct the Turing machine that decides the membership problem for language $A$ by using the Turing machine that decides the membership problem for language $B$.

**Fig. 7.12** How to construct TM that decides the membership problem for $A$ by using TM that decides the membership problem for $B$. Note that in the figure $[C]$ takes value 1 if condition $C$ is satisfied, and takes value 0 otherwise

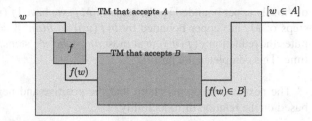

As a final comment on reducibility, we note that to make reducibility work as we intend, we have to sometimes put appropriate restrictions on reduction. No matter whether it is Turing reducibility or polynomial time reducibility, if we have $A \leq B$, solving problem $A$ is not harder than solving problem $B$. But for this to be the case, when solving problem $A$, a solution for problem $B$ must be substantially asked for. In fact, we can think of the following case which we do not intend. Consider the case where $B = \{0, 1\}$ and the reduction $f$ is such that $w \in A \Leftrightarrow f(w) = 1$. Clearly this is the case that reduction $f$ substantially decides the problem $A$ by itself and, clearly, in this case we cannot say that solving problem $A$ is not harder than solving problem $B$. For such a case not to happen, one may consider a situation where a reduction solely cannot decide problem $A$. For such situations, we may think of the case where problem $A$ is undecidable and reduction $f$ is Turing computable, or problem $A$ is not computed in polynomial time and the reduction $f$ can be computed in polynomial time.

The next proposition claims that the relation of reducibility satisfies the transitive law.

**Proposition 7.3** *If $A \leq_{\text{mapping}} B$ and $B \leq_{\text{mapping}} C$, then we have $A \leq_{\text{mapping}} C$. Similarly, if $A \leq_{\text{poly}} B$ and $B \leq_{\text{poly}} C$, then we have $A \leq_{\text{poly}} C$.*

*Proof* We begin with the part of the proof that is common to both mapping reducibility and polynomial time reducibility. Suppose that $A \leq B$ and $B \leq C$ with reductions $f$ and $g$, respectively. Then $g(f(w))$ becomes the reduction for $A \leq C$. This is because for any $w$ we have

$$A(w) = YES \quad \Leftrightarrow \quad B\big(f(w)\big) = YES \quad \text{(from } A \leq B)$$

$$\Leftrightarrow \quad C\big(g(f(w))\big) = YES \quad \text{(from } B \leq C).$$

Furthermore, if $f$ and $g$ are computed by Turing machines $M_f$ and $M_g$, respectively, then $g(f(w))$ can be computed by first converting input $w$ to $f(w)$ by $M_f$ and then feeding $f(w)$ as input to $M_g$ to obtain $g(f(w))$ as output. Let the Turing machine, denoted $M_{gf}$, be the one that computes this way. Then $M_{gf}$ computes the reduction of $A \leq_{\text{mapping}} C$.

For the case of polynomial time reducibility, suppose that $f$ and $g$ are computed by Turing machines $M_f$ and $M_g$ in $an^k$ and $bn^l$ steps, respectively, where $n$ is the length of input $w$ and $a, b, k, l$ are integers. Then, since the length of an output is obviously upper bounded by the steps of $M_f$, we have $|f(w)| \leq an^k$. Therefore, the steps of $M_g$ are upper bounded by $b|f(w)|^l \leq b(an^k)^l = ba^l n^{kl}$. Thus $M_{gf}$ computes the reduction $g(f(w))$ in at most $an^k + ba^l n^{kl}$ steps, and hence in polynomial time. This completes the proof.                                                                   $\square$

The next two theorems claim that the positive and negative results are derived based on the relation of reducibility.

**Theorem 7.4** *If B is decidable and A $\leq_{\text{mapping}}$ B holds, then A is decidable. Furthermore, if B is computable in polynomial time and A $\leq_{\text{poly}}$ B holds, then A is computable in polynomial time.*

*Proof* Let reduction of $A \leq B$ be denoted by $f$, and the Turing machine to compute $f$ be denoted by $M_f$. Let the Turing machine to compute $B$ be denoted by $M_B$. The way to compute $A$ is the same as in both the former and the latter parts of the theorem. $A$ is computed by the following Turing machine denoted $M_A$.

TM $M_A$:
Input: $w$
1.  Run TM $M_f$ on input $w$ and yield $f(w)$ as output.
2.  Run TM $M_B$ on input $f(w)$ and yield its output as output of $M_A$.

By definition, TM $M_A$ decides $A$. In an argument similar to that of Proposition 7.3, if $f$ and $B$ are computable in polynomial time, then $M_A$ decides $A$ in polynomial time.                                                                                                 $\square$

**Theorem 7.5** *If A is undecidable and A $\leq_{\text{mapping}}$ B, then B is undecidable. Furthermore, if A is not computable in polynomial time and A $\leq_{\text{poly}}$ B, then B is not computable in polynomial time.*

*Proof* The proof is by contradiction. Assume the hypotheses of the theorem and the fact that $B$ is decidable. Then, by Theorem 7.4 and by the hypothesis $A \leq_{\text{mapping}} B$, it is concluded that $A$ is decidable, obtaining a contradiction. Similarly, the latter half of the theorem is proved.                                                                    $\square$

## 7.4   Undecidability of Post Correspondence Problem

In Sect. 7.2, we verified that the halting problem is undecidable. In this section, we discuss another undecidable problem, called the Post correspondence problem (PCP). It will be proved that the PCP is undecidable by showing that the halting problem is reducible to the PCP and applying Theorem 7.5.

An instance of the PCP is given as a list of pairs of strings

$$(u_1, v_1), \ldots, (u_k, v_k),$$

**Fig. 7.13**  A match generates
the same sequence

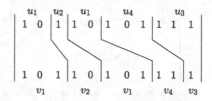

where the strings $u_1, v_1, \ldots, u_k, v_k$ as well as the length $k$ of the list are taken arbitrarily. The *Post correspondence problem* is the one to decide whether there exists a sequence of subscripts $i_1, \ldots, i_m$ such that $u_{i_1} u_{i_2} \cdots u_{i_m} = v_{i_1} v_{i_2} \cdots v_{i_m}$ for a given list of pairs of strings. That is, it is the problem to decide whether there exists a sequence of subscripts $i_1, \ldots, i_m$ such that the string obtained by concatenating strings from $\{u_1, \ldots, u_k\}$ in the order of subscripts $i_1, \ldots, i_m$ is the same as that obtained by concatenating strings from $\{v_1, \ldots, v_m\}$ in the same order. Note that repetitions are permitted in a sequence of subscripts, and hence there exists no limit on the length $m$ of the sequence in advance. When $u_{i_1} u_{i_2} \cdots u_{i_m} = v_{i_1} v_{i_2} \cdots v_{i_m}$ holds, the sequence of subscripts $i_1, i_2, \ldots, i_m$ is called a *match* of the PCP. Later, a match will refer to not only such a sequence of subscripts, but also the sequence itself $u_{i_1} u_{i_2} \cdots u_{i_m} (= v_{i_1} v_{i_2} \cdots v_{i_m})$ that corresponds to the sequence of subscripts.

*Example 7.6*  Consider the example of PCP $P'$ given as the following list of pairs of strings:

$$(10, 101), (1, 10), (111, 1), (101, 11).$$

That is, $(u_1, v_1) = (10, 101)$, $(u_2, v_2) = (1, 10)$, $(u_3, v_3) = (111, 1)$, and $(u_4, v_4) = (101, 11)$. For this example, a match is given as 12143 as Fig. 7.13 illustrates.

In order to show that the PCP is undecidable, it suffices to show that the halting problem can be reduced to the PCP because, if it is proved, then it follows from Theorem 7.5 that the PCP is undecidable. In this section, it is shown that, when TM $M$ and string $w$ are given, we can construct an instance $P$ of the PCP that somehow simulates TM $M$ on input $w$ so that TM $M$ halts on the input $w$ if and only if $P$ has a match, hence establishing that the halting problem can be reduced to the PCP.

We shall explain how to construct an instance $P$ of the PCP from TM $M$ and input $w$. Let $P$ be denoted by $\{(u_i, v_i) \mid 1 \le i \le k\}$. The point is that $P$ is constructed so that, when TM $M$ with input $w$ eventually halts, $P$ has a common sequence of configurations, each separated by symbol $\sharp$, which corresponds to the computation from the start to the halt. Assume that TM $M$ with input $w$ eventually halts. When string $w$ is written as $a_1 a_2 \cdots a_n$, the start configuration becomes $q_0 a_1 a_2 \cdots a_n$, so we put for the first pair of strings $(u_1, v_2) = (\sharp, \sharp q_0 a_1 a_2 \cdots a_n \sharp)$. As is shown later, we can assume without loss of generality that this pair appears first in the match.

Since TM $M$ with input $w$ is assumed to halt eventually, the instance $P'$ of PCP we construct has a match. Let the match be denoted by $u_{i_1} u_{i_2} \cdots u_{i_m} = v_{i_1} v_{i_2} \cdots v_{i_m}$. As shown in Fig. 7.14, the sequences $u_{i_1} u_{i_2} u_{i_3} \cdots$ and $v_{i_1} v_{i_2} v_{i_3} \cdots$ gradually extend

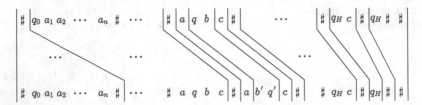

**Fig. 7.14** Match of PCP $P$

and finally both of them coincide. For example, when $\sharp aqbc$ first appears in the bottom, the same subsequence does not appear in the top at that moment. As shown in Fig. 7.14, the bottom proceeds by one configuration as compared to the top subsequently. Let us suppose that the configuration next to $aqbc$ is $ab'q'c$ because the transition function of $M$ is specified as $\delta(q, b) = (q', b', R)$. The point of constructing $P'$ is to force a match to extend in such a way that $ab'q'c\sharp$ comes after $aqbc\sharp$. To do so, we add $(\sharp, \sharp)$, $(a, a)$, $(qb, b'q')$, and $(c, c)$ as the pairs. Then, since $aqbc\sharp$ in the bottom must also appear in the top exactly at the same corresponding positions of $aqbc\sharp$ in the bottom, the pairs added force the match to extend itself as desired, as shown in the middle part of Fig. 7.14. In this way the successive configurations are added repeatedly until a halt state appears in a configuration. Then the top catches up with the bottom so that eventually the same sequence appears at both the top and the bottom. To do this, the pairs $(q_H a, q_H)$ and $(aq_H, q_H)$ are added for all symbols $a$ so that any symbol $a$ adjacent to the halt state $q_H$ disappears, where $q_H$ denotes the states such as an accept state or a reject state, which is supposed to halt. In this way, $\sharp q_H \sharp$ appears at the bottom. So, adding a pair $(q_H \sharp\sharp, \sharp)$ to the list, we can make $q_H$ disappear and the top catch up with the bottom.

Before preceding to a precise definition of the set of pairs of strings, we explain why we can assume without loss of generality the requirement that the pair $(u_1, v_1)$ appears first in a match. So let us introduce the PCP with this requirement as the *modified Post correspondence problem* (MPCP) as follows: when given a list of pairs of strings $\{(u_1, v_1), \ldots, (u_k, v_k)\}$, decide whether there exists a sequence of subscripts $i_1, i_2, \ldots, i_m$ such that $u_1 u_{i_1} u_{i_2} \cdots u_{i_m} = v_1 v_{i_1} v_{i_2} \cdots v_{i_m}$. So far we have constructed MPCP $P'$ that simulates TM $M$. In what follows we show that, in general, we can somehow modify MPCP $P'$ to obtain PCP $P$ that behaves essentially the same way as MPCP $P'$ even if we disregard the requirement that $(u_1, v_1)$ must be used first. This is done by adding new symbols to $P'$ so that a match of $P$ must start necessarily with the pair $(u_1, v_1)$, and otherwise $P'$ and $P$ manipulate strings in substantially the same way.

Taking $P'$ of Example 7.6 as an example, we explain what is described above by giving the corresponding modified $P$. To begin with, we give a match of the $P$ as shown in Fig. 7.15, which corresponds to the match of the $P'$ given in Fig. 7.13. Observe that symbol $\cent$'s newly introduced appear in every other symbol in the match. So to search for a match we focus on the remaining symbols 0 and 1 which appear between $\cent$'s. To change the situation this way, we put pairs

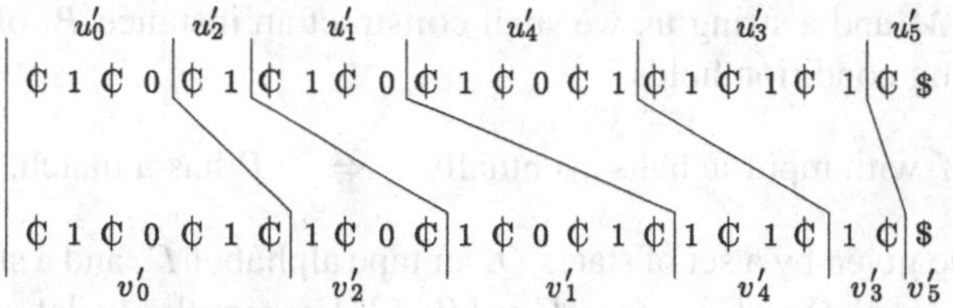

$$u'_0 \quad u'_2 \quad u'_1 \quad u'_4 \quad u'_3 \quad u'_5$$

¢ 1 ¢ 0 | ¢ 1 | ¢ 1 ¢ 0 | ¢ 1 ¢ 0 ¢ 1 | ¢ 1 ¢ 1 ¢ 1 | ¢ $

¢ 1 ¢ 0 ¢ 1 ¢ | 1 ¢ 0 ¢ | 1 ¢ 0 ¢ 1 ¢ | 1 ¢ 1 ¢ | 1 ¢ $

$$v'_0 \qquad v'_2 \qquad v'_1 \qquad v'_4 \qquad v'_3 \; v'_5$$

**Fig. 7.15** Match corresponding to that in Fig. 7.13

$(u'_i, v'_i)$ of $P$, each corresponding to $(u_i, v_i)$ of $P'$, which are specified as follows: $(u'_1, v'_1) = (\text{¢}1\text{¢}0, 1\text{¢}0\text{¢}1\text{¢})$, $(u'_2, v'_2) = (\text{¢}1, 1\text{¢}0\text{¢})$, $(u'_3, v'_3) = (\text{¢}1\text{¢}1\text{¢}1, 1\text{¢})$, and $(u'_4, v'_4) = (\text{¢}1\text{¢}0\text{¢}1, 1\text{¢}1\text{¢})$. Observe that none of these could appear as the first pair in a match because in any pair of strings one string starts with ¢ while the other string starts with a symbol other than ¢, hence any pair is unable to be the first. So, in addition to those, we add $(u'_0, v'_0) = (\text{¢}1\text{¢}0, \text{¢}1\text{¢}0\text{¢}1\text{¢})$ which could appear once as the first pair in a match. This is because $(u'_0, v'_0)$ is the only pair that could possibly start a match with the same symbol, namely ¢. To complete the collection of pairs of $P$, we further add $(u'_5, v'_5) = (\text{¢}\$, \$)$. Specifying the pairs this way we can say that, as compared to the top, the bottom proceeds in placing symbol ¢ in a match so that the top must catch up with the bottom by using the pair $(\text{¢}\$, \$)$ at the end of the match.

We are ready to summarize how to convert MPCP $P'$ to an equivalent PCP $P$. To do this, we need the following notation. For string $w = a_1 a_2 \cdots a_i$ of length $i$, let

$$\text{¢}w = \text{¢}a_1 \text{¢}a_2 \cdots \text{¢}a_i,$$

$$w\text{¢} = a_1 \text{¢}a_2 \text{¢} \cdots a_i \text{¢},$$

$$\text{¢}w\text{¢} = \text{¢}a_1 \text{¢}a_2 \text{¢} \cdots a_i \text{¢}.$$

Using this notation, from $P' = \{(u_1, v_1), \ldots, (u_k, v_k)\}$, we construct

$$P = \big\{ (\text{¢}u_1, \text{¢}v_1\text{¢}), (\text{¢}u_1, v_1\text{¢}), \ldots, (\text{¢}u_k, v_k\text{¢}), (\text{¢}\$, \$) \big\}.$$

It follows from what is described so far that the conditions that MPCP $P'$ has a match and that PCP $P$ has a match are equivalent with each other. In this way, MPCP is reduced to PCP, which is described as the following lemma.

**Lemma 7.7** *MPCP $\leq$ PCP.*

We are ready to prove that PCP is undecidable.

**Theorem 7.8** *The Post correspondence problem is undecidable.*

*Proof* In this proof, it is shown that the halting problem can be reduced to the MPCP. Since the MPCP is reduced to the PCP from Lemma 7.7, the halting problem is reduced to the PCP from Proposition 7.3. Therefore, since the halting problem is undecidable, the PCP becomes undecidable from Theorem 7.5, completing the proof.

From a TM $M$ and a string $w$, we shall construct an instance $P'$ of MPCP such that the following condition holds.

The TM $M$ with input $w$ halts eventually    $\Leftrightarrow$    $P$ has a match.

Let TM $M$ be given by a set of states $Q$, a tape alphabet $\Gamma$, and a state transition function $\delta : (Q - \{q_H\}) \times \Gamma \to Q \times \Gamma \times \{R, L\}$. For simplicity, let $q_H \in Q$ be the only halting state, and let the halting problem be the problem to decide whether the TM $M$ with input $w$ eventually goes to the state $q_H$. An instance $P$ is defined to consist of the pairs of strings from (1) to (5) as follows:

(1) Put the pair into $P'$ as the first pair

$$(\sharp, \sharp q_0 w \sharp) \in P',$$

which sets the starting configuration.

(2) Put the pairs into $P'$ which correspond to state transitions. For all $a, a', b \in \Gamma$, $q \in Q - \{q_H\}$, and $q' \in Q$, if $\delta(q, a) = (q', a', R)$, then put

$$(qa, a'q') \in P'.$$

If $\delta(q, a) = (q', a', L)$, then put

$$(bqa, q'ba') \in P'.$$

(3) Put the pairs into $P'$ to copy symbols. That is, for all $a \in \Gamma \cup \{\sharp\}$ put

$$(a, a) \in P'.$$

Furthermore, to add a blank symbol $\sqcup$ at the right end of the configurations infinitely, put

$$(\sharp, \sqcup\sharp) \in P'.$$

(4) To make symbols adjacent to $q_H$ disappear, for all $a \in \Gamma$ put

$$(aq_H, q_H), (q_H a, q_H) \in P'.$$

(5) To complete a match when $\sharp q_H \sharp$ appears in the bottom, put

$$(q_H \sharp\sharp, \sharp) \in P'.$$

Clearly, the collection of the pairs defined above forces a match to simulate $M$ with input $w$. So, if $M$ with input $w$ never halts, then the common sequence continues to extend infinitely. So in such a case we will never end up with a match which has a finite length by definition. This completes the proof.                    $\square$

In the next example, we apply the construction of the MPCP to a certain concrete TM and see how the match extends itself.

**Table 7.3** Transition function $\delta(q, a)$

| $q \backslash a$ | 0 | 1 | ⊔ |
|---|---|---|---|
| $q_0$ | $(q_1, 1, R)$ | $(q_1, 0, L)$ | $(q_1, 1, L)$ |
| $q_1$ | $(q_H, 0, L)$ | $(q_0, 0, R)$ | $(q_1, 0, R)$ |
| $q_H$ | – | – | – |

*Example 7.9* Given TM $M$ and string $w$, we construct the instance $P'$ of the MPCP. Let $w = 01$ and let the transition function of TM $M$ be given in Table 7.3.

As shown in the proof of Theorem 7.8, the pairs of $P'$ are specified as follows:
(1) To set the initial configuration, put

$(\sharp, \sharp q_0 01 \sharp)$.

(2) Corresponding to specification of the transition function, put

$(q_0 0, 1 q_1)$,
$(0 q_0 1, q_1 00)$,
$(1 q_0 1, q_1 10)$,
$(0 q_0 \sqcup, q_1 01)$,
$(1 q_0 \sqcup, q_1 11)$,
$(0 q_1 0, q_H 00)$,
$(1 q_1 0, q_H 10)$,
$(q_1 1, 0 q_0)$,
$(q_1 \sqcup, 0 q_1)$.

(3) To copy symbols and supply the blank symbol ⊔, put

$(0, 0)$,
$(1, 1)$,
$(\sqcup, \sqcup)$,
$(\sharp, \sharp)$,
$(\sharp, \sqcup \sharp)$.

(4) To make symbols adjacent to $q_H$ disappear, put

$(0 q_H, q_H)$,
$(1 q_H, q_H)$,
$(q_H 0, q_H)$,
$(q_H 1, q_H)$,
$(\sqcup q_H, q_H)$,
$(q_H \sqcup, q_H)$.

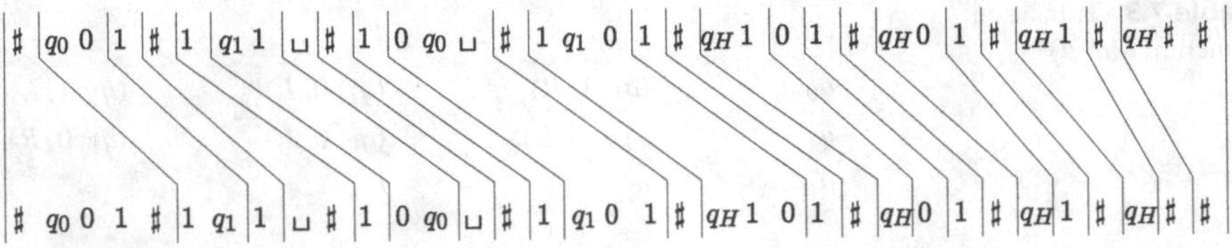

**Fig. 7.16** Match that represents moves of TM $M$

(5) To complete a match at the end, put

$$(q_H \sharp\sharp, \sharp).$$

Putting the pairs of $P'$ as above, we have the match shown in Fig. 7.16, which corresponds to the transitions of TM $M$:

$$q_0 01 \Rightarrow 1q_1 1 \Rightarrow 10q_0 \Rightarrow 1q_1 01 \Rightarrow q_H 101.$$

## 7.5   Problems

**7.1** Modify the state diagram of the universal Turing machine $U$ in Fig. 7.3 so that $U$ can simulate a nondeterministic Turing machine.

**7.2** Show that the sentence "this sentence is not correct" is contradictory.

**7.3**\*\* The halting problem is the one to decide whether TM $M$ with input $w$ eventually halts or not. Let the empty tape halting problem be the one to decide whether TM $M$ with input $\varepsilon$ eventually halts or not. Verify that the empty tape halting problem is undecidable.

**7.4** We present below a proof that derives the fact that the Post correspondence problem is decidable. Find a flaw in the proof.

Define two graphs $G_1$ and $G_2$ as

$$G_1 = (\{1, 2\}, \{(1, 2)\}),$$
$$G_2 = (\{1, 2\}, \emptyset).$$

Consider the reachability problem for these graphs with $s = 1$ and $t = 2$. Then, clearly the reachability problem returns *YES* if the input is $G_1$, and *NO* if the input is $G_2$. Furthermore, define a reduction as follows.

$$f(\langle P \rangle) = \begin{cases} \langle G_1 \rangle & \text{if PCP } P \text{ has a match,} \\ \langle G_2 \rangle & \text{otherwise.} \end{cases}$$

Then, the following equivalence relation clearly holds.

The Post correspondence problem          The reachability problem returns
returns *YES* for input *P*          $\Leftrightarrow$          *YES* for $f(\langle P \rangle)$

Hence we have

the Post correspondence problem $\leq_{\text{mapping}}$ the reachability problem.

On the other hand, since the reachability problem is decidable, the Post correspondence problem is decidable from Theorem 7.5.

# Part IV

# Complexity of Computation

# Computational Complexity Based on Turing Machines

**8**

Among the problems that can be solved in principle by Turing machines, there exists a problem that requires one to run a modern supercomputer, say, for the life time of the earth. This fact naturally leads us to classify real-world problems into two types: *tractable* problems, which can be computed in a feasible amount of time, and *intractable* problems, which cannot be computed in a feasible amount of time. In order to classify problems we employ a Turing machine as a computational model, because it turns out that the time required to run a typical computer is somehow related to that required to run an equivalent Turing machine. First, we introduce a way of measuring the time required to compute problems. The class of tractable problems, denoted by P, consists of problems that can be computed in polynomial time on a Turing machine in terms of the measure. On the other hand, we often encounter in practice a problem that can be solved somehow by checking all the possible *certificates* for the problem instance. Finally, we introduce the other type of problems such that, once an appropriate certificate is given, checking the validity can be done in polynomial time. The class of this type of problems is denoted by NP, which includes P by definition.

## 8.1 Time Complexity

Among the resources required to perform computation there exist time and memory. In this book we exclusively deal with time and introduce the notion of the time complexity of Turing machine (TM) $M$ which gives the number of steps of $M$ in terms of a function of the length of inputs to $M$. In particular, we are concerned with how fast the time complexity increases as the length of the input increases. The time complexity plays an important role throughout this final part of the book.

The *time complexity* of a Turing machine takes the form $t(n) : \mathcal{N} \to \mathcal{N}$, where $\mathcal{N}$ denotes the set of integers and $t(n)$ gives the steps in the case that the length of an input is $n$. Since there are several inputs whose length is $n$, we must specify what input we consider in defining $t(n)$. The method that we adopt throughout this book is to define $t(n)$ to be the maximum number of steps that the Turing machine takes

A. Maruoka, *Concise Guide to Computation Theory*,
DOI 10.1007/978-0-85729-535-4_8, © Springer-Verlag London Limited 2011

among all the inputs of the length $n$. This is called *worst-case analysis*. In addition to worst-case analysis, we can consider *average-case analysis*, where $t(n)$ is defined to be the average of all the steps for inputs of length $n$.

**Definition 8.1** Let $M$ be a deterministic Turing machine that halts on all inputs. The *time complexity* of $M$ is the function $t(n) : \mathcal{N} \to \mathcal{N}$ such that $t(n)$ is the maximum number of steps that $M$ takes until it halts among all inputs of length $n$. When the time complexity of $M$ is $t(n)$, we say that $M$ runs in time $t(n)$ and that $M$ is a $t(n)$ time Turing machine.

## Time Required to Check Well-Nested Parentheses

We consider the Turing machine given in Fig. 6.4 that decides if an input is well-nested and analyze the time complexity of the Turing machine.

As shown below, $M$ starts on the leftmost square. In (1), the machine moves rightward searching for " ) " first met, and crosses off the " ) " (rewrites it to $X$). Then the machine moves leftward searching for " ( " first met and crosses off the " ( " (rewrites it to $X$). In this way the machine repeats crossing off the innermost " ( " and " ) " among the symbols that have not been crossed off as long as such a pair of " ( " and " ) " remains. When such pairs of " ( " and " ) " are all crossed off, in (2), the machine checks if any " ( " or " ) " remains unmatched and accepts if no such parenthesis remains.

TM that accepts well-nested strings of parentheses:
1. Starting on the leftmost square, repeat the following as long as a pair of " ( " and " ) " described below remains.
   Move rightward searching for " ) " first met and rewrite the " ) " into $X$. Then move leftward searching for " ( " first met and rewrite the " ( " into $X$.
2. If blank symbol ⊔ is encountered while scanning rightward in 1, scan across the tape leftward until the leftmost square is encountered. If no parenthesis is found while scanning, accept the input.

Figures 8.1 and 8.2 show how the head moves when well-nested strings of parentheses are fed as input to $M$. Symbol ◉ in the figures indicates the moment when a parenthesis is rewritten to $X$. Figure 8.3 gives the state diagram of $M$.

Next we analyze the time complexity of $M$. First we derive the number of steps required when a well-nested string of the form in Fig. 8.1 is fed as input. In general, when string $( ( \cdots ( \quad ) \cdots ) )$ of length $n$ is fed as input, the number of steps required is given as

$$\frac{n}{2} + 1 + 2 + \cdots + (n-1) + n + n = \frac{n^2}{2} + 2n.$$

In this equation $n/2, 1, 2, \ldots, n-1$ are the numbers of steps required to handle the corresponding right and left parentheses and the next two $n$'s are the numbers of steps to check that the right and the left parentheses do not remain.

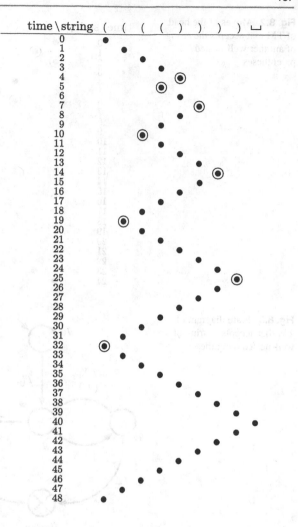

**Fig. 8.1** Moves of the head of TM that accepts the string of well-nested parentheses

On the other hand, when a string of the form ( )( ) ··· ( ) of length $n$ is fed as input, the number of steps required is given as

$$2 + 4\left(\frac{n}{2} - 1\right) + 2 + n = 3n.$$

In this equation, the first 2 is the number of steps to handle the first right and left parentheses, $4(n/2 - 1)$ is the number of steps to handle the remaining $(n/2 - 1)$ pairs, each requiring 4 steps, the next 2 is for checking that no right parentheses remain, and the final $n$ is for checking that no left parentheses remain.

We can prove that the maximum number of steps that $M$ uses on all the inputs of length $n$ is given by $(n^2/2) + 2n$, which is left as Problem 8.3$^\sharp$. In fact, as we have shown above, the input written as ( ( ··· ( ) ··· ) ) requires this number of

**Fig. 8.2** Moves of the head of TM that accepts the string of another well-nested parentheses

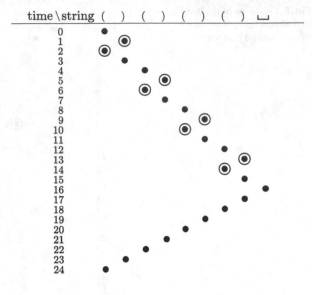

**Fig. 8.3** State diagram of TM that accepts a string of well-nested parentheses

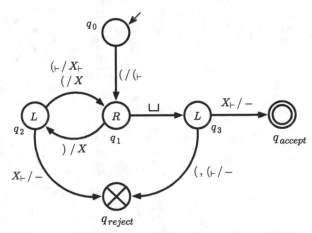

steps. Since throughout this book the time complexity is the largest number of steps on inputs of length $n$, the time complexity of TM $M$ given in Fig. 8.3 turns out to be $(n^2/2) + 2n$.

## Big-$O$ Notation

As described in Problem 8.4$^\sharp$, given a TM $M$ with time complexity $t(n)$, we can modify $M$ so as to obtain a new $M'$ with time complexity proportional to $t(n)$ with an arbitrarily small coefficient. Namely, for any $c > 1$, we can make the time complexity of the new $M'$ $t(n)/c$. This speedup can be achieved by executing $c$

steps of the original TM by one step at the cost of increasing the number of tape symbols and that of states.

On the other hand, substantial speedup beyond any small constant factor may be accomplished by introducing a Turing machine equivalent to the original one, but behaving in a completely different manner. So we introduce a notation for the time complexity that disregards the differences up to a constant factor.

The notation we introduce is not exact in expressing the time complexity in two respects. First, it focuses on the most dominating term, neglecting the remaining terms, and disregards the coefficient of the dominant term. Second, it expresses in terms of an upper bound. For example, when $f(n) = 50n^3 + 100n^2 + 500$, we express $f(n) = O(n^3)$, reading $f(n)$ is order $n^3$. In this case, we can write either $f(n) = O(n^4)$ or $f(n) = O(n^5)$. But $f(n) = O(n^2)$ is not valid. When $f(n) = 100n \log n$, we can write $f(n) = O(n \log n)$ and $f(n) = O(n^2)$. But $f(n) = O(n)$ is not valid.

We formalize the notation for the time complexity as follows. Let $\mathcal{R}^+$ denote the set of nonnegative real numbers.

**Definition 8.2** Let $f$ and $g$ be functions $f, g : \mathcal{N} \to \mathcal{R}^+$. Write $f(n) = O(g(n))$ if there exist positive integers $c$ and $n_0$ such that for all $n \geq n_0$

$$f(n) \leq cg(n).$$

As in the definition above, since in order to make the inequality $f(n) \leq cg(n)$ hold we can take a sufficiently large $c$, we can disregard the coefficient of $g(n)$ in the expression $f(n) = O(g(n))$. Furthermore, because of the restriction on the range of $n$ in terms of $n_0$, the inequality $f(n) \leq cg(n)$ does not need to hold for all $n$. If the inequality does not hold only for a finite number of $n$, say, $n_1, n_2, \ldots, n_m$, we can still make the condition of Definition 8.2 be satisfied by taking $n_0$ to be $n_m + 1$, provided that $n_1 < n_2 < \cdots < n_m$.

## Time Complexity of Turing Machines and that of Problems

The time complexity of an algorithm is the function that gives the maximum number of steps that the algorithm takes on all the inputs of length $n$. We can see this by simply identifying a Turing machine in Definition 8.2 with an algorithm. In what follows we take it for granted that the time complexity of a Turing machine and that of an equivalent algorithm are somehow related. In fact, all reasonable computational models with their complexity measure are polynomially equivalent. More precisely, two functions $f(n)$ and $g(n)$ are *polynomially equivalent* if there exist $c > 0, k > 0$, and $n_0 \geq 1$ such that $f(n) \leq cg(n)^k$ and $g(n) \leq cf(n)^k$ for any $n \geq n_0$. On the other hand, when we talk about the time complexity of a problem $P$ rather than an algorithm, we have in mind that the time complexity of a problem $P$ is somewhat the "minimum" among the time complexities of the algorithms that solve the problem $P$. But throughout this book the terminology "time complexity" associated with a problem only appears in a statement like "The time complexity of a problem $P$ is $O(g(n))$," which means that there exists an algorithm to solve the problem $P$ in $t(n)$ time such that $t(n) = O(g(n))$. So actually we do not need to worry about how to define the minimum of a certain collection of functions expressing time complexity.

As an example of the time complexity of a problem, let us consider the membership problem for a regular language given in terms of a state diagram. To solve the membership problem of a regular language, we can think of an algorithm that works as follows: beginning at the start state, the algorithm given an input just traces the moves on the state diagram along with the input and outputs YES if the state at the end of the input is an accept state, and NO otherwise. The number of steps required is obviously proportional to the length $n$ of an input. So we can conclude that, since the algorithm works in $O(n)$ time, the time complexity of the membership problem for regular languages is $O(n)$. On the other hand, the time complexity of the membership problem for a context-free language is $O(n^3)$ because the algorithm in Sect. 4.5 solves the problem in $O(n^3)$ time.

In what follows, we discuss the membership problem and analyze its time complexity. We consider that a language $L$ means the membership problem for the language $L$ as well. From the context we can see whether the language or the membership problem for the language is meant.

**Definition 8.3** Let P denote the class of languages that can be decided by a deterministic Turing machine that runs in polynomial time. Namely,

$$P = \big\{ L \mid \text{there exists a deterministic TM that decides } L \text{ in } O\big(n^k\big) \\ \text{time for an integer } k, \text{ where } k \text{ is arbitrary} \big\}.$$

Furthermore, for function $t : \mathcal{N} \to \mathcal{R}^+$, the class of languages, denoted by TIME$(t(n))$, is defined as

$$\text{TIME}\big(t(n)\big) \\ = \big\{ L \mid \text{there exists a deterministic TM that decides } L \text{ in time } O\big(t(n)\big) \big\}.$$

As described in this definition, "$t(n)$ is *polynomial time*" means that $t(n)$ is such that $t(n) = O(n^k)$ for some natural number $k$. So P defined above is said to be the class of languages that can be decided by a deterministic polynomial time TM. Our standpoint is that P corresponds to the class of problems that are solvable on actual computers in practical time. Of course, a polynomial time like, say, $n^{100}$ can never seem practical. But when actual concrete problems turn out to be solvable in polynomial time $O(n^k)$, the constant $k$ can be mostly reduced to be rather small. It should be noted here that a deterministic polynomial time Turing machine can also be used as a basis in Sect. 8.3 and Chap. 10 to characterize the class of problems that might not be solved in polynomial time.

## 8.2    Problems that Are Solvable in Polynomial Time and those that Have not Been Solved in Polynomial Time

Let us discuss a type of problem that asks whether or not, when given a graph as input, there exists a path in the graph that satisfies a certain condition. Some of the problems of this type can be solved in polynomial time, while others have not been solved in polynomial time. Since we can enumerate all the potential paths and check to see if the required condition is satisfied for each potential path, we can solve this type of problem by using brute-force search. Since the number of potential paths is usually given by $O(m^m)$ for graphs with $m$ nodes, a problem is such that the brute-force search is necessary, then the time complexity of the problem turns out to be exponential, which is beyond any polynomial, so that the problem does not seem to be solvable in a manageable time. However, if we could find some trick to avoid the brute-force search to solve this type of problem, then the problem might be solvable in polynomial time. In general, whether the time complexity of this type of problem is polynomial or beyond polynomial depends on a required condition on a path. We examine three concrete problems of this type with various time complexities.

The first problem, called the reachability problem, is to determine whether there exists a path from node $s$ to node $t$ in a graph, where $s$ and $t$ denote designated nodes. The second problem, called the Eulerian path problem, is to determine whether there exists a closed path that passes all the edges in a graph going through each edge exactly once. The third problem, called the Hamiltonian path problem, is to determine whether there exists a closed path that passes all the nodes in a graph going through each node exactly once. As will be shown shortly, the reachability problem and the Eulerian path problem can be solved in polynomial time. But so far no one knows of an algorithm that solves the Hamiltonian path problem in polynomial time. For ease of presentation, the reachability problem and the Hamiltonian path problem are given in terms of directed graphs, whereas the Eulerian path problem is stated in terms of undirected graphs.

The problems mentioned so far concerning graphs can be seen to ask whether there exists an object that satisfies a certain condition among potential candidates. It turns out that there are a large variety of this type of combinatorial problems in practice. Furthermore, the problem we have discussed so far can be seen from this framework: the membership problem for a context-free language can be interpreted to decide whether or not there exists a sequence of rules such that applying successively the rules in the sequence to the start symbol leads to a string given as input; the halting problem can be thought of as deciding whether or not there exists a permissible sequence of configurations from the start configuration up to the configuration with the halting state. The difficulty with these types of problems and hence the number of steps required to solve the problems depends on whether or not we can avoid brute-force search, or whether the number of potential candidates we must check is finite or infinite.

## Reachability Problem

The *reachability problem* is to determine whether there exists a path from node $s$ to node $t$, where $s$ and $t$ denote designated nodes. We can avoid brute-force search to solve this problem. A polynomial time algorithm to solve the problem will be shown below.

Before proceeding to the polynomial time algorithm, we consider an algorithm to solve the reachability problem by means of a brute-force search. The algorithm enumerates all the potential paths from node $s$ to node $t$

$$v_0(=s), v_1, v_2, \ldots, v_{i-1}, v_i(=t),$$

and outputs YES if at least one of such potential paths connects from node $s$ to node $t$, and outputs NO otherwise. Let $s \neq t$ and the number of the nodes in the graph be denoted by $m$. Let us estimate how many potential paths the algorithm must check. Since the number of nodes $v_1, \ldots, v_{i-1}$ except for $v_0$ and $v_i$ that the algorithm must check ranges over 0 through $m - 2$, the number of such sequences is given by

$$1 + m + m^2 + \cdots + m^{m-2} = (m^{m-1} - 1)/(m - 1).$$

So the number of potential sequences of nodes is given by $O(m^{m-2})$, which is exponential in the number of nodes in graphs. Thus the brute-force algorithm runs in exponential time in the number of nodes $m$ in a graph. The time complexity of the brute-force algorithm is exponential in the input size $n$ under any reasonable encoding of graphs. As such an encoding of graphs, let us take the adjacency matrix explained in Sect. 2.4. Recall that an adjacency matrix is such that the $(i, j)$th entry is 1 if there exists an edge from node $i$ to node $j$. Then the size of an input graph is given by $m^2$. Hence the running time of the algorithm is still exponential in the input size $n = m^2$. In a similar argument, we can conclude that the brute-force algorithm to solve the reachability problem runs in exponential time in the input size $n$ for any reasonable encoding of graphs.

To avoid the brute-force search we again consider the algorithm *REACH* that is explained in Sect. 2.7. Instead of examining all the potential paths, the algorithm *REACH* starts with the set consisting of the single node $s$ and then repeats to add a node connected by an edge from a node already found reachable until no further node is added. The algorithm outputs YES if the set of reachable nodes contains $t$ and NO otherwise; this behavior is shown in Fig. 8.4. The algorithm is rewritten as follows.

**Algorithm** *REACH*
Input: $\langle G, s, t \rangle$,
where $s$ and $t$ are the nodes of the graph $G$ and the set of edges
of $G$ is denoted by $E_G$.
**1.** $R \leftarrow \{s\}$,
   $S \leftarrow \{s\}$.
**2.** While $S \neq \emptyset$, do the following for each $v$ in $S$.

**Fig. 8.4** Explanation of behavior of algorithm *REACH*

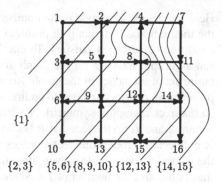

For each $(v, v')$ in $E_G$, do the following if $v' \notin R$.
$$R \leftarrow R \cup \{v'\},$$
$$S \leftarrow S \cup \{v'\}.$$
$$S \leftarrow S - \{v\}.$$
**3.** If $t \in R$, output YES,
   If $t \notin R$, output NO.

Next, we analyze the time complexity of algorithm *REACH*. Since each time stage **2** is executed at least one node is added to $R$, the number of stages executed is given by $O(m)$. Furthermore, the running time of each stage is bounded by $O(m^2)$. So the time complexity of *REACH* is given by $O(m \times m^2)$ which is a polynomial in the number of nodes $m$. Thus we can conclude that *REACH* is a polynomial time algorithm in terms of the input size $n$ with $m^2 = O(n)$.

Furthermore, if we modify algorithm *REACH* so as to access each node in $S$ only once at the first line of stage **2**, then we can improve the time complexity. Suppose that graph $G$ in input $\langle G, s, t \rangle$ is encoded as the adjacency matrix. Although we omit the details, choosing $(v, v')$ in $E_G$, checking if $v' \in R$, and updating $R$ and $S$ can be done in constant steps. Clearly, throughout stage **2** each $(v, v')$ in $E_G$ is processed only once, so the time complexity of the modified algorithm is given by $O(m^2)$, which is also given by $O(n)$.

## Eulerian Path Problem

An *Eulerian path* of a graph is a closed path that passes all the edges of the graph, going through each edge exactly once. The *Eulerian path problem* is to determine whether there exists an Eulerian path in a graph. As in the case of the reachability problem, a brute-force search to solve the Eulerian path problem requires an exponential time.

To avoid the brute-force search we verify a theorem which says that the following two conditions are equivalent with each other:
(1) A graph has an Eulerian path.
(2) A graph is connected and the degree of every node is even.

Here the degree of a node is the number of edges incident to the node. Because of the theorem, the Eulerian path problem can be solved by determining whether the condition (2) above is satisfied. To check if a graph is connected, we only have to run algorithm *REACH* for the graph and see if the set $R$ obtained constitutes the entire set of the nodes of the graph. Recall that when we consider the Eulerian path problem we restrict ourselves to undirected graphs and that the adjacency matrix for an undirected graph is symmetric, i.e., the $(i, j)$th entry is equal to the $(j, i)$th entry for any $i$ and $j$. So the fact that $R$ becomes the entire node set means that any node is connected through undirected edges to the node $s$. Furthermore, to check that the degree of every node is even, it suffices to check that any row of the adjacency matrix has an even number of 1's. Since these conditions can be checked in $O(m^2)$ time, the Eulerian path problem is solved in $O(m^2)$ time, and hence in $O(n)$ time.

Before proceeding to a proof of the theorem, we informally explain why (1) and (2) above are equivalent with each other. First, we explain that (2) implies (1). To do so, assume that condition (2) holds. Starting with an arbitrary node in the graph, we walk along a path by repeatedly visiting a new edge arbitrarily chosen as long as we encounter an unvisited edge. Then we eventually go back to the node where we started. This is because, if we became stuck at a node other than the start node, then that node would have an odd number of incident edges, thereby contradicting the assumption that the degree of every node is even. If there remains an edge not visited, then eliminate all the visited edges from the graph to obtain a new graph and repeat the process just described on the new graph: choose an arbitrary edge and walk along a path as long as we encounter an edge not visited. Iterating this process, we will end up with a graph with no edge. It is easy to form a single Eulerian path that connects all the closed paths obtained this way. Conversely, it is easy to see that condition (1) implies condition (2).

**Theorem 8.4** *The following two conditions are equivalent with each other:*
(1) *A graph has an Eulerian path.*
(2) *A graph is connected and the degree of every node is even.*

*Proof* (Proof of (1) $\Rightarrow$ (2)) Suppose that a graph $G$ has an Eulerian path. Then, $G$ is clearly connected. Take any node in $G$ and go along the Eulerian path. Every time the Eulerian path passes through the node we find two edges incident to that node. Since all the incident edges found are different from each other, we can conclude that the degree of this node is even, completing the proof of (1) $\Rightarrow$ (2).

   (Proof of (2) $\Rightarrow$ (1)) Assume that condition (2) of the theorem holds. Take a node $v$ and go forward by taking edges arbitrarily and forming a disjoint path as long as we can, where a disjoint path is one whose edges are different from each other. Then this path comes back to the original node $v$ to form a disjoint path. To verify this fact, assume that the path ends at a node $v'$ other than node $v$. If we go along the path, we encounter two edges every time we go through node $v'$. Finally, we become stuck at node $v'$. Thus, since all the edges incident to node $v'$ appear on the path, it follows that the degree of the node $v'$ is odd, contradicting condition (2). So let the closed path be denoted by $C_1$. Then we construct a new graph $G'$ by eliminating edges of

**Fig. 8.5**  Closed path $C$ is made from closed paths $C_1$ and $C_i$

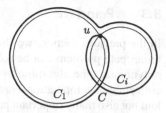

$C_1$ from $G$. This graph $G'$ may be disconnected, but the degree of any node in $G'$ is even. Hence we can apply the same argument to $G'$, obtaining a closed path $C_2$. Applying the same argument repeatedly until no edge remains, we obtain closed paths $C_1, C_2, \ldots, C_k$.

Since the original graph $G$ is connected, we have at least one closed path, say, $C_i$ that has a common node, say, $u$ with $C_1$. Then, as shown in Fig. 8.5, we can obtain a new closed path $C$ by combining these two closed paths: start at $u$; then go along $C_1$ until $u$ is encountered; then go along $C_i$ until $u$ is encountered. Thus we have $k - 1$ closed paths. Continuing this way, we finally obtain one closed path which turns out to be an Eulerian path of the original graph $G$.  $\square$

## Hamiltonian Path Problem

A *Hamiltonian path* in a graph is a closed path that passes all the nodes in the graph going through each node exactly once. The *Hamiltonian path problem* is to decide whether there exists a Hamiltonian path in a graph. Furthermore, a *Hamiltonian graph* is one that has a Hamiltonian path. For the reachability problem and the Eulerian path problem, brute-force search can be avoided so that we can obtain polynomial time algorithms to solve these problems. But, unlike these problems, so far no one knows of a polynomial time algorithm to solve the Hamiltonian problem.

The Hamiltonian path problem can be solved in exponential time by employing a brute-force search as follows. Let a graph $G$ with $m$ nodes be given. The Hamiltonian path problem can be formalized to determine whether there exists a sequence of $m$ nodes

$$v_1, v_2, v_3, \ldots, v_{m-1}, v_m,$$

that satisfies the following two conditions:

(1) All the nodes of $G$ appear in the sequence, or equivalently, all the nodes in the sequence are different from each other.
(2) All of $(v_1, v_2), (v_2, v_3), \ldots, (v_m, v_1)$ are edges of $G$.

Note that $(v_1, v_2)(v_2, v_3) \cdots (v_m, v_1)$ is defined to be a *Hamiltonian path* of $G$ exactly when these two conditions are satisfied. So to solve the Hamiltonian path problem by means of brute-force search it suffices to generate all the $m^m$ sequences of $m$ nodes and output YES if there exists a sequence that satisfies both (1) and (2), and NO otherwise.

## 8.3    P and NP

In the previous section, we showed that both the reachability problem and the Eulerian path problem can be solved in polynomial time and that no one knows any polynomial time algorithm to solve the Hamiltonian path problem. There exists a collection of computational problems including not only the Hamiltonian path problem but also many important problems in practice that no one knows how to solve in polynomial time, and hence at the present time one must essentially rely on brute-force search to solve them. These problems can be thought of as the type of problem to determine whether there exists an object that satisfies a certain condition. To formulate this type of problem, we introduce a notion of a verifier that can decide whether or not an object satisfies such a condition. Several typical problems of this type will also be given in Chap. 10.

To explain how a verifier works, we consider the Hamiltonian path problem and show how a verifier, denoted by $V$, characterizes the collection of Hamiltonian graphs. The verifier is actually a deterministic polynomial time two-tape Turing machine. When given a graph $\langle G \rangle$ on tape 1 together with a potential path $u$ on tape 2, the verifier checks if $u$ is a Hamiltonian path of $G$ in polynomial time. On the other hand, since a verifier is just a deterministic polynomial time two-tape Turing machine, the output of the verifier initially given $\langle G \rangle$ on tape 1 and $u$ on tape 2, denoted $V(\langle G \rangle, u)$, is given as follows:

$$V(\langle G \rangle, u) = \begin{cases} \text{YES} & \text{if } V \text{ given } \langle G \rangle \text{ and } u \text{ eventually goes to an accept state,} \\ \text{NO} & \text{if } V \text{ given } \langle G \rangle \text{ and } u \text{ eventually goes to a reject state.} \end{cases}$$

String $u$ on tape 2 is called a *certificate*. Using this notation, we say that verifier $V$ verifies the set of graph descriptions $\langle G \rangle$ such that there exists $u$ such that $V(\langle G \rangle, u) = $ YES. Note that the set of graph descriptions that the verifier $V$ verifies is exactly the set of graphs that has a path $u$ such that $u$ is a Hamiltonian path of $G$. By definition the set of such graph descriptions is expressed as

$$HAMPATH = \{ \langle G \rangle \mid \text{there exists a } u \text{ such that } V(\langle G \rangle, u) = \text{YES} \}.$$

Verifying existence being given a candidate might be much easier than determining existence without being given a candidate. This comes from the fact that no one knows a polynomial time algorithm for the latter, whereas the verifier can do the former in polynomial time.

In what follows we focus on the distinction between polynomial time and exponential time. So let "to solve quickly" mean "to solve in polynomial time". Likewise, let a "short string" mean a "string of polynomial length" in terms of input size $n$. Note that since a deterministic polynomial time Turing machine stops its computation in polynomial time, it can only access its tape of polynomial length.

Since we are allowed to think of any certificate $u$ and any verifier accordingly, at first glance it seems to be difficult to find a problem that cannot be verified by any verifier. But this is not the case. To explain it, think of the problem

$$\overline{HAMPATH} = \{ \langle G \rangle \mid V(\langle G \rangle, u) = \text{NO for any certificate } u \},$$

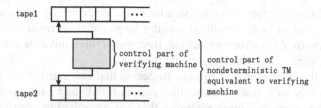

**Fig. 8.6** Verifier equivalent to nondeterministic TM

which is the complement of the set of Hamiltonian graphs. Verifying that a graph does not have any Hamiltonian graph seems to be more difficult than verifying that it has a Hamiltonian graph. Although it is not known so far that $\overline{HAMPATH}$ cannot be verified by a verifier, it seems to be difficult to find a *short* certificate that could convince that a graph does not have *any* Hamiltonian graph.

**Definition 8.5** A *verifier* is a deterministic two-tape Turing machine. A verifier $V$ verifies a language $L$ if the following two conditions are satisfied:
(1) For any $w \in L$ there exists $u \in \Sigma^*$ such that

$$V(w, u) = \text{YES.}$$

(2) For any $w \notin L$ we have

$$V(w, u) = \text{NO,}$$

for any $u \in \Sigma^*$, where $\Sigma$ is the alphabet for tape 2. The language that a verifier $V$ verifies is denoted by $L(V)$, which is expressed as

$$L(V) = \big\{w \mid \text{there exists } u \in \Sigma^* \text{ such that } V(w, u) = \text{YES}\big\}.$$

The class of languages that are verified by a verifier in polynomial time is denoted by *NP*.

Figure 8.6 illustrates a verifier schematically. A verifier is nothing more than a deterministic two-tape Turing machine that is given input $w$ on tape 1 and certificate $u$ on tape 2 when the computation starts. A verifier is the same as a Turing machine except that it is supposed to verify that $u$ is a certificate for $w$.

In Sect. 6.3 we defined that a nondeterministic Turing machine (NTM) $M$ accepts a string $w$ if there exists an accepting configuration in the computation tree determined by $M$ and $w$. We can extend this definition to apply to a nondeterministic polynomial time Turing machine by just imposing the restriction that the time complexity of an NTM is polynomial. The time complexity of an NTM $M$ is the function from the length of input $w$ to the length of the computation tree determined by $M$ and $w$, where the length of a computation tree is the maximum of the lengths of the paths from the root to the leaves of the computation tree.

Next, we prove that the condition that a language is verified by a polynomial time verifier is equivalent to the condition that the language is accepted by a nondeterministic polynomial time Turing machine. Because of this equivalence, the term NP comes from Nondeterministic Polynomial time.

To prove the equivalence, it suffices to derive that one can simulate the other. To begin with, observe that both a verifier and an NTM can be interpreted to have underlying deterministic Turing machines that are controlled by a string $u$ placed on tape 2 in the case of a verifier and by nondeterministic choices of an NTM. From this point of view, the control part together with tape 2 of a verifier illustrated in Fig. 8.6 can be interpreted to constitute the control part of an NTM.

In order to verify the equivalence, first we convert a verifier to an equivalent two-tape NTM $N_2$. The NTM $N_2$ first places a certificate on tape 2 and then simulates $V$. $N_2$ is then transformed to an equivalent single-tape NTM $N$ in a similar way to the proof of Theorem 6.8. Conversely, given an NTM $N$, we construct an equivalent verifier that uses the accepting path in the computation tree as the certificate.

**Theorem 8.6** *For a language $L$ the following hold*:

| *language L is verified by* | | *language L is accepted by a nondeterministic* |
| :-- | :-: | :-- |
| *a polynomial time verifier* | $\Leftrightarrow$ | *polynomial time Turing machine.* |

*Proof* (Proof of $\Rightarrow$) Let $V$ be a polynomial time verifier. We shall construct a nondeterministic polynomial time single-tape TM $N$ that simulates $V$. To do so we first construct a nondeterministic polynomial time two-tape TM $N_2$ and then convert $N_2$ to an equivalent nondeterministic polynomial time single-tape TM $N$. Assume that the time complexity of $V$ is $n^k$ for constant $k$. The TM $N_2$ first generates a string nondeterministically on tape 2 that corresponds to a string placed on tape 2 of verifier $V$ and then simulates $V$ as follows.

Behavior of $N_2$ equivalent to $V$:
1. Generate a string $u$ of length at most $n^k$ nondeterministically and place it on tape 2.
2. Simulate $V$ according to the string on tape 2.

Let the alphabet of tape 2 of $V$ be $\{1, \ldots, d\}$ without loss of generality. In order to generate and place a string nondeterministically on tape 2, it suffices to specify the transition function of $N_2$, denoted by $\delta_{N_2}$, as follows:

$$\delta_{N_2}(q, a, \sqcup) = \{(q, a, 1, S, R), \ldots, (q, a, d, S, R)\}$$

for every symbol $a$, where $q$ denotes the state on which $N_2$ generates and places a string on tape 2. Since $S$ above means to let the head stay in the same position, the head of tape 1 stays on the leftmost square while the head of tape 2 is writing a string. Observe that, as Problem 6.2* shows, we can introduce $S$ in addition to the directions $\{L, R\}$ without changing computational power. Clearly if $V$ verifies $L$, then $N_2$ accepts $L$. Concerning running time, $N_2$ runs in $O(n^k)$ time, hence the length of the two tapes of $N_2$ is upper bounded by $O(n^k)$. Hence each step of $N$ takes at most $O(n^k)$ time. In a similar way to the construction in the proof of Theorem 6.8 $N_2$ can be transformed to an equivalent single-tape $O(n^k \times n^k)$ $(= O(n^{2k}))$ time TM $N$. Although we need to prove that $N_2$ described above has to count length $n^k$ of $u$ in $O(n^k)$ time while generating $U$ in stage **1**, we omit its proof.

(Proof of $\Leftarrow$) Conversely, from a nondeterministic polynomial time TM $N$ we shall construct polynomial time verifier $V$ that simulates $N$. This construction is the same as the construction of an equivalent deterministic TM given in the proof of Theorem 6.9 except that $V$ simulates $N$ following the sequence of nondeterministic choices $u$ that is given as a string on tape 2. Let $d$ denote the maximum of the nondeterministic choices. That is,

$$d = \max_{q,a} \left| \delta_N(q,a) \right|,$$

where $\delta_N$ denotes the transition function of TM $N$. Then a path from the root to a leaf in a computation tree is denoted by $u \in \{1, \ldots, d\}^*$. As in the proof of Theorem 6.9, if $u$ given in tape 2 does not correspond to any valid computation, then verifier $V$ goes to a reject state. If $N$ runs in $n^k$ time, we make $V$ place the certificate of length $n^k$ on tape 2. Clearly $V$ runs in $O(n^k)$ time. $\qquad\qquad\square$

---

## 8.4    Problems

**8.1**
(1) Let $G$ be a context-free grammar. Show that $L(G) \in TIME(n^3)$ holds.
(2) Let $G$ be a regular grammar. Show that $L(G) \in TIME(n)$ holds.

**8.2** A *clique* in an undirected graph $G$ is a subgraph of $G$ such that any two nodes of the subgraph are connected. A $k$-clique is a clique consisting of $k$ nodes. Define

$$kCLIQUE = \left\{ \langle G \rangle \mid G \text{ is an undirected graph that contains a } k\text{-clique} \right\}.$$

Show the following.
(1) $3CLIQUE$ belongs to P.
(2) $\lfloor n/2 \rfloor CLIQUE$ belongs to NP where $\lfloor x \rfloor$ is the integer part of $x$ and $n$ is the number of nodes in a graph.

**8.3$^\sharp$** Prove that for an input of length $2m$ the number of steps of the Turing machine of the state diagram shown in Fig. 8.3 is at most $2m^2 + 4m$.

**8.4$^\sharp$** Let $t(n)$ be such a function that for an arbitrary $\varepsilon > 0$ there exists $n_0$ such that $n/\varepsilon \le t(n)$ holds for any $n \ge n_0$. For $k \ge 2$, prove that a $k$-tape Turing machine whose time complexity is $t(n)$ can be transformed to an equivalent $k$-tape Turing machine whose time complexity is bounded from above by $\varepsilon t(n)$ for $\varepsilon > 0$ when $n$ is sufficiently large. Note that, when the time complexity $t(n)$ is a function that increases faster than $n$ like $n \log n$ or $n^2$, the condition required in this problem is clearly satisfied.

# Computational Complexity Based on Boolean Circuits

**9**

Computation performed by a Turing machine with time complexity $t(n)$ can be thought of as being performed by a $t(n) \times t(n)$ table consisting of $t(n) \times t(n)$ cells. The $i$th row of the table transforms a configuration of the Turing machine at the $i$th step into the configuration at the next step. Each cell of the table can be regarded as a generalized gate that can be implemented by a certain number of Boolean gates under a suitable encoding. This circuit model which works as a counterpart to a Turing machine illustrates more directly how each configuration is transformed into the next configuration. By introducing this alternative circuit model, we can better understand the notion of nondeterminism discussed in Chap. 6 as well as the notion of NP-completeness which will be discussed in Chap. 10.

## 9.1 Boolean Functions and Boolean Circuits

We introduce Boolean functions and Boolean circuits, first discussing some examples and then giving definitions of these notions.

Figure 9.1 gives an example of a Boolean circuit that consists of 4 *inputs* labeled $x_1$, $x_2$, $x_3$, $x_4$ and 15 *gates* denoted by $g_1$ through $g_{15}$ together with wires connecting them. When a *Boolean value* from $\{0, 1\}$ is set to each input, the wires propagate values along them, and the gates compute their output values according to the functions associated with the gates, which are shown in Table 9.1. Given a value setting to the inputs, gate $g_1$ eventually computes an output value which is taken as an output value of the entire circuit. When a setting to the *inputs* is $(x_1, x_2, x_3, x_4)$, let the output value be denoted by $f(x_1, x_2, x_3, x_4)$. This way gate $g_1$ computes a *Boolean function* $f : \{0, 1\}^4 \to \{0, 1\}$. Gate $g_1$ is designated as the *output gate*. We say the circuit computes the function $f$ that is computed by the gate designated as the output gate.

We consider three types of gates, the OR gate, AND gate, and NOT gate, which are denoted by $\vee$, $\wedge$, and $\neg$, respectively, where $\neg$ is also denoted by $^-$. Figure 9.2 illustrates these gates, while Table 9.1 gives the functions that these gates compute. We can consider $x_1 \vee x_2$ and $x_1 \wedge x_2$ as Boolean functions with two variables, and

A. Maruoka, *Concise Guide to Computation Theory*,
DOI 10.1007/978-0-85729-535-4_9, © Springer-Verlag London Limited 2011

**Fig. 9.1** Example of
Boolean circuit

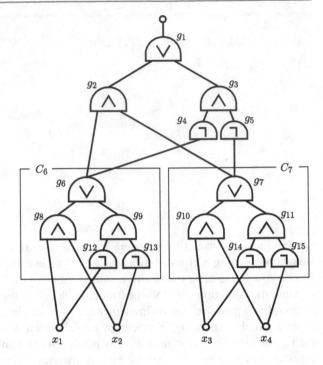

**Fig. 9.2** Gates

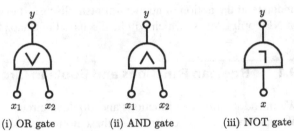

(i) OR gate          (ii) AND gate          (iii) NOT gate

**Table 9.1** Truth tables of
gates

| $x_1$ | $x_2$ | $x_1 \vee x_2$ | $x_1 \wedge x_2$ | $x$ | $\overline{x}$ |
|-------|-------|---------------|------------------|-----|----------------|
| 0     | 0     | 0             | 0                | 0   | 1              |
| 0     | 1     | 1             | 0                | 1   | 0              |
| 1     | 0     | 1             | 0                |     |                |
| 1     | 1     | 1             | 1                |     |                |

$\overline{x}$ as a Boolean function with one variable. Table 9.1 can be easily generalized to represent a Boolean function $f(x_1, \ldots, x_n)$. Such a table representing a Boolean function is called a *truth table*.

Let us explore what function the circuit in Fig. 9.1 computes. The inputs to the circuit are labeled $x_1, x_2, x_3, x_4$, which are also considered as *input variables*, and gate $g_1$ is designated as the output gate. Let us examine what values propagate along the wires from the bottom to the top. As is shown in Sect. 2.5, gate $g_8$ outputs 1 if both of its inputs $x_1$ and $x_2$ are 1. Throughout this book, when we define something by using "if and only if," we often simply use "if" without causing confusion. Similarly, gate $g_9$ outputs 1 if both of its inputs $x_1$ and $x_2$ are 0. Therefore, gate $g_6$ outputs 1 if both of $x_1$ and $x_2$ are 1, or both of $x_1$ and $x_2$ are 0. In short, gate $g_6$ outputs 1 if values of $x_1$ and $x_2$ coincide. In a similar way, gate $g_7$ outputs 1 if the values of $x_3$ and $x_4$ coincide. Therefore, gate $g_2$ outputs 1 if the values of $x_1$ and $x_2$ coincide and the values of $x_3$ and $x_4$ coincide. On the other hand, gate $g_4$ outputs 1 if the negation of "the values of $x_1$ and $x_2$ coincide" holds, that is, the values of $x_1$ and $x_2$ do not coincide. More specifically, it is when $x_1 = 1$ and $x_2 = 0$, or $x_1 = 0$ and $x_2 = 1$. Similarly, gate $g_5$ outputs 1 if the values of $x_3$ and $x_4$ do not coincide, that is, $x_3 = 1$ and $x_4 = 0$, or $x_3 = 0$ and $x_4 = 1$. Therefore, the output gate $g_1$ outputs 1 if $x_1 = x_2$ and $x_3 = x_4$, or $x_1 \neq x_2$ and $x_3 \neq x_4$. In what follows we derive the same statement by applying De Morgan's law and the distributive law to Boolean formulas.

To begin with we summarize what we derived above as follows. The Boolean circuit in Fig. 9.1 computes the Boolean function $f : \{0.1\}^4 \to \{0, 1\}$ defined as

$$f(x_1, x_2, x_3, x_4) = \begin{cases} 1 & x_1 = x_2 \text{ and } x_3 = x_4, \text{ or } x_1 \neq x_2 \text{ and } x_3 \neq x_4, \\ 0 & \text{otherwise.} \end{cases}$$

The Boolean function that gate $g_2$ computes is expressed as a Boolean formula as follows:

$$\big((x_1 \wedge x_2) \vee (\overline{x}_1 \wedge \overline{x}_2)\big) \wedge \big((x_3 \wedge x_4) \vee (\overline{x}_3 \wedge \overline{x}_4)\big).$$

By omitting $\wedge$ together with parentheses appropriately, we can rewrite the formula above to obtain

$$(x_1 x_2 \vee \overline{x}_1 \overline{x}_2)(x_3 x_4 \vee \overline{x}_3 \overline{x}_4).$$

Similarly, gate $g_3$ computes the Boolean formula

$$(\overline{x_1 x_2 \vee \overline{x}_1 \overline{x}_2})(\overline{x_3 x_4 \vee \overline{x}_3 \overline{x}_4}).$$

On the other hand, since gate $g_3$ outputs 1 if $x_1 \neq x_2$ and $x_3 \neq x_4$, the formula above might be rewritten as

$$(\overline{x}_1 x_2 \vee x_1 \overline{x}_2)(\overline{x}_3 x_4 \vee x_3 \overline{x}_4).$$

Now that we have two formulas above which should be equivalent with each other, we show that the former is transformed into the latter mechanically by using De Morgan's law

$$\overline{F \vee G} = \overline{F} \wedge \overline{G}, \qquad \overline{F \wedge G} = \overline{F} \vee \overline{G}$$

and the distributive law

$$F(G \vee H) = FG \vee FH, \qquad (F \vee G)H = FH \vee GH,$$

which hold for any formulas $F$, $G$, and $H$, and finally $\overline{x_i} x_i = 0$ for $1 \leq i \leq 4$. The rewriting goes as follows:

$$
\begin{aligned}
(\overline{x_1 x_2} \vee \overline{\overline{x_1} \overline{x_2}})(\overline{x_3 x_4} \vee \overline{\overline{x_3} \overline{x_4}}) &= (\overline{\overline{x_1 x_2}} \wedge \overline{\overline{\overline{x_1} \overline{x_2}}})(\overline{\overline{x_3 x_4}} \wedge \overline{\overline{\overline{x_3} \overline{x_4}}}) \\
&= \big((\overline{x}_1 \vee \overline{x}_2)(\overline{\overline{x}}_1 \vee \overline{\overline{x}}_2)\big)\big((\overline{x}_3 \vee \overline{x}_4)(\overline{\overline{x}}_3 \vee \overline{\overline{x}}_4)\big) \\
&= \big((\overline{x}_1 \vee \overline{x}_2)(x_1 \vee x_2)\big)\big((\overline{x}_3 \vee \overline{x}_4)(x_3 \vee x_4)\big) \\
&= (\overline{x}_1 x_2 \vee x_1 \overline{x}_2)(\overline{x}_3 x_4 \vee x_3 \overline{x}_4).
\end{aligned}
$$

Combining all the formulas obtained above, we can conclude that the output gate $g_1$ computes the formula given by

$$(x_1 x_2 \vee \overline{x}_1 \overline{x}_2)(x_3 x_4 \vee \overline{x}_3 \overline{x}_4) \vee (\overline{x}_1 x_2 \vee x_1 \overline{x}_2)(\overline{x}_3 x_4 \vee x_3 \overline{x}_4).$$

Clearly, the formula above takes value 1 if "$x_1 = x_2$ and $x_3 = x_4$" or "$x_1 \neq x_2$ and $x_3 \neq x_4$."

The *fan-out* of a gate is the number of wires going out of the gate, and the *fan-in* of a gate is the number of input wires to the gate. For the circuit given by Fig. 9.1, the fan-out of gates $g_6$ and $g_7$ is 2, and the fan-out of the remaining gates is 1. We can somehow modify the circuit in Fig. 9.1 so that all gates have fan-out 1. To do so we duplicate the subcircuits $C_6$ and $C_7$ in Fig. 9.1 to obtain an equivalent circuit shown in Fig. 9.3 with its all gates having fan-out 1. In a similar way to this example, in general a Boolean circuit can be modified to obtain an equivalent Boolean circuit such that all gates of the latter circuit have fan-out 1. Clearly any Boolean circuit with all gates having fan-out 1 can be transformed directly into an equivalent Boolean formula. So we can say that a Boolean circuit that has only gates with fan-out 1 and its corresponding Boolean formula are essentially the same with only a superficial difference in representation. A Boolean circuit generally can use an output of its subcircuit several times by sending its output to inputs to other gates, but this cannot be done in the case of Boolean formulas. Notice that circuits $C_6$ and $C_7$ in Fig. 9.3 are examples of such subcircuits.

Before ending this section, we explain notions about a Boolean circuit together with a family of Boolean circuits that corresponds to a Turing machine. *A Boolean circuit* is a collection of *gates* and *inputs* interconnected with each other through wires in such a way that no cycles are formed. One of the gates is designated as the *output gate*, which produces the output of the entire circuit. The three types of gates are given in Fig. 9.2: the OR gate, AND gate, and NOT gate. Inputs are assigned with Boolean values or Boolean variables. When Boolean variables $x_1, \ldots, x_n$ are assigned to inputs, an output gate computes a Boolean function $f(x_1, \ldots, x_n) : \{0, 1\}^n \to \{0, 1\}$ in an obvious way. We sometimes consider a

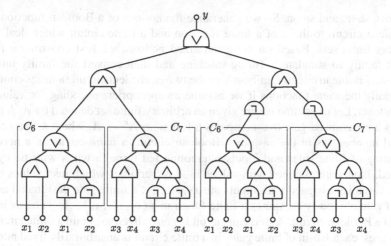

**Fig. 9.3** Boolean circuit with its all gates having fan-out 1 that is equivalent to the circuit in Fig. 9.1

Boolean circuit that has a certain number of output gates. Such a circuit with $n$ inputs computes a function $f(x_1, \ldots, x_n) : \{0, 1\}^n \to \{0, 1\}^k$ in an obvious way, where $k$ denotes the number of output gates.

When we compare a Boolean circuit and a Turing machine as a device to solve the membership problem for a language, we realize that there is a big difference between the two: any particular circuit treats inputs of a certain fixed length, whereas a Turing machine handles inputs of various lengths that are placed on the tape when the Turing machine starts to perform. A way to compensate for this difference is to introduce a family of circuits instead of a single circuit. A *circuit family* $C$ is an infinite set of circuits $\{C_0, C_1, C_2, \ldots\}$, where $C_n$ has $n$ input variables. A family $C$ *decides* a language $L$ over $\{0, 1\}$ if for any string $w$ in $\{0, 1\}^*$

$$w \in L \quad \Leftrightarrow \quad C_n(w) = 1,$$

where $n$ is the length of $w$. The language that is decided by a circuit family $\{C_0, C_1, C_2 \ldots\}$ is denoted by $L(\{C_n\})$ as in the case of Turing machines. In this chapter we shall construct a circuit family that simulates a deterministic Turing machine as well as a nondeterministic Turing machine.

The *size* of a circuit $C$ is the number of gates in the circuit and is denoted by $s(C)$. The *size complexity* of a circuit family $\{C_0, C_1, C_2, \ldots\}$ is the function $s(n)$ : $\mathcal{N} \to \mathcal{N}$ such that $s(n) = s(C_n)$.

## 9.2   Finite Functions and Finite Circuits

The object of this chapter is to construct a circuit family that simulates a Turing machine. For this purpose, it is convenient to generalize the Boolean values $\{0, 1\}$ to a finite collection of values so that we can handle values that correspond to tape symbols, state, and so on. So we generalize the notions of a Boolean function and a Boolean circuit to those of a finite function and a finite circuit which deal with arbitrary finite sets. Based on the generalized notions, we first construct a finite circuit family to simulate a Turing machine and then convert the family into an equivalent Boolean circuit family, where the two equivalent circuit families compute essentially the same functions if we assume an appropriate encoding for values in the finite set. Let us assume we are given an arbitrary finite set denoted by $A$. A *finite gate* is a generalized gate that computes a function: $A^k \to A$, where $k$ is a certain natural number. As in the case of a Boolean circuit, a *finite circuit* is a network consisting of finite gates and inputs interconnected through wires with no cycles allowed. Inputs are assigned with values in $A$ or variables which take values in $A$. When variables assigned with inputs are $x_1, \ldots, x_n$, a finite gate designated as the output gate obviously computes a finite function $f(x_1, \ldots, x_n) : A^n \to A$ as in the case of a Boolean circuit. A Boolean circuit is generalized to obtain a finite circuit in two senses: each fan-in of finite gates in a finite circuit is an arbitrarily fixed number rather than 1 or 2 in the case of Boolean gates; the collection of values that appear on wires is an arbitrarily fixed set rather than $\{0, 1\}$ in the case of a Boolean circuit.

First, we explain how to convert a finite circuit into an equivalent Boolean circuit. To do so, we assume a one-to-one correspondence between $A$ and $\{0, 1\}^l$ so that each value in $A$ can be represented by a sequence in $\{0, 1\}^l$, where $l$ is the least integer such that $2^l \geq |A|$. Assuming the one-to-one correspondence, we replace each wire and each variable in a finite circuit by a collection of $l$ binary wires and a collection of $l$ Boolean variables. Furthermore, each finite gate computing a finite function $g : A^k \to A$ is replaced with a corresponding Boolean circuit with $l$ designated output gates that computes the corresponding function $g' : \{0, 1\}^{kl} \to \{0, 1\}^l$. In the same way as we have the Boolean circuit that computes $g' : \{0, 1\}^{kl} \to \{0, 1\}^l$ corresponding to a finite gate computing $g : A^k \to A$, we have the Boolean circuit that computes $f : \{0, 1\}^{nl} \to \{0, 1\}^l$ corresponding to a finite circuit computing $f : A^n \to A$.

To explain what is described above, we use an example of a finite gate and construct a Boolean circuit equivalent to it. The finite gate in this example computes a finite function $f$ defined as follows:

$$f(z_1, z_2) = (z_1 + z_2) \quad \mod 4,$$

where $(z_1 + z_2) \mod 4$ denotes the remainder of $z_1 + z_2$ divided by 4. Let $A$ denote $\{0, 1, 2, 3\}$. Then the function $f : A^2 \to A$ is shown in Table 9.2. We consider the one-to-one correspondence such that $\{0, 1, 2, 3\}$ correspond to $\{00, 01, 10, 11\}$, respectively. That is, $z \in A$ is represented by a binary number of two digits. When this correspondence is assumed, Table 9.2 is replaced with Table 9.3, which is divided into two tables in Table 9.4, each of them representing a Boolean function. Now we proceed to construct the Boolean circuit that computes these two Boolean functions.

Figure 9.4 gives a Boolean circuit that computes the Boolean functions $f_1$ and $f_2$. We construct the circuit so that the circuit obtained corresponds to Table 9.4 rather than trying to minimize the size of the circuit. We can easily see that the four hor-

**Table 9.2** Example of a finite function

| $z_1 \backslash z_2$ | 0 | 1 | 2 | 3 |
|---|---|---|---|---|
| 0 | 0 | 1 | 2 | 3 |
| 1 | 1 | 2 | 3 | 0 |
| 2 | 2 | 3 | 0 | 1 |
| 3 | 3 | 0 | 1 | 2 |

**Table 9.3** The finite function in Table 9.2 expressed in terms of binary numbers

| $x_1 x_2 \backslash x_3 x_4$ | 00 | 01 | 10 | 11 |
|---|---|---|---|---|
| 00 | 00 | 01 | 10 | 11 |
| 01 | 01 | 10 | 11 | 00 |
| 10 | 10 | 11 | 00 | 01 |
| 11 | 11 | 00 | 01 | 10 |

**Table 9.4** Two Boolean functions $f_1$ and $f_2$ derived from the finite function in Table 9.2

| $x_1 x_2 \backslash x_3 x_4$ | 00 | 01 | 10 | 11 | $x_1 x_2 \backslash x_3 x_4$ | 00 | 01 | 10 | 11 |
|---|---|---|---|---|---|---|---|---|---|
| | $f_1$ | | | | | $f_2$ | | | |
| 00 | 0 | 0 | 1 | 1 | 00 | 0 | 1 | 0 | 1 |
| 01 | 0 | 1 | 1 | 0 | 01 | 1 | 0 | 1 | 0 |
| 10 | 1 | 1 | 0 | 0 | 10 | 0 | 1 | 0 | 1 |
| 11 | 1 | 0 | 0 | 1 | 11 | 1 | 0 | 1 | 0 |

izontal wires in the circuit correspond to the four rows in the table, respectively. Take, for example, the setting of $x_1 = 0$ and $x_2 = 1$. For this setting only the second horizontal wire from the top carries value 1, so that only the outputs from the subcircuits $C_3$ and $C_4$ propagate toward the top. On the other hand, when the setting of $x_1 = 0$ and $x_2 = 1$ is assumed, we can see from Table 9.4 that $f_1$ and $f_2$ are expressed as $\overline{x}_3 x_4 \vee x_3 \overline{x}_4$ and $\overline{x}_4$, which in turn are computed by the subcircuits $C_3$ and $C_4$, respectively. The outputs of $C_3$ and $C_4$ are fed into subcircuits $C_1$ and $C_2$, respectively. Finally, in this case $C_1$ and $C_2$ make outputs of $C_3$ and $C_4$ go through, producing them as output. Examining the other settings this way, we can see that the circuit in Fig. 9.4 computes the functions $f_1$ and $f_2$ in Table 9.4.

## 9.3 Boolean Circuit Family to Simulate Deterministic Turing Machine

Given a deterministic Turing machine (DTM) $M$, we construct a *Boolean circuit family* that decides $L(M)$. In fact, we first construct a *finite circuit family* that decides $L(M)$ under a suitable encoding, and then convert it to an equivalent Boolean circuit family.

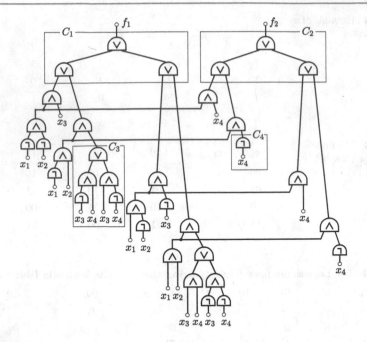

**Fig. 9.4** Boolean circuit that computes Boolean functions $f_1$ and $f_2$ in Table 9.4

**Fig. 9.5** State diagram of
TM $M$

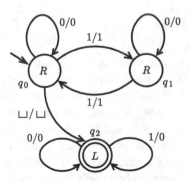

The Boolean circuit family, denoted by $\{C_n\}$, that decides $L(M)$ for a DTM $M$
is such that

$$a_1 \cdots a_n \in L(M) \quad \Leftrightarrow \quad C_n(a_1, \ldots, a_n) = 1,$$

where $C_n(a_1, \ldots, a_n)$ denotes the output of $C_n(x_1, \ldots, x_n)$ when the variables are
set as $x_1 = a_1, \ldots, x_n = a_n$. Throughout this chapter, it is supposed that a Turing
machine eventually halts for any string placed on the tape at the start.

In order to figure out intuitively how such a Boolean circuit family works, we
take the concrete DTM $M$ shown in Fig. 9.5 and discuss a finite circuit family that
simulates $M$. Figure 9.6 illustrates the successive configurations of $M$, each being

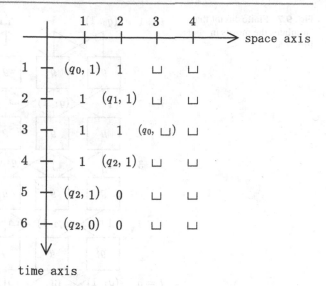

**Fig. 9.6** Moves of TM $M$ with initial tape contents 11

represented by a row. The first row $(q_0, 1)1_{\sqcup\sqcup}\cdots$ of the figure illustrates the initial configuration, where the machine is in state $q_0$ with the head on the leftmost square and with the tape content 11 on the leftmost two squares. First we construct a finite circuit illustrated in Fig. 9.7 that simulates the behavior shown in Fig. 9.6. The finite circuit consists of $5 \times 5$ finite gates connected by wires. In this circuit values propagate from the top to the bottom. We shall explain how to specify the function of the finite gates so that each configuration in Fig. 9.6 appears as a sequence of values on wires between adjacent rows, as shown in Fig. 9.7.

Let $A$ be defined as follows:

$$A = \Gamma \cup (Q \times \Gamma),$$

where $Q$ is the set of states of $M$ and $\Gamma$ is the tape alphabet. Note that a composite value, say, $(q_0, 1)$ as well as 1 and $\sqcup$ are considered as symbols in $A$. In Fig. 9.8 $(t, l)$ refers to the value corresponding to $t$ on the time axis and $l$ on the space axis. Furthermore, the gate in Fig. 9.8 is referred to as gate $(t, l)$. So, as shown in Fig. 9.8, each finite gate $(t, l)$ receives as input values $(t, l-1)$, $(t, l)$, and $(t, l+1)$, computes as output value $(t+1, l)$, and feeds that value to gates $(t+1, l-1)$, $(t+1, l)$, and $(t+1, l+1)$. This is the case with the gates attached with $g : A^3 \rightarrow A$ in Fig. 9.7. On the other hand, the function associated with the leftmost gates is denoted by $g' : A^2 \rightarrow A$. Furthermore, the output gate at the bottom left corner is associated with a function $g_{out} : A^2 \rightarrow A$. These functions will be specified shortly.

Concerning a TM $M$ to be simulated, without loss of generality we make the assumption that $M$ with time complexity $t(n)$ places its head on the leftmost square when $M$ halts. By this assumption, we can place the output gate, denoted by $g_{out}$, at the bottom left corner as shown in Fig. 9.7. This is because we modify TM $M$ so that after the $t(n)$th step the head automatically moves to the leftmost square

**Fig. 9.7** Finite circuit that simulates the moves in Fig. 9.5

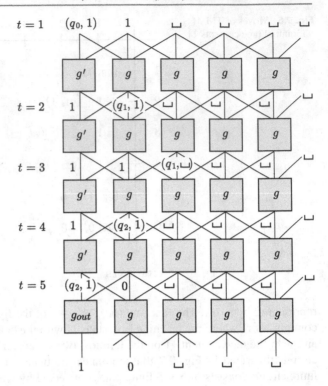

**Fig. 9.8** Typical finite gate

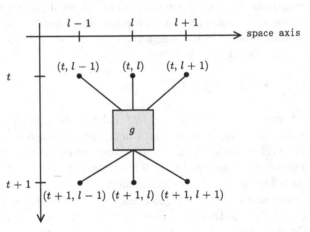

and stays there without changing the state. Although to do so we might need at most additional $t(n)$ steps, we could consider $2t(n)$ as the new time complexity and denote it by $t(n)$.

Next, we explain how to specify the functions of the gates $g$, $g'$, and $g_{out}$ so that values appear on wires, as shown in Fig. 9.7. Figure 9.9 indicates the input/output relations discussed here, which are summarized as follows:

**Fig. 9.9** Input/output specifications corresponding to the computation in Fig. 9.7

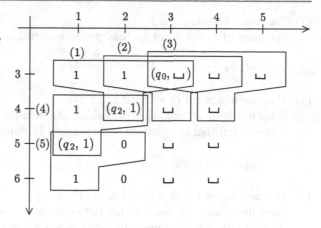

(1) $g(1, 1, (q_0, \sqcup)) = (q_2, 1)$
(2) $g(1, (q_0, \sqcup), \sqcup) = \sqcup$
(3) $g((q_0, \sqcup), \sqcup, \sqcup) = \sqcup$
(4) $g'(1, (q_2, 1)) = (q_2, 1)$
(5) $g_{\text{out}}((q_2, 1), 0) = 1$

We explain these specifications. To begin with, (1) is the specification that comes from the specification $\delta(q_0, \sqcup) = (q_2, \sqcup, L)$, where we let $\delta$ denote the transition function of $M$ to be simulated. The specification (1) indicates that $(q_0, \sqcup)$ of the input moves leftward by one square, and that the machine changes its state from $q_0$ to $q_2$. By this transition the value $(q_0, \sqcup)$ should be replaced with $\sqcup$, which is specified by specification (2) above. In a similar way, (3) and (4) are specified. As for (5), $g'$ is specified as $g'((q_2, 1), 0) = (q_2, 0)$ according to $\delta(q_2, 1) = (q_2, 0, L)$. This is because, as mentioned in Sect. 6.1, when the head tries to move to the left off the tape, the head is assumed to stay at the same leftmost square. Since in our case $M$ moves to an accept state, $g_{\text{out}}$ is specified to output 1. Recall that $q_2$ is the accept state.

As described so far, specifying a finite circuit to simulate a TM $M$ involves nothing more than specifying the input/output behavior of the gates according to the transition function of the $M$ being simulated. Observe that, since a machine moves its head position by one square, two inputs for the leftmost gates and three inputs for the remaining gates are sufficient to simulate a machine. Furthermore, in our construction, in general the finite gate illustrated in Fig. 9.8 receives three different inputs, whereas it produces the same output to the three adjacent gates.

Next, we describe how to specify the functions $g$, $g'$, and $g_{\text{out}}$ in general to simulate a TM $M$ whose transition function is denoted by $\delta$. Let the set of states be denoted by $Q$, the input alphabet be $\{0, 1\}$, and the tape alphabet be denoted by $\Gamma$. Furthermore, let

$$A = \Gamma \cup (Q \times \Gamma).$$

First we define the function $g(x, y, z) : A^3 \to A$ by considering the following five cases.

(1) The case where $x, y, z \in \Gamma$.
Since any of $x, y, z$ is not the head position, define as

$$g(x, y, z) = y.$$

(2) The case where $y \in Q \times \Gamma$ and $x, z \in \Gamma$.
Since if $y = (q, a)$ and $\delta(q, a) = (q', a', D)$, then the head at the position of $y$ is moved leftward or rightward depending on $D$, define as

$$g(x, y, z) = a'.$$

(3) The case where $x \in Q \times \Gamma$ and $y, z \in \Gamma$.
Since the head corresponds to the left input, say, $x = (q, a)$, the output corresponds to the head position, if the head moves to the right. Hence define as

$$g(x, y, z) = \begin{cases} (q', y) & \text{if } \delta(q, a) = (q', a', R), \\ y & \text{otherwise.} \end{cases}$$

(4) The case where $z \in Q \times \Gamma$ and $x, y \in \Gamma$.
In a similar way to (2) above, if $z = (q, a)$, then define as

$$g(x, y, z) = \begin{cases} (q', y) & \text{if } \delta(q, a) = (q', a', L), \\ y & \text{otherwise.} \end{cases}$$

(5) The remaining cases.
Since we do not need to take care of the cases that do not correspond to valid computation, we could define arbitrarily in the remaining cases, which include the case where two or three of $x, y, z$ take the form of $(q, a)$ as well as the case where the transition function $\delta(q, a)$ is not defined. In all the cases, define as

$$g(x, y, z) = 0.$$

Next, we proceed to define the function $g'(y, z)$. In defining it, we only need to be careful about the case where $y$ corresponds to the head position and the head on the leftmost square is about to move off the tape. In that case, by definition, the head is assumed to stay at the same leftmost square. Hence define $g'$ in terms of $g$ as follows:

$$g'(y, z) = \begin{cases} (q', a') & \text{if } y = (q, a) \text{ and } \delta(q, a) = (q', a', L), \\ g(0, y, z) & \text{otherwise.} \end{cases}$$

Note that in the second line in the definition above the first input of $g(x, y, z)$ is specified as $x = 0$, although we could specify this arbitrarily. Since we assume that, once $M$ moves to an accept state with its head on the leftmost square, the machine

**Fig. 9.10** Finite circuit $C'_n$ that simulates Turing machine $M$ with time complexity $t(n)$

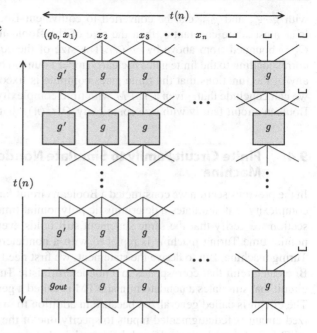

keeps its state and the head position unchanged, the $g'$ defined above is such that if $q$ is an accept state, $a, z \in \Gamma$, and $\delta(q, a) = (q, a', D)$, then $g'((q, a), z) = (q, a')$.

Finally, we proceed to defining $g_{\mathrm{out}}(y, z)$. Since the output gate has to output 1 when the machine goes to an accept state, we define

$$g_{\mathrm{out}}(y, z) = \begin{cases} 1 & \text{if } g'(y, z) = (q', a') \text{ for an accept state } q' \text{ and } a' \in \Gamma, \\ 0 & \text{otherwise.} \end{cases}$$

Using the finite functions $g$, $g'$, and $g_{\mathrm{out}}$ defined above, we construct a finite circuit consisting of $t(n) \times t(n)$ gates, shown in Fig. 9.10, that simulates a Turing machine $M$, where $t(n)$ denotes the time complexity of $M$. By the arguments so far, denoting the finite circuit by $C'_n$, we have

$$a_1 \cdots a_n \in L(M) \quad \Leftrightarrow \quad C'_n(a_1, \ldots, a_n) = 1.$$

Among the values $(q_0, x_1), x_2, \ldots, x_n, \sqcup$ fed as input to $C'_n$, only $x_1, \ldots, x_n$ are considered as input variables, whereas $q_0$ and $\sqcup$ are handled as constants. The finite circuit $C'_n$ can be converted to an equivalent Boolean circuit $C_n$ in the sense that we explained in the previous section. Thus we have a Boolean circuit family $\{C_n\}$ that simulates TM $M$.

Finally, we mention the relationship between the time complexity $t(n)$ of $M$ and the size complexity of $\{C_n\}$. As we can see from the circuit in Fig. 9.10, the number of finite gates of this circuit is given by $t^2(n)$. Hence if the finite gates associated with $g$, $g'$, and $g_{out}$ can be converted to equivalent Boolean circuits whose size is at most some constant $c$, then the size of the Boolean circuit $C_n$ equivalent to $C_n'$ is bounded from above by $ct^2(n)$. The size of the equivalent Boolean circuits corresponding to the finite gates is clearly upper bounded by some constant, because any of the functions that the finite gates compute is fixed, independent of $n$. Thus we can conclude that, given a TM $M$ with time complexity $t(n)$, we can construct a Boolean circuit family with size complexity $O(t^2(n))$ that simulates $M$.

## 9.4 Finite Circuit Family to Simulate Nondeterministic Turing Machine

In the previous section we constructed a Boolean circuit family with polynomial size complexity that simulates a deterministic polynomial time Turing machine. In this section we verify that the same statement still holds even if a deterministic polynomial time Turing machine is replaced with a nondeterministic polynomial time Turing machine. But to derive the statement, we first need to introduce a notion of a Boolean circuit that corresponds to a nondeterministic Turing machine. A Boolean circuit that simulates a nondeterministic TM is called a generalized Boolean circuit. The circuit is called generalized because in addition to ordinary inputs the generalized circuit is fed augmented inputs to specify one of the nondeterministic choices of TM. So far we have explained three models that describe nondeterminism: nondeterministic choice in a Turing machine, a certificate fed on tape 2 of a verifier, and a generalized Boolean circuit with augmented inputs. When we deal with nondeterminism, we can choose the best model among the three depending on the issue we explore.

Suppose that we want to simulate a nondeterministic TM $M$ with time complexity $t(n)$ by a somehow generalized finite circuit. The finite circuit is generalized in the sense that the circuit receives ordinary inputs $a_1, \ldots, a_n$ as well as augmented inputs $u_1, \ldots, u_{t(n)}$ which choose one of the nondeterministic choices. In that sense such a circuit is called a circuit with *input for choices*. When a nondeterministic TM or a circuit with input for choices accepts a string, it is explained as follows: for a nondeterministic TM, we use the phrase "the machine chooses a nondeterministic choice appropriately so that the machine moves to an accept state," whereas in the case of a circuit with input for choices, we use the phrase "the circuit is fed appropriately with augmented inputs so that the output gate produces 1." In summary, we shall construct a finite circuit $C_n'$ with input for choices that satisfies the following:

given input $a_1, \ldots, a_n$ there exist nondeterministic choices $u_1, \ldots, u_{t(n)}$ such that $M$ moves to an accept state
$\Leftrightarrow$
given input $a_1, \ldots, a_n$ there exist $u_1, \ldots, u_{t(n)}$ such that $C_n'(a_1, \ldots, a_n, u_1, \ldots, u_{t(n)}) = 1$.

We shall explain more details of how the finite circuit simulates $M$. Let the transition function $\delta$ of $M$ be such that

$$\delta(q, z) = \left\{ (q', a', L), (q'', a'', R) \right\}.$$

**Fig. 9.11** Finite gates that determine nondeterministic choice depending on the augmented input

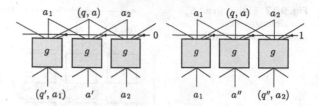

Figure 9.11 shows how to specify the finite gates to implement the specifications above. The fourth input connected to the horizontal wire carries the value to specify the next move, 0 specifying $(q', a', L)$ and 1 specifying $(q'', a'', R)$. So this fourth input is the augmented input. What is described is easily generalized to simulate the following specification:

$$\delta(q, a) = \{(q_0, a_0, D_0), \dots, (q_{j-1}, a_{j-1}, D_{j-1})\}.$$

Let $d$ denote the maximum of the number of nondeterministic choices. That is,

$$d = \max_{q,a} |\delta(q, a)|.$$

To implement the specification above, define $g : A^3 \times \{0, 1, \dots, d-1\} \to A$ in such a way that $g(x, y, z, i)$ coincides with $g(x, y, z)$ specified as in Sect. 9.3 based on $\delta(q, a) = (q_i, a_i, D_i)$. Function $g' : A^2 \times \{0, 1, \dots, d-1\} \to A$ for the leftmost gates and function $g_{\text{out}} : A^2 \times \{0, 1, \dots, d-1\} \to A$ for the output gate are defined in a similar way. Based on $g$, $g'$, and $g_{\text{out}}$ described so far, we can construct the finite circuit $C'_n$ with input for choices, as shown in Fig. 9.12, that simulates the nondeterministic $M$. Thus we have

$$M \text{ accepts } a_1 \cdots a_n \quad \Leftrightarrow \quad \begin{array}{l} \text{there exist } u_1, \dots, u_{t(n)} \in \{0, 1, \dots, d-1\} \\ \text{such that } C'_n(a_1, \dots, a_n, u_1, \dots, u_{t(n)}) = 1. \end{array}$$

Finally, as in the case of deterministic Turing machines, the finite circuit $C'_n$ can be transformed to an equivalent Boolean circuit under an appropriate encoding, which is denoted by $C_n(x_1, \dots, x_n, y_1, \dots, y_{m(n)})$. According to the encoding, a choice out of $\{0, 1, \dots, d-1\}$ is specified by a sequence in $\{0, 1\}^l$, where $l$ is the least integer such that

$$2^l \geq d.$$

Hence, each horizontal wire in Fig. 9.12 is replaced with $l$ wires carrying binary values so that $u_1, \dots, u_{t(n)}$ in Fig. 9.12 is replaced with $y_1, \dots, y_{m(n)}$ in the equivalent Boolean circuit, where

$$m(n) = l \times t(n).$$

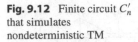

**Fig. 9.12** Finite circuit $C_n'$ that simulates nondeterministic TM

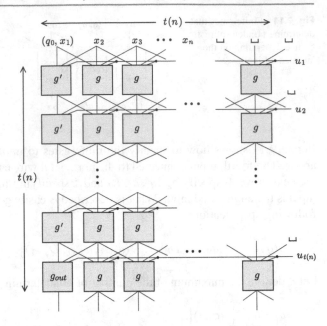

In Sects. 9.3 and 9.4, we constructed families of circuits that simulate a deterministic Turing machine and a nondeterministic one, respectively. Conversely, can we construct a Turing machine that simulates a family of circuits denoted by $\{C_n\}$? To construct an equivalent Turing machine from a family of circuits, we need a condition that there exists a Turing machine that outputs description $\langle C_n \rangle$ when given $n$ as input. In general, without the condition, the construction is impossible. To make the construction possible, the circuits $C_n$ must have a kind of uniformity. This is because an equivalent Turing machine with a fixed finite description, i.e., a description for its state diagram, must perform for all the inputs of various lengths, whereas the corresponding family of circuits $\{C_n\}$ consists of circuits $C_n$, each $C_n$ possibly being quite different from each other. In this book we omit any further discussion about the construction of the reverse direction. For further details, see Problem 9.5[#].

## 9.5    Satisfiability Problems

In this section we derive that the problem to determine whether a nondeterministic Turing machine $M$ accepts a string $w$ is reduced to the problem to determine whether a family of Boolean circuits built from $M$ and $w$ has an assignment to inputs that makes the circuit yield 1 as output. Furthermore, if the time complexity of the NTM is polynomial, then the size complexity of the corresponding family of Boolean circuits turns out to be polynomial.

We start with the Boolean circuit $C_n(x_1, \ldots, x_n, y_1, \ldots, y_{m(n)})$ in the previous section that simulates an NTM $M$. The circuit $C_n$ has as input $x_1, \ldots, x_n$ that correspond to an input to $M$ as well as $y_1, \ldots, y_{m(n)}$ that determine the nondeterministic choices of $M$, where $u_1, \ldots, u_{t(n)}$ in Fig. 9.12 correspond to $y_1, \ldots, y_{m(n)}$ with $m(n) = l \times t(n)$ by means of the encoding of $u_i$ to $\{0, 1\}^l$. Then we derive

$$M \text{ accepts } a_1 \cdots a_n \quad \Leftrightarrow \quad \begin{array}{l} \text{there exist } u_1, \ldots, u_{t(n)} \in \{0, 1, \ldots, d-1\} \\ \text{such that } C_n'(a_1, \ldots, a_n, u_1, \ldots, u_{t(n)}) = 1 \end{array}$$

$$\Leftrightarrow \quad \begin{array}{l} \text{there exist } y_1, \ldots, y_{m(n)} \in \{0, 1\} \\ \text{such that } C_n(a_1, \ldots, a_n, y_1, \ldots, y_{m(n)}) = 1. \end{array}$$

Recall that $y_1, \ldots, y_{m(n)}$ specify path $u_1 \cdots u_{t(n)}$ from the root to a leaf in the computation tree.

Among the inputs of $C_n(x_1, \ldots, x_n, y_1, \ldots, y_{m(n)})$, we substitute constants $a_1, \ldots, a_n$ to input variables $x_1, \ldots, x_n$, respectively, and leave $y_1, \ldots, y_{m(n)}$ as input variables as is. Denote the circuit obtained in this way by $C_{a_1, \ldots, a_n}(y_1, \ldots, y_{m(n)})$. A circuit is said to be *satisfiable* if there exists an assignment to inputs that makes the circuit produce 1 as output. Then the statement above is rewritten as follows:

$$\text{NTM } M \text{ accepts } a_1 \cdots a_n \quad \Leftrightarrow \quad \begin{array}{l} \text{Boolean circuit } C_{a_1, \ldots, a_n}(y_1, \ldots, y_{m(n)}) \\ \text{is satisfiable.} \end{array}$$

In this way, we can verify that the problem to determine whether an NTM accepts a string is reduced to the problem of determining whether a Boolean circuit constructed in this way is satisfiable. After giving the definition of satisfiability, we give a theorem that claims what we described above. As is stated in the next definition, the notion of satisfiability is defined for Boolean formulas as well.

**Definition 9.1** The *satisfiability problem* for Boolean circuits $C(x_1, \ldots, x_n)$ (resp. Boolean formulas $F(x_1, \ldots, x_n)$) is to determine whether there exist assignments $a_1, \ldots, a_n$ to input variables such that

$$C(a_1, \ldots, a_n) = 1 \quad (\text{resp. } F(a_1, \ldots, a_n) = 1).$$

If such assignments exist, then the Boolean circuit (resp. the Boolean formula) is defined to be *satisfiable*. The satisfiability problem for Boolean circuits (resp. Boolean formulas) is called the circuit satisfiability problem (resp. the formula satisfiability problem). The set of satisfiable Boolean circuits (resp. Boolean formulas) is denoted by *CIRCUIT-SAT* (resp. *FORMULA-SAT*), where some natural representation for circuits (resp. formulas) is assumed. *FORMULA-SAT* is also denoted by *SAT* by convention.

By the equivalence between the acceptability of a Turing machine and the satisfiability of a Boolean circuit, NTM $M$ accepts $w$ if and only if $C_w(y_1, \ldots, y_{m(n)})$ is satisfiable, namely, $C_w(y_1, \ldots, y_{m(n)}) \in CIRCUIT\text{-}SAT$, where $C_w(y_1, \ldots, y_{m(n)})$ is constructed from $M$ and $w$ as described so far. Thus we can conclude that $L(M)$ is reducible to *CIRCUIT-SAT*. Furthermore, assuming a certain natural representation for circuits, it is easy to see that the corresponding reduction is computed in polynomial time. Thus we have the following theorem.

**Fig. 9.13** Descriptions of gates

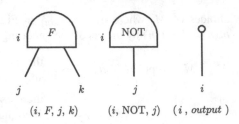

$$(i, F, j, k) \qquad (i, \mathrm{NOT}, j) \qquad (i, \text{ output })$$

**Theorem 9.2** *A language that is accepted by a nondeterministic polynomial time Turing machine is reduced to CIRCUIT-SAT in polynomial time.*

*Proof* As described in Sects. 9.4 and 9.5, let $C_w(y_1, \ldots, y_{m(n)})$ denote the Boolean circuit constructed from NTM $M$ with polynomial time complexity $t(n)$ and input string $w$, where variables $y_1, \ldots, y_{m(n)}$ correspond to variables $u_1, \ldots, u_{t(n)}$ that specify $t(n)$ nondeterministic choices out of $\{0, 1, \ldots, d-1\}$. Recall that $y_1, \ldots, y_{m(n)}$ is partitioned into $t(n)$ blocks, each corresponding to one of $u_1, \ldots, u_{t(n)}$, where $m(n) = l \times t(n)$ and $l$ is the least integer such that $2^l \geq d$.

More precisely, $C_w(y_1, \ldots, y_{m(n)})$ is constructed by replacing each finite gate in Fig. 9.12 with an equivalent Boolean circuit with $c$ gates, each horizontal wire with $l$ binary wires, and variables $x_1, \ldots, x_n$ with Boolean values $a_1, \ldots, a_n$, respectively, where $w = a_1 \cdots a_n$. So $C_w(y_1, \ldots, y_{m(n)})$ has $c \times t^2(n)$ gates and at most $2c \times t(n)$ inputs from the top and $l \times t(n)$ inputs from the right-hand side. Denote the total number of gates and inputs by $q(n)$, where

$$q(n) \leq c \times t^2(n) + 2c \times t(n) + l \times t(n).$$

From the construction of the circuit, NTM $M$ accepts $w$ if and only if $C_w(y_1, \ldots, y_{m(n)})$ is satisfiable. So we have the reduction from $L(M)$ to *CIRCUIT-SAT*.

Denote the reduction from $L(M)$ to *CIRCUIT-SAT* by $f$. The $f$ maps string $w$ to circuit $C_w(y_1, \ldots, y_{m(n)})$. So to complete the proof it suffices to show that $f$ is computed by a deterministic polynomial time TM which we denote by $M_f$.

To do this, we explain briefly how to describe circuit $C_w(y_1, \ldots, y_{m(n)})$. First, denote the gates and inputs by integers 1 through $q(n)$. The description of $C_w(y_1, \ldots, y_{m(n)})$ is a list of descriptions of gates in $C_w$, where the descriptions of the gates are illustrated in Fig. 9.13. For example, $(i, \mathrm{OR}, j, k)$ means that the $i$th gate is an OR gate and has two inputs from $j$ and $k$, where $j$ is either input variable $x_j$ or the $j$th gate, and similarly for $k$. $F$ in Fig. 9.13 denotes either OR or AND. A similar explanation applies for $(i, \mathrm{NOT}, j)$. Finally, $(i, output)$ means that the $i$th gate is designated as the output gate.

Finally, we explain briefly how the deterministic polynomial time TM $M_f$ computes the description of $C_w(y_1, \ldots, y_{m(n)})$ from string $w$. Boolean circuit $C_w(y_1, \ldots, y_{m(n)})$ is constructed like the circuit in Fig. 9.12, but with each finite gate being replaced by a copy of the equivalent Boolean circuit, and each horizontal wire being replaced by $l$ binary wires. We associate integers 1 through $q(n)$

with the gates and inputs in the circuit in a certain systematic way. So the equivalent Boolean circuit $C_w(y_1, \ldots, y_{m(n)})$ has a highly repetitive structure and is built from many nearly identical subcircuits. These subcircuits differ only in the integers associated with their gates and inputs in a certain systematic way. Since the circuit has such a uniform structure, we can make $M_f$ compute the description $\langle C_w(y_1, \ldots, y_{m(n)}) \rangle$. Furthermore, since the integers used to denote the gates and inputs are upper bounded by $q(n) \le c \times t^2(n) + 2c \times t(n) + l \times t(n)$, the length of binary numbers to denote the integers is given by $\log q(n) = O(\log n)$. This is true because, since $t(n)$ is a polynomial, and hence is denoted by, say, $n^k$ for a positive integer $k$, and the most dominant term of $c \times t^2(n) + 2c \times t(n) + l \times t(n)$ is $c \times t^2(n)$, we have

$$
\begin{aligned}
\log\left(c \times t^2(n)\right) &= \log\left(c \times n^{2k}\right) \\
&= O(\log n).
\end{aligned}
$$

Thus we can construct the deterministic TM $M_f$ that computes the description of $C_w(y_1, \ldots, y_{m(n)})$ from string $w$ in polynomial time.    □

## 9.6  Reducing Circuit Satisfiability Problem to Formula Satisfiability Problem

In this section we verify that the circuit satisfiability problem is reduced to a formula satisfiability problem in polynomial time by transforming a Boolean circuit to a Boolean formula that is equivalent to the circuit in the sense that the satisfiability is preserved. Thus we have $CIRCUIT\text{-}SAT \le_{\text{poly}} SAT$, where $SAT$ denotes $FORMULA\text{-}SAT$. We have derived $L \le_{\text{poly}} CIRCUIT\text{-}SAT$ for any $L \in \text{NP}$ in the previous section. Therefore, by the transitivity law we have

$$
L \le_{\text{poly}} SAT
$$

for any $L \in \text{NP}$, which means that computation of a nondeterministic polynomial time Turing machine can be simulated somehow by a Boolean formula.

As described in Sect. 9.1, a Boolean circuit can generally be transformed to an equivalent Boolean formula by first merely duplicating any subcircuit that is involved in computing outputs of a gate with fan-out more than 1 so as to obtain an equivalent circuit consisting of only gates with fan-out 1. A circuit consisting of only gates with fan-out 1 can be transformed to an equivalent Boolean formula in an obvious way. But, since we must repeat the duplication of such a subcircuit until there exists no gate with fan-out larger than 1, the size of the resulting circuit might possibly be exponential in the size of the original circuit. Hence a Turing machine that converts a Boolean circuit to such an equivalent Boolean formula needs exponential steps because merely writing down an equivalent Boolean formula requires exponential steps. So we must seek another way to convert a circuit to an equivalent formula.

**Fig. 9.14** Example of
Boolean circuit

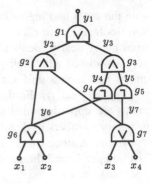

**Table 9.5** Truth table of the
formula $F_{g_1}$

| $y_1$ | $y_2$ | $y_3$ | $y_1 \vee \overline{y}_2$ | $y_1 \vee \overline{y}_3$ | $\overline{y}_1 \vee y_2 \vee y_3$ | $F_{g_1}$ |
|---|---|---|---|---|---|---|
| 0 | 0 | 0 | 1 | 1 | 1 | 1 |
| 0 | 0 | 1 | 1 | 0 | 1 | 0 |
| 0 | 1 | 0 | 0 | 1 | 1 | 0 |
| 0 | 1 | 1 | 0 | 0 | 1 | 0 |
| 1 | 0 | 0 | 1 | 1 | 0 | 0 |
| 1 | 0 | 1 | 1 | 1 | 1 | 1 |
| 1 | 1 | 0 | 1 | 1 | 1 | 1 |
| 1 | 1 | 1 | 1 | 1 | 1 | 1 |

Let $C$ denote an arbitrary Boolean circuit, where $C$ has input variables $x_1, \ldots, x_n$ and gates $g_1, \ldots, g_s$. We shall construct an equivalent Boolean formula, denoted by $F(C)$, that somehow simulates $C$. As augmented variables of $F(C)$ we introduce variables $y_1, \ldots, y_s$, each corresponding to the output of a gate in $C$. The point of constructing $F(C)$ is that for each gate $g$ in $C$ we prepare a formula $F_g$ that guarantees that the input/output relation of the gate is valid. For example, take gate $g_1$ in Fig. 9.14. Using the augmented variables $y_1, y_2, y_3$, we define the formula $F_{g_1}$ as follows:

$$F_{g_1} = (y_1 \vee \overline{y}_2)(y_1 \vee \overline{y}_3)(\overline{y}_1 \vee y_2 \vee y_3).$$

As Table 9.5 indicates, $F_{g_1} = 1$ means that the input/output relation of gate $g_1$ is valid, i.e.,

$$y_1 = y_2 \vee y_3.$$

Furthermore, let $F_{\text{out}}$ be a formula that guarantees that the output gate produces 1 as output. For the circuit in Fig. 9.14, $F_{\text{out}}$ is simply $y_1$. In general, the entire formula $F(C)$ is defined as

$$F(C) = F_{g_1} \wedge F_{g_2} \wedge \cdots \wedge F_{g_s} \wedge F_{\text{out}}.$$

Then we have

$$C \text{ is satisfiable} \quad \Leftrightarrow \quad F(C) \text{ is satisfiable},$$

which will be proved in the proof of Lemma 9.4.

Let the disjunction of $F_1, \ldots, F_m$ mean formula $F_1 \vee \cdots \vee F_m$, and let the conjunction of $F_1, \ldots, F_m$ mean formula $F_1 \wedge \cdots \wedge F_m$. In general, a Boolean variable or a negated Boolean variable is called a *literal*. A disjunction of literals is called a *clause*, and a conjunction of clauses is called a formula in *conjunctive normal form*. Note that $F_{g_1}, \ldots, F_{g_s}$, and $F_{\text{out}}$ as well as $F(C)$ take the form of formulas in conjunctive normal form.

**Example 9.3** Let the circuit in Fig. 9.14 be denoted by $C$. Let gates $g_1, \ldots, g_7$ and variables $y_1, \ldots, y_7$ be specified as indicated in Fig. 9.14. Then $F(C)$ is given as follows:

$$
\begin{aligned}
F(C) = {}& (y_1 \vee \overline{y}_2)(y_1 \vee \overline{y}_3)(\overline{y}_1 \vee y_2 \vee y_3) \quad (F_{g_1}) \\
& \wedge (y_2 \vee \overline{y}_6 \vee \overline{y}_7)(\overline{y}_2 \vee y_6)(\overline{y}_2 \vee y_7) \quad (F_{g_2}) \\
& \wedge (y_3 \vee \overline{y}_4 \vee \overline{y}_5)(\overline{y}_3 \vee y_4)(\overline{y}_3 \vee y_5) \quad (F_{g_3}) \\
& \wedge (y_4 \vee y_6)(\overline{y}_4 \vee \overline{y}_6) \quad (F_{g_4}) \\
& \wedge (y_5 \vee y_7)(\overline{y}_5 \vee \overline{y}_7) \quad (F_{g_5}) \\
& \wedge (y_6 \vee \overline{x}_1)(y_6 \vee \overline{x}_2)(\overline{y}_6 \vee x_1 \vee x_2) \quad (F_{g_6}) \\
& \wedge (y_7 \vee \overline{x}_3)(y_7 \vee \overline{x}_4)(\overline{y}_7 \vee x_3 \vee x_4) \quad (F_{g_7}) \\
& \wedge y_1 \quad (F_{\text{out}}).
\end{aligned}
$$

Let's take a look at $F_{g_1} = (y_1 \vee \overline{y}_2)(y_1 \vee \overline{y}_3)(\overline{y}_1 \vee y_2 \vee y_3)$. The condition $F_{g_1} = 1$ claims that the following three conditions must hold simultaneously: if $y_1 = 0$, then $y_2$ must be 0; if $y_1 = 0$, then $y_3$ must be 0; if $y_1 = 1$, then at least one of $y_2$ and $y_3$ must be 1. As Table 9.5 indicates, these conditions guarantee that the variables $y_1$, $y_2$, $y_3$ satisfy the input/output relation of gate $g_1$. Similarly, we have formulas $F_g$ for the other gates $g$. Furthermore, $F_{\text{out}} = 1$ clearly claims that the output gate $g_1$ produces 1 as output. $F(C) = 1$ gives the condition concerning variables $x_1, \ldots, x_4$ that the output gate $g_1$ yields 1 as output.

Generalizing the arguments so far, we explain how to derive the formulas $F_g$ for the three gates in general.

Let $G$ denote the output variable of a gate, and $H$ denote the formula that expresses the output of the gate in terms of the input variables of the gate. For example, for gate $g_1$ in Fig. 9.14, $G$ is $y_1$ while $H$ is $y_2 \vee y_3$. On the other hand, the input/output relation of the gate is expressed as the condition $G = H$. As Table 9.6 shows, this condition is written as the formula

$$(G \vee \overline{H})(\overline{G} \vee H).$$

**Table 9.6** Truth table of $G = H$

| $G$ | $H$ | $G = H$ |
| --- | --- | --- |
| 0 | 0 | 1 |
| 0 | 1 | 0 |
| 1 | 0 | 0 |
| 1 | 1 | 1 |

The formula takes the value 1 on the first and fourth rows in the truth table of Table 9.6. This is because the only rows of the table such that formula $(G \vee \overline{H})(\overline{G} \vee H)$ is not allowed to take the value 1 are the second and the third ones. In general, let the output variable be denoted by $y$, i.e., $G = y$, and let the input variables be denoted by $y'$ and $y''$. Then for the OR gate, the AND gate, and the NOT gate, the formula $H$ becomes $y' \vee y''$, $y' \wedge y''$, and $\overline{y'}$, respectively. By substituting these $G$ and $H$ into $(G \vee \overline{H})(\overline{G} \vee H)$, we have the formulas $F_g$ as follows:

*In case of OR gate*

$$
\begin{aligned}
(G &\vee \overline{H})(\overline{G} \vee H) \\
&= \left(y \vee \overline{y' \vee y''}\right)\left(\overline{y} \vee y' \vee y''\right) \\
&= \left(y \vee \left(\overline{y'} \wedge \overline{y''}\right)\right)\left(\overline{y} \vee y' \vee y''\right) \quad \text{(by De Morgan's law)} \\
&= \left(y \vee \overline{y'}\right)\left(y \vee \overline{y''}\right)\left(\overline{y} \vee y' \vee y''\right) \quad \text{(by the distributive law)}.
\end{aligned}
$$

*In case of AND gate*

$$
\begin{aligned}
(G &\vee \overline{H})(\overline{G} \vee H) \\
&= \left(y \vee \overline{y' \wedge y''}\right)\left(\overline{y} \vee \left(y' \wedge y''\right)\right) \\
&= \left(y \vee \left(\overline{y'} \vee \overline{y''}\right)\right)\left(\overline{y} \vee \left(y' \wedge y''\right)\right) \quad \text{(by De Morgan's law)} \\
&= \left(y \vee \overline{y'} \vee \overline{y''}\right)\left(\overline{y} \vee y'\right)\left(\overline{y} \vee y''\right) \quad \text{(by the distributive law)}.
\end{aligned}
$$

*In case of NOT gate*

$$
(G \vee \overline{H})(\overline{G} \vee H) = \left(y \vee \overline{\overline{y'}}\right)\left(\overline{y} \vee \overline{y'}\right) = \left(y \vee y'\right)\left(\overline{y} \vee \overline{y'}\right).
$$

From the arguments so far we have the next lemma.

**Lemma 9.4** *For any Boolean circuit* $C$

$$
C \text{ is satisfiable} \quad \Leftrightarrow \quad F(C) \text{ is satisfiable}.
$$

*Proof* Assume that a Boolean circuit $C$ has $s$ gates $g_1, \ldots, g_s$ and $n$ input variables $x_1, \ldots, x_n$. The formula $F(C)$ constructed from $C$ has $n$ variables $x_1, \ldots, x_n$ and $s$ augmented variables $y_1, \ldots, y_s$.

(Proof of $\Rightarrow$) Suppose that assignment $a_1, \ldots, a_n$ for $x_1, \ldots, x_n$ satisfies $C$. Then the outputs of gates $g_1, \ldots, g_s$ are determined accordingly. Let these outputs be denoted by $b_1, \ldots, b_s$, respectively. Obviously, assigning $a_1, \ldots, a_n, b_1, \ldots, b_s$ for $x_1, \ldots, x_n, y_1, \ldots, y_s$, respectively, satisfies $F(C)$.

(Proof of $\Leftarrow$) Suppose that assignment $a_1, \ldots, a_n, b_1, \ldots, b_s$ for $x_1, \ldots, x_n$, $y_1, \ldots, y_s$ satisfies $F(C)$. Then if we assign $a_1, \ldots, a_n$ for $x_1, \ldots, x_n$, the outputs of gates $g_1, \ldots, g_s$ become $b_1, \ldots, b_s$, respectively, and the output gate yields 1 as output.                                                                                                $\square$

**Theorem 9.5** *CIRCUIT-SAT is polynomial time reducible to SAT.*

*Proof* Suppose that we describe a circuit as explained in the proof of Theorem 9.2. The reduction $f$ in the statement of the theorem maps description $\langle C \rangle$ to formula $F(C)$. Description $\langle C \rangle$ takes the form of a list of descriptions for the gates in $C$. For example, if the description of gate $g_i$ is $(i, \text{OR}, j, k)$, then it suffices to yield $(y_i \vee \overline{y}_j)(y_i \vee \overline{y}_k)(\overline{y}_i \vee y_j \vee y_k)$ for the gate as a part of $F(C)$, where $j$ and $k$ are assumed to denote the corresponding gates. Similarly, if the description is $(i, output)$, there corresponds formula $y_i$. In this way we can construct a deterministic TM $M_f$ that successively generates portions of $F(C)$ as it reads description $\langle C \rangle$ formula $F(C)$. Thus the reduction is computed in $O(n)$ time, where $n$ represents the length of description $\langle C \rangle$.                                                                $\square$

By observing the formulas $F_g$ for gates, we see that formula $F(C)$ is the conjunction of clauses of size smaller than or equal to 3, where the size of a clause is the number of literals in the clause. We shall show that such $F(C)$ can be transformed to a formula that is the conjunction of clauses of size exactly 3. To do so, it suffices to show that clauses of size 1 or 2 can be replaced with equivalent clauses of size just 3. Clause $x \vee y$ is replaced with $(x \vee y \vee z) \wedge (x \vee y \vee \overline{z})$, relying on the following equation:

$$x \vee y \vee (z \wedge \overline{z}) = (x \vee y \vee z) \wedge (x \vee y \vee \overline{z}).$$

Since $z \wedge \overline{z} = 0$, we have $x \vee y = x \vee y \vee (z \wedge \overline{z})$. Similarly, $x$ is replaced with $(x \vee y \vee z) \wedge (x \vee y \vee \overline{z}) \wedge (x \vee \overline{y} \vee z) \wedge (x \vee \overline{y} \vee \overline{z})$ relying on the equation

$$x \vee (y \wedge \overline{y}) \vee (z \wedge \overline{z})$$
$$= x \vee \big((y \vee z) \wedge (y \vee \overline{z}) \wedge (\overline{y} \vee z) \wedge (\overline{y} \vee \overline{z})\big)$$
$$= (x \vee y \vee z) \wedge (x \vee y \vee \overline{z}) \wedge (x \vee \overline{y} \vee z) \wedge (x \vee \overline{y} \vee \overline{z}).$$

If a formula in conjunctive normal form has only clauses of size $m$, then the formula is called an $m$cnf-formula. Let

$$mSAT = \{\langle F \rangle \mid F \text{ is a satisfiable } m\text{cnf-formula}\}.$$

**Theorem 9.6** *For any language L in NP, L is polynomial time reducible to 3SAT.*

*Proof* By Theorem 9.2, we have for any $L \in \mathrm{NP}$

$$L \leq_{\mathrm{poly}} CIRCUIT\text{-}SAT.$$

On the other hand, since by Theorem 9.5 any Boolean circuit is reducible to a formula in conjunctive normal form that has only clauses whose size is less than or equal to 3, we have

$$CIRCUIT\text{-}SAT \leq_{\mathrm{poly}} (1SAT \cup 2SAT \cup 3SAT).$$

Furthermore, from the arguments described in the paragraph just before the theorem, we have

$$(1SAT \cup 2SAT \cup 3SAT) \leq_{\mathrm{poly}} 3SAT.$$

Thus, since $\leq_{\mathrm{poly}}$ is transitive by Proposition 7.3, we have

$$L \leq_{\mathrm{poly}} 3SAT,$$

completing the proof of the theorem.                                          □

## 9.7    Problems

**9.1** Let $g(x, y, z)$ denote the function of the gates of the finite circuit that simulates the TM $M$ given in Fig. 9.5. Give the value of $g(x, y, z)$ when one of $x, y, z$ is from $\{q_0, q_1, q_2\} \times \{0, 1, \sqcup\}$ and the remaining two are from $\{0, 1, \sqcup\}$.

**9.2** In general, can you construct a finite circuit that simulates a deterministic two-tape Turing machine?

**9.3**\*\* Let $F$ denote the satisfiable formula $(x_1 \vee x_2)(x_2 \vee x_3)(x_1 \vee x_3)(x_2 \vee \overline{x}_3)(\overline{x}_1 \vee x_3)$. Find a clause of the form $(s \vee t)$ such that $F \wedge (s \vee t)$ is not satisfiable, where $s$ and $t$ are literals.

**9.4**\* In a similar way to Example 9.3, give the formula $F(C)$, where Boolean circuit $C$ is given as follows. Furthermore, find an assignment to variables that satisfies $F(C)$.

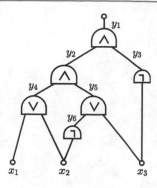

**9.5**[#]  In Chap. 9 we construct a family of Boolean circuits that simulate a deterministic TM. Conversely, in general can you construct a Turing machine that simulates a family of Boolean circuits $\{C_n\}$? If it is not possible, give conditions that make it possible.

# NP-Completeness    10

There is a special type of problem in NP, called NP-complete. An NP-complete problem is the very hardest one in the sense that any problem in NP polynomially is reduced to an NP-complete problem: for any problem $P$ in NP we have $P \leq_{\text{poly}} P_0$, where $P_0$ is an NP-complete problem arbitrarily chosen. Recall that $P \leq_{\text{poly}} P_0$ means that $P$ is not harder than $P_0$ to solve. One remarkable feature of NP-complete problems is that they are somewhat equivalent in hardness to solve. Consequently, if any single NP-complete problem is intractable, then all NP-complete problems are intractable. On the other hand, if any single NP-complete problem is tractable, then all NP-complete problems, hence all problems in NP, are tractable.

## 10.1  NP-Completeness

Before introducing the notion of NP-completeness, we present a couple of concrete examples of NP-complete problems.

*Satisfiability problem*  This problem is to decide whether a Boolean formula is satisfiable.

*Hamiltonian path problem*  This problem is to decide whether a directed graph has a Hamiltonian path, i.e., a closed path that goes through all the nodes exactly once.

*Subset sum problem*  This problem is to decide whether a collection of natural numbers $S$ together with a natural number $t$ is such that there exists a subcollection of $S$ that sums up to $t$.

Recall that a Boolean formula $F(x_1, \ldots, x_n)$ is satisfiable if there exists an assignment $(a_1, \ldots, a_n)$ to $(x_1, \ldots, x_n)$ such that $F(a_1, \ldots, a_n) = 1$. Take, for example, formula $(\overline{x}_1 \vee x_2)(x_1 \vee \overline{x}_2)(\overline{x}_1 \vee \overline{x}_2)(x_1 \vee x_3)$. This formula is satisfiable because assignment $(0, 0, 1)$ satisfies the formula. On the other hand, $x_1(\overline{x}_1 \vee x_2)(\overline{x}_1 \vee x_3)(\overline{x}_2 \vee \overline{x}_3)$ is not satisfiable. Furthermore, recall that a problem is also given in terms of a language. The satisfiability problem is considered as the membership problem of the language $\{\langle F \rangle \mid \text{formula } F \text{ is satisfiable}\}$. We can also think of the languages that correspond to the remaining problems described above.

We define the notion of NP-completeness based on polynomial time reducibility introduced in Definition 7.2.

A. Maruoka, *Concise Guide to Computation Theory*,
DOI 10.1007/978-0-85729-535-4_10, © Springer-Verlag London Limited 2011

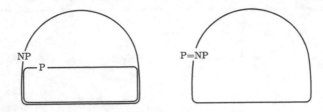

**Fig. 10.1**  The two possibilities of P $\subsetneq$ NP and P = NP

**Definition 10.1**  A language $B$ is NP-complete if it satisfies the following two conditions:
(1)  $B \in$ NP
(2)  For any $A \in$ NP, $A \leq_{poly} B$

There are thousands of important problems in practice that are NP-complete. In the next section, we shall prove that the three problems above are NP-complete.

By definition, either P $\subsetneq$ NP or P = NP holds. The question of whether P $\subsetneq$ NP or P = NP is one of the important unsolved problems in computer science, and is often referred to as the P vs. NP problem. The situation is illustrated schematically in Fig. 10.1.

Many researchers believe that P $\subsetneq$ NP. By definition, if any NP-complete problem $B$ belongs to P, then we have P = NP. This is because, if $B \in$ P, then for any $A \in$ NP we have $A \leq_{poly} B$, which, together with $B \in$ P, implies $A \in$ P by Theorem 7.4, thereby verifying P = NP. So, if P $\subsetneq$ NP, then any NP-complete problem does not belong to P. Hence, although we have not been successful in proving P $\subsetneq$ NP, verifying that a problem is NP-complete actually gives us a strong plausibility that the problem cannot be solved in polynomial time. As mentioned above, to prove P = NP, it suffices to take any NP-complete problem $B$ arbitrarily and prove that $B \in$ P. From what has been described so far, no matter whether the truth is P $\subsetneq$ NP or P = NP, any NP-complete problem turns out to be the most difficult problem in NP from the standpoint of time complexity.

## 10.2  NP-Complete Problems

In this section we begin by proving that the satisfiability problem is NP-complete, and then verify that the Hamiltonian path problem and the subset sum problem are polynomial time reducible to the satisfiability problem, thereby concluding that the remaining two problems are also NP-complete.

## Satisfiability Problem

The formula satisfiability problem is simply called the satisfiability problem and is described as

$$SAT = \big\{ \langle F \rangle \mid F \text{ is a satisfiable Boolean formula} \big\}.$$

Furthermore, by restricting formulas to the ones in conjunctive normal form with all the clauses having exactly three literals, we have 3$SAT$ defined as follows:

$$3SAT = \big\{ \langle F \rangle \mid F \text{ is a satisfiable 3cnf-formula} \big\}.$$

**Theorem 10.2** *SAT and* 3*SAT are NP-complete.*

*Proof* We shall verify conditions (1) and (2) in Definition 10.1 for $SAT$ and 3$SAT$.

For (1): Although we must cope with some details, we can verify that there exists a polynomial time verifier that, when descriptions of formula $\langle F \rangle$ and assignment $(a_1, \ldots, a_n)$ are given, accepts if $(a_1, \ldots, a_n)$ satisfies $F$. Thus $SAT$ and 3$SAT$ belong to NP.

For (2): Theorem 9.6 claims that condition (2) holds for 3$SAT$. Since obviously $3SAT \subseteq SAT$, condition (2) also holds for $SAT$ by definition. $\qquad\square$

In order to prove that a language $L$ is NP-complete based on Definition 10.1, we must verify condition (2) as well as condition (1). But, actually condition (2) could be replaced with the condition that $L' \leq_{\text{poly}} L$ for an appropriately chosen language $L'$ that has been already proven to be NP-complete. This fact is summarized as the next theorem.

**Theorem 10.3** *Let $L'$ be any NP-complete language. Then, if a language $L$ satisfies the following two conditions, $L$ is NP-complete.*
(1) $L \in \text{NP}$
(2) $L' \leq_{\text{poly}} L$

*Proof* Due to Definition 10.1, it suffices to verify condition (2) in the definition. Let $L''$ denote an arbitrary language in NP. Then, since $L'$ is NP-complete, we have

$$L'' \leq_{\text{poly}} L',$$

which, together with condition (2) in the theorem, implies

$$L'' \leq_{\text{poly}} L.$$

by the transitivity of $\leq_{\text{poly}}$, completing the proof. $\qquad\square$

On the other hand, as Problem $10.4^{\sharp}$ shows, there exists a polynomial time algorithm that decides $2SAT$, where $2SAT$ is the set that consists of satisfiable formulas in conjunctive normal form with all its clauses having two literals. The algorithm that decides $2SAT$ roughly works as follows. Suppose that a 2cnf-formula contains clauses $(x_1 \vee \overline{x}_2)$. This clause implies that $(x_1 = 0 \Rightarrow x_2 = 0)$ and $(x_2 = 1 \Rightarrow x_1 = 1)$. The algorithm determines in polynomial time whether or not the entire collection of such conditions implied by the clauses of the formula is consistent, thereby producing YES/NO accordingly. Note that, since $3SAT$ is NP-complete and $2SAT$ is polynomial time computable, there might be a big difference in computational complexity between $3SAT$ and $2SAT$.

## Hamiltonian Path Problem

The Hamiltonian path problem is associated with the language described as follows:

$$HAMPATH = \big\{\langle G \rangle \mid \text{directed graph } G \text{ has a Hamiltonian path}\big\}.$$

We shall prove that $HAMPATH$ is NP-complete. By Theorem 10.3 we verify the fact by constructing a polynomial time reduction from $3SAT$ to $HAMPATH$. The reduction is such that a satisfiable formula is converted to a graph that contains a Hamiltonian path. To explain the idea of how the reduction works, we take formula

$$F = (x_1 \vee \overline{x}_2)(\overline{x}_1 \vee \overline{x}_3)(x_2 \vee x_3),$$

and show how to construct a directed graph from this formula. The graph constructed from the formula $F$ is given in Fig. 10.2. In our case, since the original formula is satisfiable, the resulting graph has a Hamiltonian path which is denoted by the bold line in Fig. 10.2. The satisfying assignment $(x_1, x_2, x_3) = (1, 1, 0)$ specifies how to draw the Hamiltonian path. Although the formula is not a 3cnf-formula, we use it because the size of the formula and the corresponding graph is appropriate to illustrate the key idea behind the reduction. Let's take a closer look at how it works concerning our example.

The corresponding graph has three diamond structures corresponding to the three variables, respectively, and three nodes corresponding to the three clauses, respectively. The diamond structure associated with variable $x_i$ is sometimes simply called diamond $x_i$, and the node associated with clause $c_i$ is denoted by $c_i$. The three diamonds have the same structure, as illustrated in Fig. 10.3. The diamonds have a row of nodes $v_1, \ldots, v_8$ together with the edges between them so that the subpaths through all the nodes on the row are formed in both directions.

The connections between the nodes $c_1, c_2, c_3$ and the diamonds $x_1, x_2, x_3$ indicate what variables each clause consists of and whether these variables appear positively or negatively in the clauses. To show this, in each diamond structure the pairs of nodes $\{v_2, v_3\}$, $\{v_4, v_5\}$, and $\{v_6, v_7\}$ are assumed to correspond to the clauses $c_1$, $c_2$, and $c_3$, respectively, as shown in Fig. 10.3. Consider the node $c_1$. Since

**Fig. 10.2** Formula $(x_1 \vee \overline{x}_2)(\overline{x}_1 \vee \overline{x}_3)(x_2 \vee x_3)$ reduces to this graph

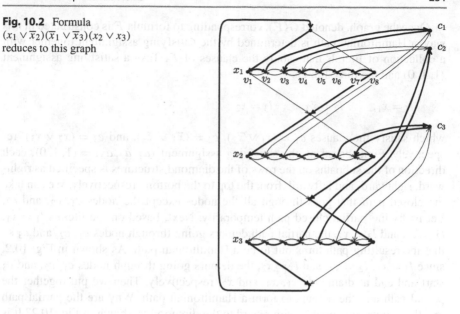

**Fig. 10.3** Nodes arranged on diamond structure

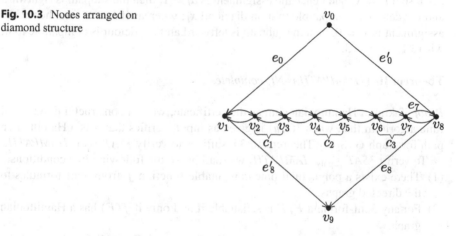

$c_1 = (x_1 \vee \overline{x}_2)$, node $c_1$ is connected with diamonds $x_1$ and $x_2$, each through the two edges. The directions of the two edges between node $c_1$ and the diamonds are specified according to the general rule described as follows: if a variable without negation appears in a clause, the directions of the two edges are specified so as to be compatible with the rightward subpath on the row; whereas if a negated variable appears, the directions are specified so as to be compatible with the leftward subpath. In a similar way, the other edges between diamond $x_i$ and node $c_j$ are drawn for $1 \le i, j \le 3$.

Once the graph, denoted $G(F)$, corresponding to formula $F$ is constructed in this way, a Hamiltonian path is determined by the satisfying assignment together with a collection of literals that satisfy the clauses of $F$. Take a satisfying assignment $(1, 1, 0)$ and the three literals

$$l_1 = x_1, \qquad l_2 = \overline{x}_3, \qquad l_3 = x_2,$$

which satisfy the clauses $c_1 = (x_1 \vee \overline{x}_2)$, $c_2 = (\overline{x}_1 \vee \overline{x}_3)$, and $c_3 = (x_2 \vee x_3)$, respectively. First, based on the satisfying assignment $(a_1, a_2, a_3) = (1, 1, 0)$, each direction of the subpaths on the rows of the diamond structures is specified as rightward, rightward, and leftward, from the top to the bottom, respectively. We can take the closed path that goes through all the nodes except the nodes $c_1$, $c_2$, and $c_3$. Let us fix the partial closed path temporarily. Next, based on the literals $l_1 = x_1$, $l_2 = \overline{x}_3$, and $l_3 = x_2$, the partial path detours going through nodes $c_1$, $c_2$, and $c_3$ so that the resulting path turns out to be a Hamiltonian path. As shown in Fig. 10.2, since $l_1 = x_1$, $l_2 = \overline{x}_3$, and $l_3 = x_2$, the detours going through nodes $c_1$, $c_2$, and $c_3$ start and end at diamonds $x_1$, $x_3$, and $x_2$, respectively. Then we put together the partial path and the detours to form a Hamiltonian path. Why are the partial path and the detours compatible with regard to the direction as shown in Fig. 10.2? It is because if $l_j = x_i$ and hence the assignment is $a_i = 1$, then the subpath is rightward and the detour is compatible to it on diamond $x_i$; whereas if $l_j = \overline{x}_i$ and hence the assignment is $a_i = 0$, then the subpath is leftward and the detour is compatible to it, where $1 \leq i, j \leq 3$.

**Theorem 10.4** *HAMPATH is NP-complete.*

*Proof* Taking a Hamiltonian path as the certificate, we can construct a polynomial time algorithm that, when $\langle G, u \rangle$ is given as input, verifies that $u$ is a Hamiltonian path for graph $G$. So by Theorem 10.3 it suffices to verify $3SAT \leq_{\text{poly}} HAMPATH$.

To verify $3SAT \leq_{\text{poly}} HAMPATH$, we shall prove the following two conditions.
(1) There exists a polynomial time computable function $f$ from 3cnf-formulas to the directed graphs.
(2) For any 3cnf-formula $F$, $F$ is satisfiable if and only if $f(F)$ has a Hamiltonian graph.

Let $F(x_1, \ldots, x_n)$ denote an arbitrary 3cnf-formula. Generalizing the construction of Fig. 10.2, we define a directed graph $G(F)$ as described below. The reduction $f$ is defined as $f(F) = \langle G(F) \rangle$.

Let a 3cnf-formula $F(x_1, \ldots, x_n)$ have $k$ clauses $c_1, \ldots, c_k$. The graph $G(F)$ has $n$ diamond structures corresponding to the $n$ variables and $k$ nodes corresponding to the $k$ clauses. The $i$th diamond structure corresponding to variable $x_i$ is sometimes called diamond $x_i$, and the node corresponding to clause $c_i$ is denoted by $c_i$. Each diamond structure has $2k + 2$ nodes $v_1, v_2, \ldots, v_{2k+2}$ on its row, as shown in Fig. 10.5. The pairs of nodes $\{v_2, v_3\}, \ldots, \{v_{2k}, v_{2k+1}\}$ are assumed to correspond to the $k$ clauses $c_1, \ldots, c_k$, respectively. Edges between node $c_j$ and diamond $x_i$ are drawn for $1 \leq i \leq n$ and $1 \leq j \leq k$ as follows: if $x_i$ appears in clause $c_j$, then edges

$(v_{2j}, c_j)$ and $(c_j, v_{2j+1})$ are added between node $c_j$ and diamond $x_i$; if $\overline{x}_i$ appears in clause $c_j$, then edges $(v_{2j+1}, c_j)$ and $(c_j, v_{2j})$ are added between node $c_j$ and diamond $x_i$. As described above, the entire graph $G(F)$ is obtained by generalizing the graph in Fig. 10.2.

If we use any natural encoding of graphs, we can obtain an algorithm that computes the function $f(F)$ in polynomial time, although we must cope with many details.

Proof of "only if" part of (2): Let $F(x_1, \dots, x_n)$ denote an arbitrary 3cnf-formula with $k$ clauses $c_1, \dots, c_k$. Assume that $F(x_1, \dots, x_n)$ is satisfied by assignment $(a_1, \dots, a_n)$. Furthermore, fix $k$ literals $l_1, \dots, l_k$ arbitrarily that satisfy clauses $c_1, \dots, c_k$ under the assignment, respectively. First, based on $(a_1, \dots, a_n)$, construct a partial closed path that goes through all the nodes except $c_1, \dots, c_k$: if $a_i = 1$, go to the right on the row in diamond $x_i$; if $a_i = 0$, go to the left on the row in diamond $x_i$. Next, based on the literals $l_1, \dots, l_k$, we add $k$ detours to the partial closed path to cover the remaining nodes $c_1, \dots, c_k$: if $l_j = x_i$, the detour from diamond $x_i$ is $(v_{2j}, c_j)$ and $(c_j, v_{2j+1})$ and hence compatible with the subpath to the right; if $l_j = \overline{x}_i$, the detour from diamond $x_i$ is $(v_{2j+1}, c_j)$ and $(c_j, v_{2j})$ and hence compatible with the subpath to the left. The point is that, whether $l_j$ is $x_i$ or $\overline{x}_i$, the detour is compatible with the partial path already fixed. The reason is that, if $l_j = x_i$ ($l_j = \overline{x}_i$), and hence $x_i = 1$ ($x_i = 0$) satisfies clause $c_j$, then the assignment is such that $a_i = 1 (a_i = 0)$, which implies the subpath goes to the right (left), which in turn is compatible with the detour $(v_{2j}, c_j)$ and $(c_j, v_{2j+1})$ $((v_{2j+1}, c_j)$ and $(c_j, v_{2j}))$. Thus, since the partial path and the detours are consistent with regard to direction, we can obtain a Hamiltonian path of $G(F)$ by combining them.

Proof of "if" part of (2): Let us assume that we are given any 3cnf-formula $F$ and construct graph $G(F)$ from it as described so far. Assume that $G(F)$ has a Hamiltonian path $P$. Our object is to specify assignment $(a_1, \dots, a_n)$ from the Hamiltonian path $P$ such that the assignment $(a_1, \dots, a_n)$ satisfies $F$. A Hamiltonian path is called *standard* if, whenever the path leaves from a diamond, it goes to a node associated with a clause and gets back to the same diamond. Before proceeding to a proof of the fact that any Hamiltonian path of $G(F)$ is standard, we show that a standard Hamiltonian path of $G(F)$ specifies an assignment $(a_1, \dots, a_n)$ that satisfies the formula $F$. As in the first half of this proof, let $(a_1, \dots, a_n)$ denote the assignment such that, if the standard Hamiltonian path $P$ goes to the right on the row in diamond $x_i$, set $a_i = 1$, and if it goes to the left, set $a_i = 0$ for $1 \le i \le n$. Then any detour on $P$ between $x_i$ and node $c_j$ guarantees that assignment $x_i = a_i$ satisfies the clause $c_j$. This is because if $a_i = 1$, then $x_i$ appears in clause $c_j$, whereas if $a_i = 0$, then $\overline{x}_i$ appears in clause $c_j$ from the construction of $G(F)$. Furthermore, since $P$ is a Hamiltonian path, every clause has such an assignment that satisfies the clause. Thus we can conclude that assignment $(a_i, \dots, a_n)$ satisfies the formula $F$.

Finally, we verify that any Hamiltonian path $P$ of graph $G(F)$ is standard. Let us take diamond $x_i$ together with nodes $v_{2j-1}, v_{2j}, \dots, v_{2j+2}$ and edges $(v_{2j}, c_j)$ and $(c_j, v_{2j+1})$, as shown in Fig. 10.4. Assume that the Hamiltonian path $P$ goes through the edge $e_5$ in the figure. The next fact says that, if we go along $P$ backward, we have only two possibilities, $e_1 e_3 e_5$ and $e_1 e_7 e_8 e_5$, as part of $P$. Clearly, if the path goes to the left, a similar fact holds.

**Fig. 10.4** Edges around detour

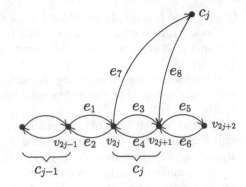

**Fig. 10.5** Diamond structure

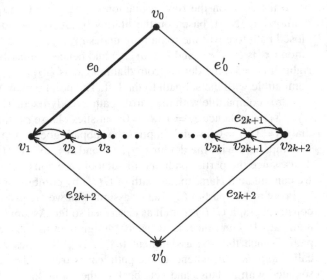

**Fact 1** If a Hamiltonian path $P$ goes through edge $e_5$, the part of $P$ is either $e_1e_3e_5$ or $e_1e_7e_8e_5$.

*Proof* Since $P$ goes through every node exactly once, $P$ that goes through $e_5$ cannot pass $e_4$. So, when $P$ passes node $v_{2j}$, there are only the following three cases: $e_1e_3$, $e_1e_7$, and $e_1e_2$. But $e_1e_2$ of these cannot be a part of a Hamiltonian path. Thus, there are only two ways that connect the remaining cases with $e_5$, that is, $e_1e_3e_5$ and $e_1e_7e_8e_5$.                                                                                □

**Fact 2** The nodes of any diamond other than nodes $c_1, \ldots, c_k$ appear in the order of $v_1, v_2, \ldots, v_{2k+2}$ or $v_{2k+2}, v_{2k+1}, \ldots, v_1$ in any Hamiltonian path.

*Proof* As shown in Fig. 10.5, assume that a Hamiltonian path $P$ goes through edge $e_0$. The only edges that go into $v_{2k+2}$ on $P$ are $e'_0$ or $e_{2k+1}$. On the other hand, since $P$ passes $v_0$ through $e_0$, $e'_0$ cannot be used on $P$. So the edge that goes to $v_{2k+2}$ on $P$ is $e_{2k+1}$. Thus, by applying Fact 1 repeatedly, the nodes of the row appear in the order of $v_1, v_2, \ldots, v_{2k+2}$. In a similar way, if $P$ goes through edge $e'_0$, the order obtained must be $v_{2k+2}, v_{2k+1}, \ldots, v_1$. □

Since it immediately follows from Facts 1 and 2 that any Hamiltonian path of graph $G(F)$ is standard, the proof is completed. □

## Subset Sum Problem

The *subset sum problem* is to determine whether there exists a subcollection $T$ of a collection of numbers $S$ that sums up to a target number $t$. So an instance of the subset sum problem is a pair of a collection of numbers $S$ and a target $t$. Let *SUBSET-SUM* denote the language corresponding to this problem. We shall prove that the subset sum problem is NP-complete by verifying

$$3SAT \leq_{\text{poly}} SUBSET\text{-}SUM.$$

To explain the idea of the reduction, we take a cnf-formula

$$F = (x_1 \vee x_2 \vee x_3)(x_1 \vee \overline{x}_2)(x_2 \vee \overline{x}_3)(x_3 \vee \overline{x}_1)$$

and give the corresponding subset sum problem instance. Although the formula is not a 3cnf-formula, we use it because the size of the formula and the corresponding subset sum problem instance is appropriate to illustrate the idea. In short, the reduction is done by arithmetization. The formula is converted to the subset sum problem instance $(S, t)$, where

$$S_1 = \{a_1, b_1, a_2, b_2, a_3, b_3\},$$
$$S_2 = \{f_1, g_1, f_2, g_2, f_3, g_3, f_4, g_4\},$$
$$S = S_1 \cup S_2,$$
$$t = 1113333.$$

The numbers in $S$ are presented in decimal notation and given in Table 10.1. The digits of the decimal numbers are at most 7 and are denoted by $x_1, x_2, x_3, c_1, c_2, c_3$, and $c_4$, as shown in Table 10.1. Let us express the clauses of $F$ as follows:

$$c_1 = x_1 \vee x_2 \vee x_3, \qquad c_2 = x_1 \vee \overline{x}_2, \qquad c_3 = x_2 \vee \overline{x}_3, \qquad c_4 = x_3 \vee \overline{x}_1.$$

To explain how the numbers in the table relate to formula $F$, the table is divided into three regions $A$, $B$, and $C$, as shown in Table 10.1. In region $A$, each digit $x_i$ is 1 only in numbers $a_i$ and $b_i$. To explain region $B$, we take literal $x_1$. Since literal $x_1$ appears in clauses $c_1 = x_1 \vee x_2 \vee x_3$ and $c_2 = x_1 \vee \overline{x}_2$, the last four digits $c_1$,

**Table 10.1** Instance of subset sum problem

| | $x_1$ | $x_2$ | $x_3$ | $c_1$ | $c_2$ | $c_3$ | $c_4$ | |
|---|---|---|---|---|---|---|---|---|
| $a_1$ | 1 | 0 | 0 | 1 | 1 | 0 | 0 | |
| $b_1$ | 1 | 0 | 0 | 0 | 0 | 0 | 1 | |
| $a_2$ | | 1 | 0 | 1 | 0 | 1 | 0 | $B$ |
| $b_2$ | | 1 | 0 | 0 | 1 | 0 | 0 | |
| $a_3$ | | | 1 | 1 | 0 | 0 | 1 | |
| $b_3$ | | | 1 | 0 | 0 | 1 | 0 | |
| $f_1$ | | | | 1 | 0 | 0 | 0 | |
| $g_1$ | | | | 1 | 0 | 0 | 0 | |
| $f_2$ | | | | | 1 | 0 | 0 | |
| $g_2$ | | | | | 1 | 0 | 0 | $C$ |
| $f_3$ | | | | | | 1 | 0 | |
| $g_3$ | | | | | | 1 | 0 | |
| $f_4$ | | | | | | | 1 | |
| $g_4$ | | | | | | | 1 | |
| $t$ | 1 | 1 | 1 | 3 | 3 | 3 | 3 | |

($S_1$ comprises rows $a_1$ through $b_3$; $S_2$ comprises rows $f_1$ through $g_4$; region $A$ covers the $x$-columns.)

$c_2, c_3, c_4$ of $a_1$ are 1100. On the other hand, since literal $\overline{x}_1$ appears only in clause $c_4 = x_3 \vee \overline{x}_1$, the last four digits of $b_1$ are 0001. In this way each column in region $B$ indicates what literals each clause comprises. In region $C$, each digit $c_i$ is 1 only in numbers $f_i$ and $g_i$. Finally, digits $x_1$, $x_2$, and $x_3$ of $t$ are 1, while digits $c_1, c_2, c_3$, and $c_4$ of $t$ are 3.

Since each column in the table has at most five 1's, a carry to the next higher digit never happens in any digit when the numbers in $S$ are added. Obviously, it is so when numbers in any subset $T$ of $S$ is added. Thus, for any subset $T$, the sum of the numbers in $T$ is equal to $t$ means that, for every digit, the sum of 1's in the digit of the numbers in $T$ is equal to the digit of $t$. With this fact, we shall explain what the requirement $\sum_{i \in T} i = t$ means. For digits $x_1$, $x_2$, and $x_3$, exactly one of $a_i$ or $b_i$ is selected. We interpret that $a_i \in T$ means setting $x_i = 1$, while $b_i \in T$ means setting $\overline{x}_i = 1$. In this way, the rows selected in region $A$ specify a Boolean assignment to variables $x_1$, $x_2$, and $x_3$. It turns out that the problem to choose an assignment that satisfies $F$ is reduced to the problem to select a subset $T'$ of $\{a_1, b_1, \ldots, a_3, b_3\}$ such that exactly one of $a_i$ and $b_i$ is chosen for $1 \le i \le 3$ and that the number of 1's in digit $c_i$ of the chosen numbers is 1, 2, or 3 for $1 \le i \le 4$. Finally, the numbers $f_1$, $g_1, \ldots, f_4, g_4$ in region $C$ raise the total numbers of digit $c_i$ from 1 or 2 up to 3 so that the sum of the numbers chosen this way is equal to the target number 3.

**Theorem 10.5** *SUBSET-SUM is NP-complete.*

*Proof* We shall show that *SUBSET-SUM* $\in$ NP and $3SAT \le_{\text{poly}} SUBSET\text{-}SUM$. By taking a subset as the certificate, we can verify that *SUBSET-SUM* $\in$ NP.

Let $F$ be a 3cnf-formula. Let $F$ have variables $x_1, \ldots, x_n$ and $k$ clauses. To verify $3SAT \le_{\text{poly}} SUBSET\text{-}SUM$, we need to specify an instance of the subset sum problem, denoted by $(S, t)$, such that the reduction $f$ converts $F$ to $(S, t)$. We can specify this $(S, t)$ by generalizing the construction given by Table 10.1. Let $S$ consist of $2n + 2k$ decimal numbers, each having at most $n + k$ digits. As in the case of Table 10.1, these numbers are denoted by $a_1, b_1, \ldots, a_n, b_n, f_1, g_1, \ldots, f_k, g_k$. It is clear from the example shown in Table 10.1 how to specify these numbers $a_1, b_1, \ldots, a_n, b_n, f_1, g_1, \ldots, f_k, g_k$ together with $t$. Let the $n + k$ digits be denoted by $x_1, \ldots, x_n, c_1, \ldots, c_k$ as in Table 10.1.

Assume that $F$ is satisfied by an assignment $(x_1, \ldots, x_n) \in \{0, 1\}^n$. Based on this assignment, we shall specify $T \subseteq S$ that adds up to $t$. Let $a_i \in T$ if $x_i = 1$ and $b_i \in T$ if $x_i = 0$ for $1 \leq i \leq n$. Furthermore, define number $U$ as $U = \sum_{a_i \in T} a_i + \sum_{b_i \in T} b_i$. Then clearly every $x_i$ digit of $U$ is 1, where $1 \leq i \leq n$. Since the assignment $(x_1, \ldots, x_n)$ satisfies $F$, every $c_i$ digit of $U$ takes on a value of 1 through 3, where $1 \leq i \leq k$. Depending on the value of the $c_i$ digit of $U$, add the numbers $f_i$ and $g_i$ to $T$ as follows for $1 \leq i \leq k$: let $f_i \in T$ and $g_i \in T$ if the value is 1; let $f_i \in T$ if the value is 2; none of $f_i$ and $g_i$ is added if the value is 3. Let $T$ consist of only the numbers specified so far. Clearly $T$ as specified is the solution for the instance $(S, t)$.

Conversely, assume that $T$ is a solution for the instance $(S, t)$. Since on every digit $S$ summed up is at most five, a carry never occurs by adding any subset of $S$. Hence on every digit the sum of numbers in $T$ is equal to the corresponding digit of $t$. Since exactly one of $a_i$ and $b_i$ belongs to $T$ for $1 \leq i \leq n$, we can put the assignment $(x_1, \ldots, x_n) \in \{0, 1\}^n$ as follows: if $a_i \in T$, set $x_i = 1$; if $b_i \in T$, set $x_i = 0$.

Put $U = \sum_{a_i \in T} a_i + \sum_{b_i \in T} b_i$. Since $T$ is a solution to $(S, t)$, every digit of $U$ is equal to or more than 1. Then, since every clause of $F$ contains a literal taking on value 1, formula $F$ is satisfied by the assignment $(x_1, \ldots, x_n)$.

Finally, we can construct a polynomial time algorithm that computes the reduction $f$ defined as $f(\langle F \rangle) = \langle S, t \rangle$. $\qquad \square$

There are thousands of NP-complete problems; some of which are studied in the problems of this chapter.

## 10.3 Problems

**10.1**[#] Let a directed graph $G = (V, E)$ and an integer $k$ together with a function $g : E \to \mathcal{N}$ be given, where $g(e)$ indicates a cost of edge $e$. The *traveling salesman problem* is to determine whether there exists a Hamiltonian path whose cost is equal to or smaller than $k$, where the cost of a path is the sum of the cost of the edges on the path. Prove that the traveling salesman problem is NP-complete.

**10.2**[#] A *clique* of an undirected graph $G = (V, E)$ is a subgraph of $G$ in which every two nodes are connected by an edge. A $k$-clique is a clique consisting of $k$ nodes. Put

$$CLIQUE = \{\langle G, k \rangle \mid G \text{ is an undirected graph that has a } k\text{-clique}\}.$$

Prove that *CLIQUE* is NP-complete.

**10.3**$^\sharp$  An *independent set* of an undirected graph $G = (V, E)$ is a subset of $V$ in which any two nodes are not connected by an edge. A *cover* of $G = (V, E)$ is a subset $A$ of $V$ such that for any $(u, v) \in E$ at least one of $u$ and $v$ belongs to $A$. Furthermore, put

> *INDEPENDENT SET*
>
> $\quad = \{\langle G, k \rangle \mid k \geq 1$ and $G$ has an independent set of $k$ nodes$\}$,
>
> *COVER*
>
> $\quad = \{\langle G, k \rangle \mid k \geq 1$ and $G$ is a cover of $k$ nodes$\}$.

Letting graph $G = (V, E)$ and its complement graph $\overline{G} = (V, \overline{E})$ be as shown below, answer the following questions, where $\overline{E} = V \times V - E$.

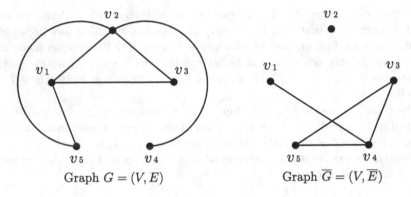

Graph $G = (V, E)$                Graph $\overline{G} = (V, \overline{E})$

(1) Give all the 3-cliques of $G$, all the independent sets consisting of three nodes of $\overline{G}$, and all the covers consisting of two nodes of $\overline{G}$.
(2) Prove that, in general, $A \subseteq V$ is a clique of graph $G = (V, E)$ if and only if $A \subseteq V$ is an independent set of $\overline{G} = (V, \overline{E})$.
(3) Prove that, in general for $G = (V, E)$, $A \subseteq V$ is an independent set if and only if $V - A$ is a cover.
(4) Prove that both *INDEPENDENT SET* and *COVER* are NP-complete.

**10.4**$^\sharp$  Give an algorithm that solves 2*SAT* in polynomial time.

**10.5**$^{**}$  Define an undirected version of the Hamiltonian problem as follows:

$$UHAMPATH = \{\langle G \rangle \mid \text{an undirected graph } G \text{ has a Hamiltonian path}\}.$$

Prove that *UHAMPATH* is NP-complete.

# Part V

# Conclusions and Exercise Solutions

# Solutions

11

Solving problems is essential to obtain deeper insights into those topics which might not be learned merely through reading the materials included in the book. In particular, to receive the greatest benefit, the readers are urged to try to solve the problems before reading the solutions included in this chapter. However, note that these exercises range from elementary drills to hard-to-solve problems; that is, there is a mix of elementary and advanced topics. The readers are advised to consult the marks assigned to the problems to understand how difficult they are and what characteristics the problems have.

---

## 11.1   Chapter 2

2.1 $2^n$.

2.2 $n^m$.

2.3 Let $R$ be a relation on set $S$ that satisfies the three laws. For each element $x$, let $R[x]$ denote the collection of the elements that are in relation $R$ with $x$. That is,

$$R[x] = \{y \mid x R y\}.$$

Let us assume that $R[x] \cap R[y] \neq \emptyset$. Then there exists $z$ such that $z \in R[x]$ and $z \in R[y]$. Then we have $x R z$ and $y R z$, which implies $z R y$ by the symmetric law. Hence we have $x R y$ by the transitivity law. On the other hand, for any $y' \in R[y]$ we have $x R y$ and $y R y'$ which implies $x R y'$ which in turn implies $R[x] \supseteq R[y]$. Similarly, by interchanging the roles of $x$ and $y$, we have $R[y] \supseteq R[x]$. Thus, if $R[x] \cap R[y] \neq \emptyset$, then $R[x] = R[y]$. On the other hand, by the reflexive law, every element $x$ belongs to at least one collection, i.e., $x \in R[x]$. Thus, we conclude that an equivalence relation $R$ derives the partition $\{R[x] \mid x \in S\}$.

Conversely, let $P$ be a partition. Define relation $R$ as follows:

$$x R y \quad \Leftrightarrow \quad \text{there exists } B \in P \text{ such that } x \in B \text{ and } y \in B.$$

Clearly, the relation $R$ becomes an equivalence relation that satisfies the three laws.

A. Maruoka, *Concise Guide to Computation Theory*,
DOI 10.1007/978-0-85729-535-4_11, © Springer-Verlag London Limited 2011

2.4  This proof only establishes the statement that if there exists $y$ such that $xRy$ holds, then $xRx$ holds. So $xRx$ holds only in such a case.

2.5  We shall prove the statement for the case of directed graphs. Similarly, the statement can be proved in the case of undirected graphs.

Assume that there exists a path from node $s$ to node $t$ of length equal to or larger than $n$, which is expressed as

$$s = v_0 \rightarrow v_1 \rightarrow \cdots \rightarrow v_{m-1} \rightarrow v_m = t,$$

where $m \geq n$. Since more than or equal to $n + 1$ nodes appear on the path, at least two of them are the same. Let these nodes be $v_i$ and $v_j$ where $i < j$. Then,

$$v_0 \rightarrow v_1 \rightarrow \cdots \rightarrow v_i \rightarrow v_{j+1} \rightarrow \cdots \rightarrow v_m$$

also becomes a path. If the length of the path obtained this way is equal to or larger than $n$, the above argument can be applied repeatedly until the length of the resulting path is equal to or less than $n - 1$, establishing the statement.

2.6
$$\overline{F \vee G \vee H} = \overline{(F \vee G) \vee H} = \overline{(F \vee G)} \wedge \overline{H} = (\overline{F} \wedge \overline{G}) \wedge \overline{H}$$
$$= \overline{F} \wedge \overline{G} \wedge \overline{H},$$
$$\overline{F \wedge G \wedge H} = \overline{(F \wedge G) \wedge H} = \overline{(F \wedge G)} \vee \overline{H} = (\overline{F} \vee \overline{G}) \vee \overline{H}$$
$$= \overline{F} \vee \overline{G} \vee \overline{H}.$$

Let us interpret formula $F \vee G \vee H$ as saying "at least one of $F$, $G$, and $H$ takes the value 1." The first of De Morgan's laws can be interpreted as the negation of "at least one of $F$, $G$, and $H$ takes the value 1" is equivalent to "all of $F$, $G$, and $H$ take the value 0." Similarly, the second of De Morgan's laws can be interpreted as the negation of "all of $F$, $G$, and $H$ take the value 1" is equivalent to "at least one of $F$, $G$, and $H$ takes the value 0." We can generalize De Morgan's law to the case of an arbitrary number of variables. The generalized De Morgan's law can be proved by mathematical induction, but it is omitted here.

2.7  The induction step from the case of $n = 1$ to the case of $n = 2$ is invalid. The induction step is valid only when the two groups for which the statement is assumed to be true must have at least one member in common.

## 11.2   Chapter 3

3.1  (a)  Strings in which every 0 is immediately followed by at least one 1.

(b)  Strings $w$ such that $N_1(w) - N_0(w) \equiv 1 \pmod 3$.

(c)  Strings in which no $a$ appears or $aa$ appears exactly once as a substring.

(d)  Strings that begin with an $a$ and end with an $a$, or begin with a $b$ and end with a $b$.

(e) Strings in which every $a$ appears more than or equal to two times consecutively.

(f) Strings in which $a$ appears in every odd numbered position, or $b$ appears in every even numbered position.

3.2 (1)

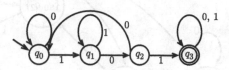

(2)

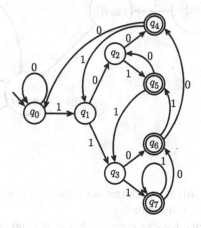

(3)

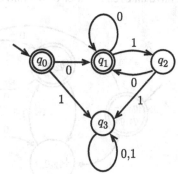

3.3 (1)

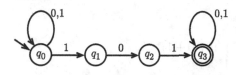

(2)

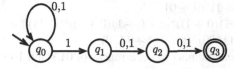

3.4

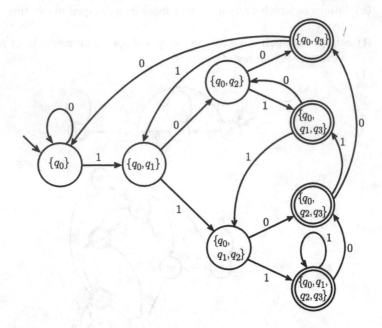

Note that this state diagram is identical to that of (2) in Problem 3.2 except for the names of the states.

3.5 (1)  Strings that have an even number of 0's and an odd number of 1's.
     (2)

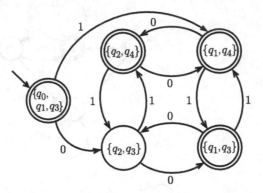

3.6 (1)  $(0+1)^*101(0+1)^*$
     (2)  $(0+11^*00)^*$
     (3)  $00(0+1)^*11$
     (4)  $01(0+1)^*01+01$
     (5)  $((0+1)(0+1))^* + ((0+1)(0+1)(0+1))^*$
     (6)  $(0+11^*00)^*(\varepsilon+11^*+11^*0)$
3.7 (1)  The total number of occurrences of 01's and 10's in the strings is even.

(2) The condition that strings of $\{01, 10\}$ appear an even number of times is equivalent to saying that the pairs of strings from $\{01, 10\}$ appear an arbitrary number of times, each being interleaved by strings in $(00+11)^*$. Thus a regular expression for the strings is expressed as $(00+11)^*((01+10)(00+11)^*(01+10)(00+11)^*)^*$, which is equivalent to $(00+11+(01+10)(00+11)^*(01+10))^*$.

(3)   (i)  Adding a new start state and a new accept state

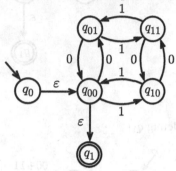

(ii)  Deleting $q_{01}$

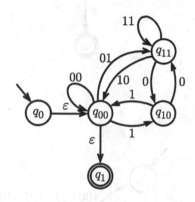

(iii)  Deleting $q_{10}$

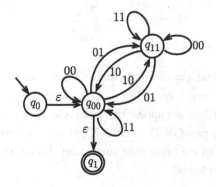

(iv) Replacing edges and regular expressions labeled appropriately

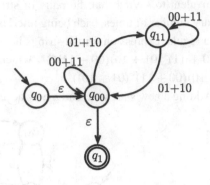

(v) Deleting $q_{11}$

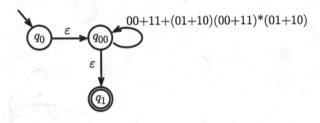

(vi) Deleting $q_{00}$

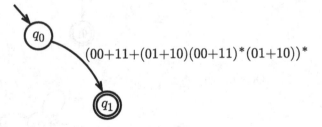

3.8 ($\Rightarrow$) is clear. ($\Leftarrow$) Suppose that string $w$ does not have 10 as a substring. Then $w$ is expressed as $w_1 2 w_2 \cdots 2 w_m$, where $w_1, \ldots, w_m$ are strings of $\{0, 1\}^*$. Since 10 does not appear in any $w_i$, clearly $w_i \in 0^* 1^*$.

3.9 From the proof of Theorem 3.22, $m$ in the theorem is taken to be the number of states of the finite state automaton. Therefore, from (1) of the theorem the statement holds.

3.10

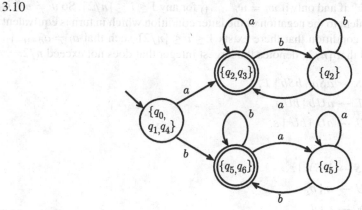

3.11

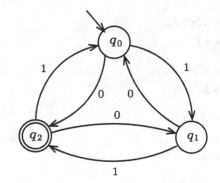

## 11.3 Chapter 4

4.1

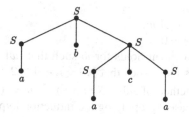

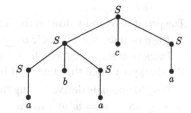

$$S \Rightarrow SbS \Rightarrow abS \Rightarrow abScS$$
$$\Rightarrow abacS \Rightarrow abaca$$

$$S \Rightarrow ScS \Rightarrow SbScS$$
$$\Rightarrow abScS \Rightarrow abacS \Rightarrow abaca$$

4.2 Let $G_1 = (\Gamma_1, \Sigma_1, P_1, S_1)$, $L_1 = L(G_1)$, $G_2 = (\Gamma_2, \Sigma_2, P_2, S_2)$, and $L_2 = L(G_2)$ such that $\Gamma_1 \cap \Gamma_2 = \emptyset$. Put $G = (\Gamma_1 \cup \Gamma_2 \cup \{S\}, \Sigma_1 \cup \Sigma_2, P_1 \cup P_2 \cup \{S \to S_1, S \to S_2\}, S)$. Then, clearly, $L(G) = L_1 \cup L_2$.

4.3 $S \to aSa \mid bSb \mid a \mid b \mid \varepsilon$

4.4 (1) $w = w^R$ if and only if $w_i = w_{n-(i-1)}$ for any $1 \le i \le \lceil n/2 \rceil$. So $w \ne w^R$ is equivalent to the negation of the latter condition which in turn is equivalent to the condition that there exists $1 \le i \le \lceil n/2 \rceil$ such that $a_i \ne a_{n-(i-1)}$. Recall that $\lceil n/2 \rceil$ denotes the largest integer that does not exceed $n/2$.

(2)
$$S \to aSa \mid bSb \mid T,$$
$$T \to aUb \mid bUa,$$
$$U \to aU \mid bU \mid \varepsilon.$$

4.5

(1)
$$S \to USV \mid \varepsilon,$$
$$U \to a \mid b,$$
$$V \to c \mid d,$$

(2)
$$S \to aSc \mid T,$$
$$T \to bT \mid \varepsilon,$$

(3)
$$S \to aSd \mid T,$$
$$T \to bTc \mid \varepsilon,$$

(4)
$$S \to ASB \mid \varepsilon,$$
$$A \to a \mid \varepsilon,$$
$$B \to b,$$

(5)
$$S \to aSbB \mid \varepsilon,$$
$$B \to b \mid \varepsilon,$$

(6)
$$S \to TU,$$
$$T \to aTb \mid \varepsilon,$$
$$U \to bUc \mid \varepsilon.$$

4.6 The proof is by induction on the length $n$ of $w$.

Base of induction: It is clear when $n = 0$.

Induction step: Suppose that the statement holds for $w$ such that $|w| \le n$. We shall prove that the statement holds for $w$ such that $|w| = n + 1 \ge 2$.

($\Rightarrow$) String $w$ is derived using the method of either $S \Rightarrow (S) \overset{*}{\Rightarrow} w = (w')$ or $S \Rightarrow SS \Rightarrow w = w'w''$. In the first case, by applying the induction hypothesis to $S \overset{*}{\Rightarrow} w'$, it is clear that $w = (w')$ satisfies the conditions (1) and (2). Similarly, in the second case, by applying the induction hypothesis to $S \overset{*}{\Rightarrow} w'$ and $S \overset{*}{\Rightarrow} w''$, $w = w'w''$ satisfies the conditions (1) and (2).

($\Leftarrow$) Suppose that $w$ satisfies the conditions (1) and (2). We consider two cases: the case where there exists a proper prefix $w'$ of $w$ such that $N_(w') - N_)(w') = 0$; the remaining case where $N_((w') - N_)(w') \ge 1$ for any proper prefix $w'$ of $w$, where a proper prefix $w'$ of $w$ of length $n$ is a prefix of $w$ whose length is 1 through $n-1$. In the former case, $w$ is expressed as $w'w''$, and both $w'$ and $w''$ satisfy the conditions (1) and (2). Since $S \overset{*}{\Rightarrow} w'$ and $S \overset{*}{\Rightarrow} w''$ by the induction hypothesis, we have $S \Rightarrow SS \Rightarrow w'w'' = w$. In the latter case, $w$ is expressed as $(w')$ and $w'$ satisfies the conditions (1) and (2). Thus, since $S \overset{*}{\Rightarrow} w'$ by the induction hypothesis, we have $S \Rightarrow (S) \overset{*}{\Rightarrow} (w') = w$.

4.7 (1) Strings of even length, including the empty string.

(2)
$$S \to aT \mid bT \mid \varepsilon,$$
$$T \to aS \mid bS.$$

4.8 (1) Strings that take the form $a^i b^j$ with $i + j \geq 1$.

(2)
$$S \to aS \mid T \mid a,$$
$$T \to bT \mid b.$$

4.9 (1) The language $L$ expressed as $L = \{w \mid N_a(w) = N_b(w)\}$.

(2) Let $L_{+a} = \{w \mid N_a(w) = N_b(w) + 1\}$ and $L_{+b} = \{w \mid N_b(w) = N_a(w) + 1\}$. Let's denote the context-free grammars with the rules given in the problem, but with the start symbols $S$, $A$, and $B$ represented by $G$, $G_A$, and $G_B$, respectively. We shall establish that $L(G) = L$, $L(G_A) = L_{+a}$, and $L(G_B) = L_{+b}$. More specifically, we shall prove that

For any $w \in \Sigma^*$, $\quad w \in L \quad \Leftrightarrow \quad w \in L(G)$,

For any $w \in \Sigma^*$, $\quad w \in L_{+a} \quad \Leftrightarrow \quad w \in L(G_A)$,

For any $w \in \Sigma^*$, $\quad w \in L_{+b} \quad \Leftrightarrow \quad w \in L(G_B)$,

where $\Sigma = \{a, b\}$. The proof is done by induction on the length of $w$.

(Base) Concerning the string of length zero, i.e., the empty $\varepsilon$, we have by definition the followings:

$$\varepsilon \in L(G), \qquad \varepsilon \notin L(G_A), \qquad \varepsilon \notin G(B),$$

$$\varepsilon \in L, \qquad \varepsilon \notin L_{+a}, \qquad \varepsilon \notin L_{+b}.$$

(Induction step) Assume that the three equivalences hold for any string of length smaller than or equal to $n$. Let $w$ be an arbitrary string of length $n + 1 \geq 1$. We shall prove that the three equivalences hold for string $w$.

($\Leftarrow$) Assume that $w \in L(G)$. Then $w$ is derived in either of the two ways:

$$S \Rightarrow aB \overset{*}{\Rightarrow} aw_B(= w),$$

$$S \Rightarrow bA \overset{*}{\Rightarrow} bw_A(= w).$$

In the former case, we have by induction hypothesis $w_B \in L_{+b}$, hence $aw_B \in L$. Similarly, in the latter case, we have $w_A \in L_{+a}$, hence $bw_A \in L$.

Next assume that $w \in L(G_A)$. Then $w$ is derived in either of the two ways:

$$A \Rightarrow aS \overset{*}{\Rightarrow} aw_S(= w),$$

$$A \Rightarrow bAA \overset{*}{\Rightarrow} bw_A A \overset{*}{\Rightarrow} bw_A w'_A(= w).$$

In the former case, we have by induction hypothesis $w_S \in L$, hence $aw \in L_{+a}$. Similarly, in the latter case, we have $w_A \in L_{+a}$ and $w'_A \in L_{+a}$, hence $bw_A w'_A \in L_{+a}$. In a similar way, if we assume $w \in L(G_B)$, then we can conclude that $w \in L_{+b}$.

($\Rightarrow$) Assume that $w \in L$. Then we have the following two cases:

$$w = aw_B \quad \text{with } w_B \in L_{+b},$$

$$w = bw_A \quad \text{with } w_A \in L_{+a}.$$

In the former case, we have by induction hypothesis $B \overset{*}{\Rightarrow} w_B$, hence $S \Rightarrow aB \overset{*}{\Rightarrow} aw_B$. Thus $w = aw_B \in L(G)$. In the latter case, we can similarly prove $w = bw_A \in L(G)$. Next assume that $w \in L_{+a}$. Then we have the following two cases:

$$w = aw_S \quad \text{with } w_S \in L,$$

$$w = bw_A w'_A \quad \text{with } w_A \in L_{+a} \text{ and } w'_A \in L_{+a}.$$

In the former case, we have by induction hypothesis $w_S \in L(G)$, hence $A \Rightarrow aS \overset{*}{\Rightarrow} aw_S$, hence $w = aw_S \in L(G_A)$. Similarly, in the latter case, we have

$$A \Rightarrow bAA \overset{*}{\Rightarrow} bw_A A \overset{*}{\Rightarrow} bw_A w'_A,$$

hence $w = bw_A w'_A \in L(G_A)$. In a similar way, if we assume $w \in L_{+b}$, then we can conclude that $w \in L(G_B)$.

4.10 (1)

$$S \to aSbS \mid bSaS \mid \varepsilon.$$

(2) We only give a proof idea to show that $L(G) \subseteq L$ and $L \subseteq L(G)$, omitting the precise proof.

(Proof idea for $L(G) \subseteq L$) Applying a rule in (1) adds an $a$ and a $b$, or the empty string. Hence any string eventually generated has an equal number of $a$'s and $b$'s.

(Proof idea for $L \subseteq L(G)$) Any string $w \in L$ can be written as $aw'bw''$ or $bw'aw'$ for appropriately chosen $w' \in L$ and $w'' \in L$. We can see this fact by drawing a graph as shown in Fig. 4.9 for any $w \in L$. By regarding these $w'$ and $w''$ as new $w$, the argument will be repeated until we have the empty string. By applying $S \to aSbS$ and $S \to bSaS$ in the cases of $aw'bw''$ and $bw'aw''$, respectively, and applying $S \to \varepsilon$ in the case that the empty string is obtained, we can generate the original string $w$.

**Fig. 11.1** Graph associated with *aaabbbabaaaa*

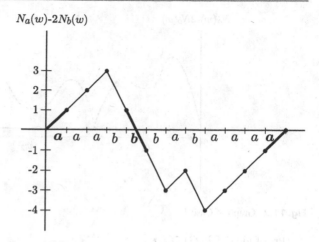

4.11 The substitution rules are given as follows:

$$S \rightarrow aSaSb \mid bSaSa \mid aSbSa \mid SS \mid \varepsilon.$$

In order to verify that the context-free grammar specified by the rules above generates the language in question, we introduce the graph associated with a string in a similar way to Fig. 4.9. But this time the vertical axis represents $N_a(w) - 2N_b(w)$, where $N_x(w)$ is the number of $x$'s that appear in $w$. To figure out how a graph is associated with a string, we give an example in Fig. 11.1. The graph associated with string *aaabbbabaaaa* consists of 13 nodes. What the bold symbols and edges in the figure mean will be explained shortly. The height of a node is simply defined to be the coordinate of the vertical axis. This definition extends to an edge. For example, edge $((3, 3), (4, 1))$ goes from height 3 to height 1. Similarly, the symbol associated with the edge, namely the leftmost $b$, goes from height 3 to height 1. Clearly, string $w$ satisfies the condition that $N_a(w) = 2N_b(w)$ if the graph associated with $w$ begins at the origin and ends on the horizontal axis.

**Statement** Let $G$ be the context-free grammar specified by the substitution rules above. Then the grammar $G$ generates the language $\{w \in \{a, b\}^* \mid N_a(w) = 2N_b(w)\}$.

(Proof idea of Statement)
Let

$$L = \{w \in \{a, b\}^* \mid N_a(w) = 2N_b(w)\}.$$

(Proof idea of $L(G) \subseteq L$)
Whenever terminals are generated by applying one of the substitution rules, two $a$'s and one $b$ are generated. Thus we have $L(G) \subseteq L$.

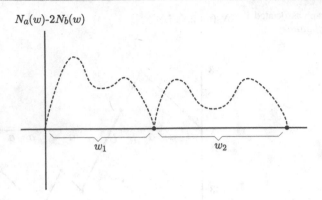

**Fig. 11.2**   Graph of Case 1

(Proof idea of $L(G) \supseteq L$)

Let Condition C be specified as follows:

Condition C: $N_a(w) = 2N_b(w)$.

To prove that $L(G) \supseteq L$ we shall verify the Fact: any string $w$ satisfying Condition C is devided into substrings, each corresponding to one of the righthand parts of the rules so that the substrings associated with nonterminal $S$ satisfies Condition C. In a rigorous argument to establish $L(G) \supseteq L$ based on the Fact, we use mathematical induction on the number of steps, assuming induction hypothesis that a substring associated with $S$ can be derived from the start symbol $S$. Take, for example, string *aaabbbabaaaa* and the graph associated with it. In this case, the string is devided into 5 substrings based on $aSbSa$: the bold symbols in Fig. 11.1 correspond to the terminals and the remaining parts left, namely, *aab* and *babaaa*, are substrings associated with nonterminal $S$. The point of the argument is that the height of the start point of the substring *aab* is the same as that of the end point of the substring. This is also the case with the other substring *babaaa*.

To establish the Fact we consider two cases depending on the graph associated with string $w$.

*Case 1*   There exists a node on the graph other than the origin and the end point that is placed on the horizontal axis.

In this case, $w$ is written $w_1 w_2$ such that both $w_1$ and $w_2$ satisfy Condition C. This is because, as shown in Fig. 11.2, we can divide the string into two substrings where the graph touches the horizontal axis so that both substrings satisfy Condition C. In this case, we can verify

$$S \Rightarrow SS \overset{*}{\Rightarrow} w_1 S \overset{*}{\Rightarrow} w_1 w_2 .$$

Note that either of the parts of the graph corresponding to the substrings might be below the horizontal axis.

*Case 2*   Any node on the graph other than the origin and the end point is not placed on the horizontal axis.

Case 2 is further divided into three cases.

**Fig. 11.3**  Graph of Case 2.1

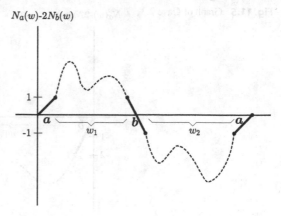

**Fig. 11.4**  Graph of Case 2.2

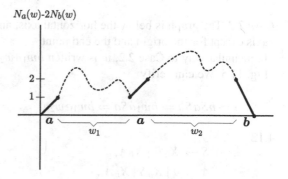

*Case 2.1* An edge of the graph crosses the horizontal axis, but not at any node. More specifically, no node on the graph other than the origin and the end point has height 0.

Figure 11.1 gives a concrete example of this case. More generally, the graph of this case is given in Fig. 11.3, because the graph touches the horizontal axis, but not at any node. In this case, we can verify

$$S \Rightarrow aSbSa \overset{*}{\Rightarrow} aw_1bSa \overset{*}{\Rightarrow} aw_1bw_2a.$$

*Case 2.2* The graph is above the horizontal axis, and does not touch the horizontal axis except for the origin and the end point.

It is easy to see that in this case string $w$ starts with $a$ and ends with $b$. Furthermore, $w$ has $a$ from height 1 to height 2. Thus the graph is drawn as in Fig. 11.4. String $w$ is written $aw_1aw_2b$, and the graph associated with it is drawn as in Fig. 11.4. Thus we can verify

$$S \Rightarrow aSaSb \overset{*}{\Rightarrow} aw_1aSb \overset{*}{\Rightarrow} aw_1aw_2b.$$

It should be noticed that we can take any $a$ in $w$ from height 1 to height 2 as the second $a$ in $aw_1aw_2b$.

**Fig. 11.5** Graph of Case 2.3   $N_a(w)\text{-}2N_b(w)$

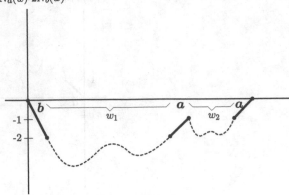

*Case 2.3* The graph is below the horizontal axis, and does not touch the horizontal axis except for the origin and the end point.

In a similar way to Case 2.2, $w$ is written $bw_1aw_2a$, and the graph is drawn as in Fig. 11.5. We can verify

$$S \Rightarrow bSaSa \stackrel{*}{\Rightarrow} bw_1aSa \stackrel{*}{\Rightarrow} bw_1aw_2a.$$

4.12
$$S \to X_aB \mid X_bA,$$
$$A \to a \mid X_aS \mid X_bA',$$
$$A' \to X_bA,$$
$$B \to b \mid X_bS \mid X_aB',$$
$$B' \to BB,$$
$$X_a \to a,$$
$$X_b \to b.$$

4.13 (1) Apply Theorem 4.12 to string $a^mb^mc^m$, where $m$ is the number described in the theorem. Then $a^mb^mc^m$ is expressed as $uvxyz$. By the condition $|vxy| \le m$ the substring $vxy$ does not touch the region of $c^m$ or that of $a^m$. In the former case, if $uv^2xy^2z$ is expressed as $a^ib^jc^m$, then at least one of $i > m$ or $j > m$ holds by the condition that $|vy| \ge 1$ so that $a^ib^jc^m$ does not belong to the language of (1). On the other hand, in the latter case, if $uxz$ is expressed as $a^mb^jc^k$, then at least one of $m > j$ or $m > k$ holds so that $a^mb^jc^k$ does not belong to the language. In any case, a contradiction occurs.

   (2) In a similar way to (1), apply Theorem 4.12 to string $a^mb^mc^m$, which is expressed as $uvxyz$. Since the substring $vxy$ touches at most two of the three regions of $a^m$, $b^m$, and $c^m$, the numbers of occurrences of $a$, $b$, and $c$ in $uv^2xy^2z$ are not equal to each other, resulting in a contradiction.

4.14 (1) Let $L = \{ww \mid w \in \{a, b\}^*\}$. Apply Theorem 4.12 to string $a^m b^m a^m b^m \in L$, where $m$ is the number given in the theorem. Then $a^m b^m a^m b^m$ is written as $uvxyz$. String $uxy$ obtained by deleting the portions of $v$ and $y$ belongs to $L$ and hence is written as $w'w'$ for some $w' \in \{a, b\}^*$. Since $|vxy| \le m$, either of the first $a^m$ or the last $b^m$ in $a^m b^m a^m b^m$ remains in $uxy$ without being deleted. Without loss of generality, assume that the last $b^m$ remains. On the other hand, since the length of $uxy$ is at least $3m$, the last $b^m$ is contained in the last half of $w'w'$. Hence $b^m$ must also appear in the first half of $w'w'$ so that both $b^m$ in $a^m b^m a^m b^m$ must remain without being deleted. On the other hand, since $|vy| \ge 1$ and $|vxy| \le m$, $uxy$ is obtained from $uvxyz$ by deleting a certain number of $a$'s in one of the $a^m$ regions of $a^m b^m a^m b^m$. This contradicts the fact that $uxy$ is written as $w'w'$.

(2) Let $\Sigma = \{a, b\}$. Then, $\Sigma^* - L$ is generated by

$$S \to AB \mid BA \mid A \mid B,$$

$$A \to CAC \mid a,$$

$$B \to CBC \mid b,$$

$$C \to a \mid b.$$

It is easy to see that the grammar generates (i) all the strings whose lengths are odd, and (ii) strings that are in the form of $xayubv$ or $ubvxay$ where $|x| = |y|$ and $|u| = |v|$. On the other hand, the strings that belong to $\{a, b\}^* - \{ww \mid w \in \{a, b\}^*\}$, namely, the strings that cannot be expressed as $ww$ are either (i) strings whose length are odd or (ii)' strings that are in the form of $xauybv$ or $xbuyav$ where $|x| = |y|$ and $|u| = |v|$. This is because, if $w \ne w'$ and $|w| = |w'|$, then there exist two symbols that appear in the same position of $w$ and $w'$, but are different from each other. On the other hand, clearly the conditions (ii) and (ii)' are equivalent with each other.

(3) It is clear because the complement $L$ of the context-free language $\Sigma^* - L$ is not a context-free language.

4.15

| Row No./Column No. | 1 | 2 | 3 | 4 |
|---|---|---|---|---|
| 4 | $\{S, A, C\}$ | | | |
| 3 | $\{S\}$ | $\{S, A, B, C\}$ | | |
| 2 | $\{S\}$ | $\{A, C\}$ | $\{B\}$ | |
| 1 | $\{A\}$ | $\{A\}$ | $\{B, C\}$ | $\{A\}$ |
| | $a$ | $a$ | $b$ | $a$ |

## 11.4   Chapter 5

5.1 (1)

$$L = \{x0y \in \{0, 1\}^* \mid |x| = |y|\}$$

(2)

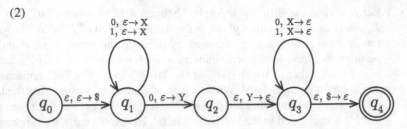

**Fig. 11.6** State diagram of solution of Problem 5.1

(3) We give four types of rules in the proof of Theorem 5.8 in (3.1), (3.2), (3.3) and (3.4), respectively, where states $q_0, q_1, \ldots, q_4$ are denoted by $0, 1, \ldots, 4$, respectively.

(3.1)
$$A_{13} \to 0A_{13}0 \mid 0A_{13}1 \mid 1A_{13}0 \mid 1A_{13}1,$$

$$A_{13} \to 0A_{22},$$

$$A_{04} \to A_{13}.$$

(3.2) The rules of the type $A_{pq} \to A_{pr}A_{rq}$, where $p, q, r$ are over all combinations of three states from $\{0, 1, \ldots, 4\}$. Hence, the total number of these rules is $125 (= 5^3)$.

(3.3) For each $q \in \{0, 1, \ldots, 4\}$,
$$A_{qq} \to \varepsilon.$$

(3.4)
$$S \to A_{04}.$$

(4)
$$S \to 0S0 \mid 0S1 \mid 1S0 \mid 1S1 \mid 0.$$

When the PDA accepts a string, it pushes repeatedly a certain number of times and then pops the same number of times. So in fact all the rules of the type (2.2) are not necessary, and the only rule of the type (2.3) that we need is $A_{22} \to \varepsilon$. Taking these facts into account, it is easy to see that the rules of (2) is equivalent to the rules of (3).

5.2 $\{w \in \{a, b\}^* \mid w = a^i b^j c^k$ for some $i \leq 0$, $j \leq 0$, $k \leq 0$, and the length of $w$ is odd and the symbol in its center is $b\}$

5.3

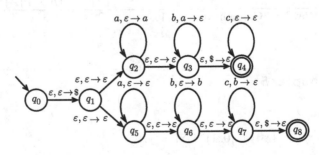

5.4

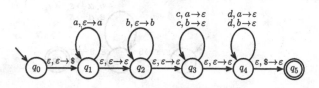

5.5

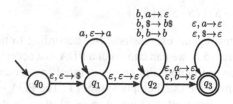

5.6  (1)  $\{w \in \{a, b\}^* \mid N_a(w) = N_b(w)\}$.

(2)  Add the two edges $q_1 \xrightarrow{b,\varepsilon \to b} q_2$ and $q_1 \xrightarrow{b,a \to \varepsilon} q_2$ to the PDA in (1).

5.7  (1)  The proof is by induction on the length of string $w$. Let us denote the right-hand condition by $C$.

(Base) It is clear because $\varepsilon \in L(G)$ and $N_a(\varepsilon) - N_b(\varepsilon) = 0$.

(Induction step) Suppose that the statement is true for strings of length smaller than or equal to $n \geq 1$.

($\Rightarrow$) Let $w \in L(G)$ be such that $|w| = n + 1$. Then a derivation of $w$ is either $S \Rightarrow aSbS \overset{*}{\Rightarrow} aw'bw'' = w$ with $S \overset{*}{\Rightarrow} w'$ and $S \overset{*}{\Rightarrow} w''$ or $S \Rightarrow aS \overset{*}{\Rightarrow} aw' = w$ with $S \overset{*}{\Rightarrow} w'$. In the former case, since $w'$ and $w''$ satisfy the condition $C$ by the induction hypothesis, $w = aw'bw''$ satisfies $C$. On the other hand, in the latter case, since $w'$ satisfies $C$, $w = aw'$ also satisfies $C$.

($\Leftarrow$) Let $w$ with $|w| = n + 1$ satisfy the condition $C$. We consider the two cases: $N_a(w') - N_b(w') \geq 1$ for any prefix $w'$ of $w$ of length longer than or equal to 1; there exists a prefix $w'$ of $w$ longer than or equal to 1 such that $N_a(w') - N_b(w') = 0$. In the former case, $w$ is written as $aw'$ such that $w'$ satisfies the condition $C$. Then $w' \in L(G)$ by the induction hypothesis, so we have $S \Rightarrow aS \overset{*}{\Rightarrow} aw' = w$. On the other hand, in the latter case, $w$ is written as $aw'bw''$ such that $w'$ and $w''$ satisfy the condition $C$. Then, since $w' \in L(G)$ and $w'' \in L(G)$ by the induction hypothesis, we have $S \Rightarrow aSbS \overset{*}{\Rightarrow} aw'bw'' = w$.

(2)  We give a PDA that accepts $L(G)$, which is simpler than the one constructed in the way described in the proof of Theorem 5.6.

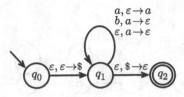

5.8  Based on Definition 3.15 we have a context-free grammar

$$S \to a \mid b \mid \tilde{\varepsilon} \mid \emptyset,$$

$$S \to (S + S) \mid (S \cdot S) \mid (S^*),$$

which generates regular expressions. So, according to the construction in the proof of Theorem 5.6, we can construct a PDA that accepts regular expressions. In this solution we denote the empty symbol in regular expressions by $\tilde{\varepsilon}$ to distinguish it from the usual empty symbol $\varepsilon$. The PDA is given as in Fig. 5.15 by specifying, as labels of the type $\{a, a \to \varepsilon\}$ of $q_2 \to q_2$, $a, a \to \varepsilon$, $b, b \to \varepsilon$, $\tilde{\varepsilon}, \tilde{\varepsilon} \to \varepsilon$, $\emptyset, \emptyset \to \varepsilon$, $+, + \to \varepsilon$, $\cdot, \cdot \to \varepsilon$, $*, * \to \varepsilon$, $($ , $( \to \varepsilon$, and $)$, $) \to \varepsilon$, and, similarly, as labels of the type $\{\varepsilon, A \to u\}$, $\varepsilon, S \to (S + S)$, $\varepsilon, S \to (S \cdot S)$, $\varepsilon, S \to (S^*)$, $\varepsilon, S \to a$, $\varepsilon, S \to b$, $\varepsilon, S \to \tilde{\varepsilon}$, and $\varepsilon, S \to \emptyset$.

## 11.5  Chapter 6

6.1

$$\{\vdash wwX \mid w \in \{0, 1\}^*\}.$$

6.2  Let $(q', a', D) \in \delta(q, a)$ be denoted by $q \xrightarrow{a/a',D} q'$. In general, to simulate a three-way model by a two-way model, every transition $q \xrightarrow{a/a',D} q'$ of staying put in the three-way model is replaced by $q \xrightarrow{a/a',R} q'' \xrightarrow{*,L} q'$ by introducing a different new state $q''$, where $*$ means that a transition is made automatically without changing any content of the tape.

6.3  (1)

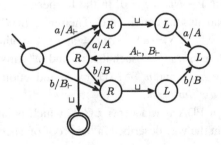

(2)

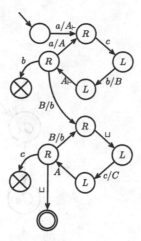

Note that the head of the TMs above moves off the tape leftward in certain cases such as *a* for (1) and *aabc* for (2).

6.4

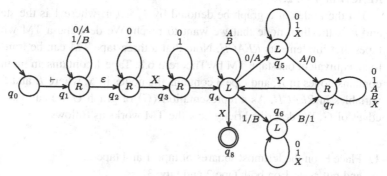

Note that when the TM accepts ⊢ *uyvXy*, it moves its head from the portion of *y* to the portion of *v* making transition $q_1 \to q_2$. Furthermore, note that the TM checks if the identical substring *y* also appears somewhere in the portion of *w* by moving the head to the left and making transitions between states $q_4$, $q_5$, $q_6$, and $q_7$.

6.5

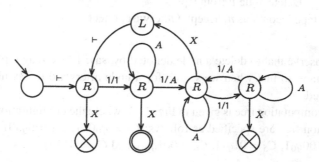

**Fig. 11.7** Computation tree
of solution of Problem 6.7

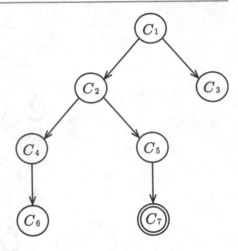

6.6 We explain roughly how to make the TM work so as to perform algorithm
*REACH* in Sect. 2.7.

Let the nodes of a graph be denoted by $1, \ldots, n$, where 1 is the start node
and $n$ is the final node that we want to reach. We describe a TM with three
tapes that implements *REACH*. Note that a three-tape TM can be transformed
to an equivalent one-tape TM by Theorem 6.8. Tape 1 contains an input, tape 2
contains nodes in $R$, and tape 3 contains the nodes in $S$, where $R$ and $S$ are the
variables in *REACH*. As a representation $\langle G \rangle$ of graph $G$ we can use a list of
edges of $G$. We describe briefly how the TM works as follows:

1. Place $\vdash$ on the leftmost squares of tape 1 and tape 2,
   and put node 1 on both tape 2 and tape 3.
2. Do the following as long as there remains a node on tape 3.
   Let such a node be denoted by $v$.
       Do the following as long as there remains an edge expressed as $(v, u)$
       for any node $u$.
           If tape 2 does not contain node $u$, add it to both tape 2 and tape 3.
       Delete node $v$ from tape 3.
3. If tape 2 contains $n$, accept. Otherwise, reject.

Observe that to delete a node denoted by, say, 12 we may replace it by $XX$
on a tape so that when we search for a node afterward the symbol $X$ will be
skipped.

6.7 The computation tree is given in Fig. 11.7, where the configurations in the computation tree are specified as follows: $C_1 = q_0001$, $C_2 = 0q_001$, $C_3 = 1q_101$,
$C_4 = 00q_01$, $C_5 = 01q_11$, $C_6 = 001q_0\sqcup$, and $C_7 = 011q_2\sqcup$.

## 11.6   Chapter 7

7.1 We shall modify the deterministic universal TM $U$ in Fig. 7.3 to obtain the state diagram of a nondeterministic universal TM $U'$. In the state diagram in Fig. 7.3, the machine searches for a 5-tuple, say, $(q, a, q', a', D)$ that coincides with the current state $q$ and the current symbol $a$ by moving its head to the right from the leftmost square of the machine description region. To obtain $U'$, we modify $U$ to make $U'$ search for such a 5-tuple nondeterministically. Suppose that $U'$ is in state $q_6$ with its head on the leftmost square in the machine description region. We make $U'$ skip symbol $X$ an appropriate number of times to get to a 5-tuple. To do so, add transition $q_6 \xrightarrow{X} q_6$ to the state diagram in Fig. 7.3, so that $U'$ guesses the 5-tuple matched by repeating transition $q_6 \xrightarrow{X} q_6$ an appropriate number of times and then making transition $q_6 \xrightarrow{X} q_1$. Although we do not need to check whether the content of the current state and symbol coincides with the first part of the 5-tuple selected in this nondeterministic version, we leave the part of the diagram to check the coincidence as it is so that the modification of the state diagram becomes small. On the other hand, we add all the transitions of the type $q \xrightarrow{a/a} q$ that are omitted in the deterministic machine Fig. 7.3 by convention.

7.2 Let the sentence "this sentence is false" be denoted by $S$. If $S$ is true, then what $S$ means is that $S$ is false. On the other hand, if $S$ is false, then it follows that $S$ is true. Thus in either case we have a contradiction, which is summarized as follows:

| Truth-value of sentence $S$ | What sentence $S$ means |
|---|---|
| True | False |
| False | True |

7.3 To establish the statement it suffices by Theorem 7.5 to show that the halting problem is mapping reducible to the empty tape halting problem. To do this we introduce reduction $f$ as follows. The Turing machine $M_f$ that computes $f$ works as follows: given $\langle M, w \rangle$ as input, $M_f$ yields as output description $\langle M_w, \varepsilon \rangle$, where the Turing machine $M_w$ first writes $w$ on its tape, then moves its head to the leftmost square, and finally simulates TM $M$. That is, $f(\langle M, w \rangle) = \langle M_w, \varepsilon \rangle$. Clearly, the halting problem is mapping reducible to the empty tape halting problem by means of the reduction $f$. More specifically, according to the given string $w = a_1 \cdots a_n$, $M_w$ writes $w$ through transitions $q_0 \xrightarrow{\sqcup/a_{1\vdash}, R} q_1 \xrightarrow{\sqcup/a_2, R} q_2 \xrightarrow{\sqcup/a_3, R} \cdots \xrightarrow{\sqcup/a_n, R} q_n$, where in general $a_{i\vdash}$ means that symbol $a_i$ is placed on the leftmost square. Then, to simulate $M$, $M_w$ makes transition $q_n \xrightarrow{a_{1\vdash}/a_{1\vdash}, S} q_s$. Note that $q_0$ and $q_s$ are the start state of $M_w$ and that of $M$, and that concerning head movement we follow the convention of Problem 6.2*. For the description of $\langle M_w \rangle$ we can use what is described in Sect. 7.1.

7.4 For the proof to be valid, there must be a Turing machine that computes the reduction $f$. This is not true in this case.

## 11.7  Chapter 8

8.1 We only explain a basic idea of how to construct a Turing machine for each case.
 (1) Construct a Turing machine that simulates *MEN* in Sect. 4.5. As explained in that section, the time complexity is $O(n^3)$.
 (2) A deterministic finite automaton that accepts $L(G)$ can be considered as a deterministic TM that accepts $L(G)$ in $n$ steps.

8.2 (1) Let the nodes of graph $G$ be denoted by $1, \ldots, n$ and $\langle G \rangle$ be denoted by a list of edges of $G$, where the length of the list is $O(n^2)$. We can construct a deterministic TM that determines if $\langle G \rangle \in 3CLIQUE$ as follows: enumerating all 3-tuples $\{i, j, k\}$ of the nodes, check if all the three edges $(i, j)$, $(j, k)$, and $(i, k)$ are contained in the list of edges of $G$: if there exists at least one such 3-tuple, accept; otherwise, reject. The number of steps required is $O(n^3)$ to enumerate all 3-tuples and $O(n^2)$ to check each 3-tuple. Thus the time complexity of the TM is $O(n^5)$.
 (2) Let $m = \lfloor n/2 \rfloor$. In a similar way to (1) above, we can construct a nondeterministic TM that chooses $m$ nodes nondeterministically in $O(m)$ time and check if the $m$ nodes chosen form a clique in $O(m^2 \times n^2)$ time. Clearly the nondeterministic TM accepts $\lfloor n/2 \rfloor CLIQUE$ in $O(m) + O(m^2 \times n^2) = O(m^4)$ time.

8.3 We shall show that the TM shown in Fig. 8.3 decides whether an input string is well-nested or not in at most $2m^2 + 4m$ steps. We shall verify it only when an input is well-nested, because when an input is not well-nested the TM goes to the reject state at some step during the computation to accept well-nested parentheses.

**Lemma** *Let the TM be given as input a string of length $n = 2m$. Starting at state $q_0$ or $q_1$ with the head on the leftmost square, the TM replaces all the parentheses in a well-nested sequence to $X$'s (including $X_\vdash$) in at most $2m^2$ steps.*

Before proceeding to the proof of the lemma, we verify the statement provided that the lemma is proved. When a well-nested string of length $2m$ parentheses is given as input, the TM replaces the input to a string of $X$'s in at most $2m^2$ steps, then moves its head to the right until it reads the blank symbol for the first time in at most $2m$ steps, and finally moves its head back until it reads symbol $X_\vdash$ in $2m$ steps. So the total number of steps is at most $2m^2 + 2m + 2m = 2m^2 + 4m$.

*Proof of the Lemma* The proof is by induction on the length of an input.
 (Base) When $m = 1$, the only well-nested string is ( ) and the TM replaces it with $XX$ in two steps.
 (Induction step) Assume that the lemma holds for well-nested strings of length at most $2m$. We shall prove the lemma for well-nested strings $w$ of length $2m + 2$.
 *Case 1* $w$ is written as $(w')$ for some well-nested string $w'$.

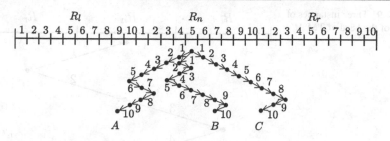

**Fig. 11.8** Three instances of traces of the head of TM $M$

By the induction hypothesis, since the TM replaces $w'$ to the string of $X$'s beginning and ending with its head somewhere in $w'$ in at most $2m^2$ steps, the TM replaces $(w')$ with the same situation in at most $1 + 2m^2 + 2m + (2m + 1) = 2(m + 1)^2$ steps, where the first 1 is to move from $($ to the leftmost of $w'$, the third $2m$ is to move from some position in $w'$ to the rightmost of $(w')$, and finally the last $(2m + 1)$ is to move from the rightmost to the leftmost of the entire string $(w')$.

*Case 2* $w$ is written as $w'w''$ for some well-nested strings $w'$ and $w''$.

Let the lengths of $w'$ and $w''$ be denoted by $2m_1$ and $2m_2$, respectively, where $m = m_1 + m_2$. In a way similar to Case 1, the TM replaces $w'w''$ to the string of $X$'s in at most $2m_1^2 + 2m_1 + 2m_2^2 \le 2(m_1 + m_2)^2$, where $2m_1$ is to move the head rightward from the position when all the parentheses in $w'$ are replaced to $X$'s to the leftmost square of $w''$. Note that when we apply the lemma for a certain string, we assume that the TM starts with its head on the leftmost of the string and ends on one of the squares where a well-nested parenthesis is placed. □

8.4 Let $\varepsilon$ be an arbitrary positive number. For an arbitrary $t(n)$ time $k$-tape TM $M$, we shall construct an equivalent $\varepsilon t(n)$ time $k$-tape TM $M'$. The idea of the construction is that $m$ transitions of $M$ are performed through 6 transitions of $M'$ so that the time complexity of $M'$ becomes $6t(n)/m$. If we take $m$ such that $\varepsilon > 6/m$, then the time complexity of $M'$ is at most $\varepsilon t(n)$.

To implement the idea, $M'$ collects the information concerning $m$ transitions of $M$ and updates the state, tape contents, and head position of $M'$ accordingly in 6 steps. Let us call $m$ consecutive transitions of $M$ a stage and $m$ consecutive squares of a tape a region. Taking a region of $M$ as a square, $M'$ simulates a stage of $M$ in 6 transitions. Figure 11.8 illustrates three cases $A$, $B$, and $C$ of how the head moves during a stage over the squares of three regions $R_l$, $R_n$, and $R_r$ of a tape, each consisting of 10 squares ($m = 10$), where the stage starts with the head on the fifth square of region $R_n$. These three are typical cases used just for explanation. Each head of the $k$ tapes moves in a way illustrated in the figure. Let the set of states of $M$ be denoted by $Q$ and the tape alphabet be $\Gamma$. Then the set of states of $M'$ is taken to be $Q \times \{1, \ldots, m\}^k \times \Gamma^{3km}$ and the tape alphabet as $\Gamma' = \Gamma^m \cup \Gamma$, where $\{1, \ldots, m\}^k$ is to specify positions within a region for $k$ tapes and $\Gamma^{3km}$ is to denote the contents of the three regions $R_l$, $R_n$, and $R_r$ of $k$ tapes. Figure 11.9 illustrates how the head of $M'$ moves to simulate the three

**Fig. 11.9** Three instances of traces of the head of TM $M'$

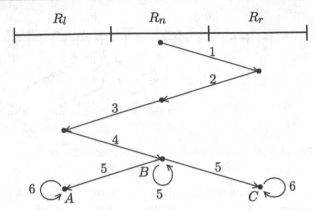

cases of Fig. 11.8 in at most 6 steps (6 steps for cases $A$ and $C$ and 5 steps for case $B$), respectively. As illustrated in Fig. 11.9, in any case of $A$, $B$, and $C$, $M'$ collects the information of the contents of regions $R_l$, $R_n$, and $R_r$ of every tape during the first 4 steps and stores the information as $(a_1, \ldots, a_{3km}) \in \Gamma^{3km}$. Then $M'$ updates its state based on $q$, $(j_1, \ldots, j_k)$, and $(a_1, \ldots, a_{3km})$ according to the transition function of $M$, which corresponds to the next $m$ transitions of $M$. Let the updated state be denoted by $(q', j'_1, \ldots, j'_k, a'_1, \ldots, a'_{3km})$. Finally $M'$ updates the contents of regions $R_l$, $R_n$, and $R_r$ of every tape according to $a'_1, \ldots, a'_{3km}$ together with its head positions in at most 2 further steps, as shown in Fig. 11.9. Observe that the position of the $i$th head of $M$ is the $j'_i$th square of the region where the $i$th head of $M'$ is placed. This is how $M'$ simulates $m$ steps of $M$ in at most 6 steps. For ease of explanation, we assume that $M$ adopts the two-way model, whereas $M'$ adopts the three-way model. These models are explained in Problem 6.1. In the solution of the problem it is shown that, in general, a TM in the three-way model can be simulated by a TM in the two-way model.

In order that $M'$ is equivalent to $M$, $M'$ must receive the same strings as $M$. For a string $a_1 \cdots a_n$ fed as input in tape 1, $M'$ converts it to $(a_1, \ldots, a_m)(a_{m+1}, \ldots, a_{2m}) \cdots (a_{(\lceil n/m \rceil - 1)m+1}, \ldots, a_n, \sqcup, \ldots, \sqcup)$ of length $\lceil n/m \rceil$ in tape 2, where $\lceil n/m \rceil$ denotes the smallest integer larger than or equal to $n/m$. To do so, $M'$ reads each group of $m$ consecutive symbols in tape 1 and writes the corresponding $m$-tuple in tape 2. After converting this way, $M'$ neglects the portion in tape 1 where $a_1 \cdots a_n$ was written, taking the remaining part newly as the entire tape, and moves the head of tape 2 to the leftmost square. $M'$ does this in $n + \lceil n/m \rceil$ steps. Then, interchanging the roles of tape 1 and tape 2, $M'$ starts to simulate $M$ as described above. The condition that the number of tapes is larger than or equal to two is not only to make the statement general, but also to make it possible to prepare the sequence of $m$-tuples in $O(n)$ steps by using the two tapes. From the arguments so far, the number of steps of $M'$ is given as $6\lceil t(n)/m \rceil + n + \lceil n/m \rceil$. On the other hand, since $\lceil x \rceil < x + 1$, we have $6\lceil t(n)/m \rceil + n + \lceil n/m \rceil < 6t(n)/m + n + n/m + 7 =$

$6t(n)/m + (1 + 1/m + 7/n)n$. Let $\varepsilon > 0$ be an arbitrary number. Let $m$ be such that $6/m \leq \varepsilon/2$. Furthermore, let $n_0$ be such that $(18/\varepsilon)n \leq t(n)$ for any $n \geq n_0$. Then we have $6t(n)/m + (1 + 1/m + 7/n)n \leq (6/m)t(n) + 9n \leq (6/m)t(n) + (\varepsilon/2)(18/\varepsilon)n \leq (\varepsilon/2)t(n) + (\varepsilon/2)t(n) \leq \varepsilon t(n)$.

## 11.8  Chapter 9

9.1 For $q \in \{q_0, q_1, q_2\}$ and $a, x, y, z \in \{0, 1, \sqcup\}$, $g$ is specified as follows:

$$g((q, a), y, z) = \begin{cases} (q_1, y) & \text{if } (q, a) \text{ is } (q_0, 1) \text{ or } (q_1, 0), \\ (q_0, y) & \text{if } (q, a) \text{ is } (q_0, 0) \text{ or } (q_1, 1), \\ y & \text{otherwise}, \end{cases}$$

$$g(x, (q, a), z) = \begin{cases} 0 & \text{if } (q, a) \text{ is } (q_2, 1), \\ a & \text{otherwise}, \end{cases}$$

$$g(x, y, (q, a)) = \begin{cases} (q_2, y) & \text{if } q = q_2 \text{ or } (q, a) = (q_0, \sqcup), \\ y & \text{otherwise}. \end{cases}$$

9.2 The way of constructing the circuit in Sect. 9.3 cannot be applied to a two-tape TM. The reason is that, when the two heads of the TM are apart, a finite gate that receives three inputs from the previous row cannot compute an output to represent update. So in order to construct a finite circuit to simulate a two-tape TM in the way described in Sect. 6.2, we must transform the TM to an equivalent one-tape TM for which we construct the equivalent circuit.

9.3 $\bar{x}_2 \vee \bar{x}_3$.

9.4

$$(y_1 \vee \bar{y}_2 \vee \bar{y}_3)(\bar{y}_1 \vee y_2)(\bar{y}_1 \vee y_3)$$
$$\wedge (y_2 \vee \bar{y}_4 \vee \bar{y}_5)(\bar{y}_2 \vee y_4)(\bar{y}_2 \vee y_5)$$
$$\wedge (y_3 \vee x_3)(\bar{y}_3 \vee \bar{x}_3)$$
$$\wedge (y_4 \vee \bar{x}_1)(y_4 \vee \bar{x}_2)(\bar{y}_4 \vee x_1 \vee x_2)$$
$$\wedge (y_5 \vee \bar{y}_6)(y_5 \vee \bar{x}_3)(\bar{y}_5 \vee y_6 \vee x_3)$$
$$\wedge (y_6 \vee x_2)(\bar{y}_6 \vee \bar{x}_2)$$
$$\wedge y_1,$$
$$(y_1, y_2, y_3, y_4, y_5, y_6, x_1, x_2, x_3) = (1, 1, 1, 1, 1, 1, 1, 0, 0).$$

9.5 In order to construct TM $M$ such that "$a_1 \cdots a_n \in L(M) \Leftrightarrow C_n(a_1, \ldots, a_n) = 1$" from a family $\{C_n\}$ of circuits, the family $\{C_n\}$ must be *uniform*, where $\{C_n\}$ is called uniform if there exists a Turing machine that outputs $\langle C_n \rangle$ when $1^n$ is given as input. Observe that $\{C_n\}$ in Sects. 9.3 and 9.4 is uniform. But in general a family $\{C_n\}$ arbitrarily given is not necessarily uniform.

Furthermore, if $\{C_n\}$ is uniform, we can construct TM $M$ equivalent to $\{C_n\}$, which is roughly described as follows: given $a_1 \cdots a_n$ as input, compute $\langle C_n \rangle$, where $\langle C_n \rangle$ is assumed to be a list of items such as $(i, F, j, k)$ as described in the proof of Theorem 9.2; replace variables $x_1, \ldots, x_n$ in $\langle C_n \rangle$ with $a_1, \ldots, a_n$, respectively; compute outputs of each gate in $\langle C_n \rangle$ starting with inputs $a_1, \ldots, a_n$; accept or reject depending on the value of the output gate.

## 11.9 Chapter 10

10.1 The traveling salesman problem (*TSP* for short) belongs to NP because it can be verified in polynomial time that a sequence of nodes given as a certificate satisfies the condition for *TSP*. Furthermore, if we define the cost function $g$ as $g(e) = 1$ for every edge $e$ and the integer $k$ to be the number of nodes of a graph, then the traveling salesman problem turns out to be exactly the Hamiltonian path problem. Hence we have $HAMPATH \leq_{\text{poly}} TSP$.

10.2 *CLIQUE* belongs to NP because it can be verified in polynomial time that a collection of $k$ nodes given as a certificate becomes a $k$-clique, where an instance of *CLIQUE* is written as $(G, k)$. Next, we shall prove $3SAT \leq_{\text{poly}}$ *CLIQUE*. To explain the idea of the reduction, we take an example: formula $F = (x_1 \vee x_2 \vee x_3)(\overline{x}_1 \vee \overline{x}_2 \vee x_3)(x_1 \vee x_2 \vee \overline{x}_3)$ and the graph $G_F$ shown below, which corresponds to the formula. For each clause of the formula, there corresponds a group of three nodes, each corresponding to a literal of the clause. Label each node with its corresponding literal. So if $F$ has $k$ clauses, $G_F$ has $k$ such groups. Between any two nodes from the same group, no edge is drawn. On the other hand, between any two nodes from different groups, an edge is drawn except for the case that the two nodes have contradictory literals like, say, $x_3$ and $\overline{x}_3$. Then as shown in the figure, the two 3-cliques drawn by bold lines specify satisfying assignments of $x_1 = 1$, $x_2 = 0$, and $x_3 = 0$, and $x_2 = 1$ and $x_3 = 1$.

Generalizing the construction, let $G_F$ denote the graph of $k$ collections of three nodes that corresponds to a 3cnf-formula $F$ with $k$ clauses of three literals, where 3cnf means cnf, each clause consisting of exactly 3 literals. We shall prove that $F$ is satisfiable if and only if $G_F$ has a $k$-clique.

($\Rightarrow$) Assume that $F$ is satisfiable, and denote its satisfying assignment by $T \subseteq \{x_1, \overline{x}_1, \ldots, x_n, \overline{x}_n\}$, where $x_i \in T$ means $x_i = 1$ and $\overline{x}_i \in T$ means $x_i = 0$. Since $T$ satisfies $F$, each clause of $F$ has at least one literal in $T$. If we choose such a literal from each clause, the chosen nodes form a $k$-clique in $G_F$ because the literals associated with the chosen nodes do not contain contradictory literals.

($\Leftarrow$) Assume that $G_F$ has a $k$-clique. Since there are no edges between nodes in the same group, each group has a node of the $k$-clique. Let assignment $T$ consist of literals associated with the nodes of the $k$-clique. Since $T$ does not contain contradictory literals, $T$ satisfies $F$.

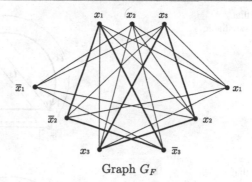

Graph $G_F$

10.3 (1)

3-cliques of $G$: $\{v_1, v_2, v_3\}$ and $\{v_1, v_2, v_5\}$,
Independent sets of $\overline{G}$: $\{v_1, v_2, v_3\}$ and $\{v_1, v_2, v_5\}$,
Covers of $\overline{G}$: $\{v_4, v_5\}$ and $\{v_3, v_4\}$.

(2) It is clear because nodes $v$ and $u$ are connected by an edge in $G = (V, E)$ if and only if $v$ and $u$ are not connected by an edge in $\overline{G} = (V, \overline{E})$.

(3) The statement follows from the following claims for graph $G = (V, E)$.

$A \subseteq V$ is an independent set

$\Leftrightarrow$  for any edge $(v, u)$ in $G$, (i) $v \in A$ and $u \in V - A$,

or (ii) $v \in V - A$ and $u \in A$, or (iii) $v \in V - A$ and $u \in V - A$

$\Leftrightarrow$   $V - A$ is a cover.

(4) It is clear that *INDEPENDENT-SET* $\in$ NP and *NODE-COVER* $\in$ NP. If we define reduction $f$ as $f((G, k)) = (\overline{G}, k)$, we have *CLIQUE* $\leq_{\text{poly}}$ *INDEPENDENT-SET* from (2). On the other hand, since *CLIQUE* is an NP-complete problem from Problem 10.2$^\sharp$, *INDEPENDENT-SET* is NP-complete by Theorem 10.2.

Furthermore, if we define reduction $f$ as $f((G, k)) = (G, |V| - k)$, we have *INDEPENDENT-SET* $\leq_{\text{poly}}$ *NODE-COVER* from (3). On the other hand, since *INDEPENDENT-SET* is an NP-complete problem, *NODE-COVER* is NP-complete.

10.4 For literal $s$, let $s$ mean specifying as $s = 1$, and let $\overline{s}$ mean specifying as $\overline{s} = 1$. Then the condition $(s \vee t) = 1$ is equivalent to "$\overline{s} \Rightarrow t$ and $\overline{t} \Rightarrow s$," which is described as the following truth table and graph.

Truth table of $\overline{s} \Rightarrow t$ and $\overline{t} \Rightarrow s$

| $s$ | $t$ | $\overline{s} \Rightarrow t$ | $\overline{t} \Rightarrow s$ |
|-----|-----|------------------------------|------------------------------|
| 0 | 0 | 0 | 0 |
| 0 | 1 | 1 | 1 |
| 1 | 0 | 1 | 1 |
| 1 | 1 | 1 | 1 |

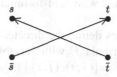

Graph that expresses
condition $s \vee t = 1$

**Fig. 11.10** Graph that
indicates that $F_1$ is satisfiable

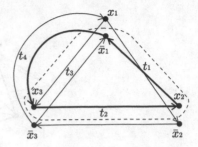

**Fig. 11.11** Graph that
indicates that $F_2$ is not
satisfiable

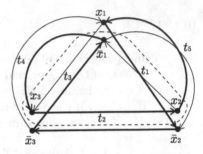

Let us consider 2cnf-formulas $F_1$ and $F_2$ described below. Note that $F_1$ is satisfiable and $F_2$ is not. Figures 11.10 and 11.11 give graphs showing the conditions to satisfy the clauses of $F_1$ and $F_2$, respectively. Let $t_1 = \overline{x}_1 \vee \overline{x}_2$, $t_2 = x_2 \vee \overline{x}_3$, $t_3 = x_3 \vee \overline{x}_1$, $t_4 = x_1 \vee x_3$, and $t_5 = \overline{x}_2 \vee x_1$

$$F_1 = (\overline{x}_1 \vee \overline{x}_2)(x_2 \vee \overline{x}_3)(x_3 \vee \overline{x}_1)(x_1 \vee x_3),$$

$$F_2 = (\overline{x}_1 \vee \overline{x}_2)(x_2 \vee \overline{x}_3)(x_3 \vee \overline{x}_1)(x_1 \vee x_3)(\overline{x}_2 \vee x_1).$$

$F_1$ is satisfied by the assignment $(x_1, x_2, x_3) = (0, 1, 1)$ which is also denoted by $\{\overline{x}_1, x_2, x_3\}$, which is surrounded by the dotted line. In Fig. 11.10 all the paths that start at nodes in $\{\overline{x}_1, x_2, x_3\}$ are drawn with bold lines. Similarly, in Fig. 11.11 all such paths are drawn with bold lines. But in this figure there exists path $\overline{x}_1 \rightarrow x_3 \rightarrow x_2 \rightarrow x_1$, which means that $\overline{x}_1 = 1$ cannot be a part of an assignment that satisfies $F_2$. This is because both $\overline{x}_1 = 1$ and $x_1 = 1$ are required for the assignment. On the other hand, no such contradictory situation occurs in the case of Fig. 11.10 for $F_1$. The arguments above can be generalized as follows: the problem of determining whether a 2cnf-formula is satisfiable can be computed in polynomial time by constructing the graph associated with 2cnf-formula and checking whether any contradiction can be deduced.

Let us define the directed graph $G(F) = (V_F, E_F)$ that corresponds to a 2cnf-formula $F$ with variables $x_1, \ldots, x_n$, where $V_F = \{x_1, \overline{x}_1, \ldots, x_n, \overline{x}_n\}$ and $E_F = \{(\overline{s}, t), (\overline{t}, s) \mid (s \vee t) \text{ is a clause of } F\}$. Furthermore, denote an assignment by $T = \{l_1, \ldots, l_n\}$ such that $l_i = x_i$ or $\overline{x}_i$ for $1 \leq i \leq n$, where $l_i = x_i$ means $x_i = 1$ and $l_i = \overline{x}_i$ means $x_i = 0$. For the assignment $T$, let $E(T) = T \cup \{s \mid \bar{t} \in T$ and $(\bar{t}, s) \in E_F\}$. In the case of Fig. 11.10, $E(\{\bar{x}_1, x_2, x_3\}) = \{\bar{x}_1, x_2, x_3\}$, and in the case of Fig. 11.11, $E(\{\bar{x}_1, x_2, x_3\}) = \{x_1, \bar{x}_1, x_2, \bar{x}_2, x_3, \bar{x}_3\}$.

**Fact 1** Let $T$ be an assignment. $F$ is satisfied by $T$ if and only if $E(T) = T$.

*Proof* ($\Rightarrow$) The proof is by contradiction. Suppose that $F$ is satisfied by $T$ and that $E(T) \neq T$. Then there exists clause $(s \vee t)$ in $F$ such that $\bar{s} \in T$, $t \notin T$, and $t \in E(T)$. Then $T$ does not satisfy $(s \vee t)$, and hence does not satisfy $F$. This contradicts the assumption that $T$ satisfies $F$.

($\Leftarrow$) Suppose that $E(T) = T$. We shall prove that $T$ satisfies an arbitrary clause $(s \vee t)$ in $F$. Since it is clear when $s \in T$, assume that $\bar{s} \in T$. Then, since $(\bar{s}, t) \in E_F$, we have $t \in E(T)$. On the other hand, since $E(T) = T$, we have $t \in T$, which implies that $T$ satisfies $(s \vee t)$. □

By the definition of $G(F)$, if there exists a path $\bar{s} \rightarrow t$, then there exists a path $s \leftarrow \bar{t}$. So we have the next fact.

**Fact 2** If there is a path $s_1 \rightarrow s_2 \rightarrow \cdots \rightarrow s_m$, there is also a path $\bar{s}_1 \leftarrow \bar{s}_2 \leftarrow \cdots \leftarrow \bar{s}_m$.

**Fact 3** Let $F$ be a 2cnf-formula composed of clauses, each consisting of literals associated with different variables. Then the following two conditions are equivalent with each other: (1) $F$ is satisfiable; (2) for every variable $x$ $G(F)$ has at most one of a path from $x$ to $\bar{x}$ and that from $\bar{x}$ to $x$.

*Proof* ((1) $\Rightarrow$ (2)) The proof is by contradiction. Suppose that $T$ is an assignment that satisfies $F$. Furthermore, suppose that there exists variable $x_i$ such that in $G(F)$ there exist both a path from $x_i$ to $\bar{x}_i$ and one from $\bar{x}_i$ to $x_i$. Since $x_i \in T$ or $\bar{x}_i \in T$, suppose without loss of generality that $x_i \in T$ and $\bar{x}_i \notin T$. Then there exists an edge $(\bar{t}, s)$ on the path such that $\bar{t} \in T$, $s \notin T$ and such that $(t \vee s)$ is a clause of $F$. But, since $T$ is such that $\bar{t} \in T$ and $\bar{s} \in T$, $T$ does not satisfy clause $(t \vee s)$, hence resulting in a contradiction.

((2) $\Rightarrow$ (1)) Assume (2) in the Fact. We shall construct $T \subset \{x_1, \bar{x}_1, \ldots, x_n, \bar{x}_n\}$. Starting with $T = \emptyset$, we add a variable to $T$ as described below repeatedly until $T$ has literals associated with all the variables. While there exists a variable $x_i$ such that $x_i \notin T$ and $\bar{x}_i \notin T$, choose $t \in \{x_i, \bar{x}_i\}$ such that there exists no path from $t$ to $\bar{t}$, and add all the literals on the paths that start at literal $t$. Let the set constructed this way be denoted by $T$. By the construction, $T$ is clearly such that, if $\bar{t} \in T$ and $(\bar{t}, s)$ is an edge of $G(F)$, then $s \in T$. Thus, since we have $E(T) = T$, $T$ satisfies $F$ by Fact 1. □

To summarize, the satisfiability problem for 2cnf-formulas can be solved by constructing graph $G(F)$ for 2cnf-formula $F$ and then checking if condition (2) of Fact 3 holds for $G(F)$. To check condition (2) we can use the polynomial time algorithm for the reachability problem in Sect. 8.2. Thus the satisfiability problem for 2cnf-formulas is computable in polynomial time.

10.5 Since it is easy to see that $UHAMPATH \in$ NP we shall show that $HAMPATH \leq_{poly} UHAMPATH$. We define reduction $f$ that takes as input a directed graph $G$ and produces as output an undirected graph $G' = f(G)$. To do so for each node $v$ in $G$, we associate three nodes denoted by $v^{in}$, $v^{mid}$, and $v^{out}$. The set of nodes of $G'$ consists of these nodes. For edges of $G'$ we define undirected edges between $v^{in}$ and $v^{mid}$, between $v^{mid}$ and $v^{out}$, and finally between $u^{out}$ and $v^{in}$ if $u \to v$ is a directed edge of $G$. The set of edges of $G'$ consists of these edges. Clearly each $v^{mid}$ is connected by edges only to $v^{in}$ and $v^{out}$. Then, any Hamiltonian path in $G'$ has the following two properties.

**Property 1** For any $v$, nodes denoted by $v^{in}$, $v^{mid}$, $v^{out}$ appear consecutively.

**Property 2** The only possible edges between $\{v^{in}, v^{mid}, v^{out}\}$ and $\{u^{in}, u^{mid}, u^{out}\}$ with different nodes $v$ and $u$ are $(u^{out}, v^{in})$ and $(v^{out}, u^{in})$.

Next, we shall verify that directed graph $G$ has a Hamiltonian path if and only if undirected graph $G'$ has a Hamiltonian path. It is clear from the definition of $G'$ that if $G$ has a Hamiltonian path

$$u_1, u_2, \ldots, u_k, \tag{1}$$

then $G'$ has a Hamiltonian path

$$u_1^{in}, u_1^{mid}, u_1^{out}, u_2^{in}, u_2^{mid}, u_2^{out}, \ldots, u_k^{out}, \tag{2}$$

where $G$ has edge $(u_k, u_1)$ and $G'$ has edge $(u_k^{out}, u_1^{in})$.

Conversely, suppose that $G'$ has a Hamiltonian path $P'$. Then by Properties 1 and 2 $P'$ takes the form of (2). Thus by the definition of $G'$, $G$ has edges $(u_1, u_2), (u_2, u_3), \ldots, (u_{k-1}, u_k)$, and $(u_k, u_1)$ so that $G$ has a Hamiltonian path. Finally, the reduction $f$ can be computed in polynomial time, completing the proof.

# Concluding Remarks

12

This book aims at the concise presentation of fascinating results chosen from three fields, which together constitute the theory of computation: automata and language theory, computability theory, and complexity theory. The author will be grateful if this book helps readers find it pleasurable to explore those fields. The theory of computation incorporates not only aspects of science, but also intimate connections with engineering practices that might be useful to businesses. Among such typical examples are public key cryptography, which obviates risks in sending a key to someone far away, the concept of relational databases, which enables expression of enormous amounts of data in tabular form, thereby facilitating updating and searching for data, and an algorithm to compute linear programming efficiently, which solves many practical optimization problems. The database industry, for example, has grown to a tens-of-billion-dollar scale. I hope that the theory of computation, which has such an extensive effect on business, will develop along a relevant and healthy direction in the future.

Many people helped me in many ways to publish this book. Takashi Yokomori of Waseda University advised me to publish this book. Osamu Watanabe read the manuscript and offered both encouragement and valuable advice. Satoshi Okawa and Yoshifumi Sakaki read an early version of the manuscript, uncovered numerous errors, and informed me on how to improve the manuscript. Eiji Takimoto read carefully through the manuscript, found errors, and pointed out ideas of how to revise the manuscript. Kazuyuki Amano read the manuscript in a surprisingly short period of time and pointed out how to make the book accessible and readable. I am indebted to Keita Iwasaka, Shingo Kawabata, Kosuke Kimura, and Katsuhiro Tashiro for their valuable comments on how to revise the manuscript, thereby making it easy to read from a student's perspective. I am also appreciative of Kazuyoshi Funamizu for producing so many beautifully drawn illustrations, and my secretaries Fumi Ozima and Kazuko Takimoto for typing large volumes of the manuscript. I am also indebted to employees of Saiensu-sha Co. Ltd.: Nobuhiko Tajima for giving me generous encouragement and Ayako Suzuki for careful proofreading. I wish to thank my three teachers, Zen'iti Kiyasu, Namio Honda, and Masayuki Kimura, who helped me shape my style of thought and enabled me to complete the writing

A. Maruoka, *Concise Guide to Computation Theory*,
DOI 10.1007/978-0-85729-535-4_12, © Springer-Verlag London Limited 2011

of this book. Without the help of all of those described above, I could never have written this book. Finally, my thanks go to my wife, Reiko, for her continuous support and assistance, and to my son, Atsushi, and daughter, Tamae, for continuously inspiring me to write this book.

# Bibliography

In writing this book, I consulted [1–3] and revised some parts of the material from those books to make them accessible as parts of this book. The proof of the equivalence of context-free grammars and pushdown automata, as well as the proof of the NP completeness for Hamiltonian path problem and subset sum problem are based on the contents of [1]. Solutions to Problems $8.4^{\sharp}$ and $10.2^{\sharp}$ come from [2]. The construction of the universal Turing machine in Sect. 7.1 is from [3]. Furthermore, in discussing the general aspect of the theory of computation in computer science, I refer to [4–6].

Many other books other than those described above deal with the theory of computation. Regarding the broader introductory topics of computer science that include the theory of computation as their theoretical core, several books have been published that are easy to understand, even for those who are not that familiar with computers. Biermann [8] is such an instructive book, although it is quite a large volume; Kawai [9] and Watanabe [10] are introductory books in this regard, but it is concisely organized for ease of reading; Wilf [11] and Iwama [12] are also introductory books, but are somewhat focused on theoretical aspects; Inagaki [13] is another such book, which provides an account of computation as well as communication.

For those readers who want to study further the topics of this book, but who need more advanced material or more comprehensive details, many useful books are available. Among them, Sipser [1] is a fascinating one which emphasizes throughout the intuitive bases underlying ideas. Papadimitriou [2] is remarkable for its depth of advanced arguments and its wide coverage of the topics. For that reason, this book is a must for readers who want to start research in these fields. Hopcroft et al. [14] is a comprehensive standard textbook that has been revised several times over the years. Its recent version emphasizes practical applications more than did the earlier versions. Savage [15] is another comprehensive book that particularly addresses the Boolean circuit model as well as the lower bounds of complexity in terms of circuit size. Arora and Barak [16] describes recent achievements of complexity theory. Kasai [17] adopts a Turing machine as its computational model, expressing its behavior in terms of a certain programming language introduced in the book. It gives arguments in detail, but succinctly. Koren [18] covers only the material described in Parts II and III of this book; it comprises lectures numbered 1 through 39 together with supplementary lectures, each being more or less self-contained.

A. Maruoka, *Concise Guide to Computation Theory*,
DOI 10.1007/978-0-85729-535-4, © Springer-Verlag London Limited 2011

Numerous books partially address the material contained in this book. Watanabe [19] covers Parts III and IV; Kobayashi [20] covers Part IV; Kasai and Toda [21] covers Parts III and IV as well as parallel computation and approximate computation, which this book does not cover; Moriya [22] deals with Part II in detail; Tomita and Yokomori [23] and Yoneda et al. [24] explain the material of Parts II and III mainly using examples; Takeuti [25] specifically examines how to solve the P vs. NP problem; and Leeuwen [26] is a sort of encyclopedia of computer science, describing various topics written by authors who contributed to establishing their respective fields.

Designing good algorithms is a fundamental common objective that is related somehow with every topic in this book. So many books have been published on how to design algorithms: Knuth [27, 28] give thorough arguments on designing algorithms; Hopcroft and Ullman [29], read extensively, has imparted a large influence on the research on designing and analyzing algorithms; Ibaraki [30] is a good textbook on algorithms; and Hirata [31] is both concise and easy to read.

The books, including others than those described above, that are referred to in this book are listed as follows.

# References

1. M. Sipser, Introduction to the Theory of Computation, *Course Technology Ptr, Florence*, 2005.
2. C. Papadimitriou, Computational Complexity, *Addison-Wesley, Reading*, 1994.
3. M. Minsky, Computation: Finite and Infinite Machines, *Prentice-Hall, New York*, 1967.
4. J. von Neumann, Theory of Self-Reproducing Automata, *University of Illinois Press, Champaign*, 1966.
5. L. Valiant, Circuit of the Mind, *Oxford University Press, London*, 1994.
6. J. Hartmanis, Observations About the Development of Theoretical Computer Science, *Annals of the History of Computing*, Vol. 3, No. 1, pp. 42–51, 1981.
7. G. Johnson, All Science Is Computer Science, *The New York Times*, March 25, 2001.
8. A. Biermann, Great Ideas in Computer Science, *MIT Press, Cambridge*, 1993.
9. S. Kawai, Computer Science, *Tokyo University Press, Tokyo*, 1995 (in Japanese).
10. O. Watanabe, Computer Science as a Culture, *Saiensu-sha, Tokyo*, 2001 (in Japanese).
11. H. Wilf, Algorithms and Complexity, *AK Peters, Wellesley*, 2002.
12. K. Iwama, Automaton-Languages and Computation Theory, *Corona Publishing, San Antonio*, 2003 (in Japanese).
13. Y. Inagaki, Communication and Computation, *Iwanami-Shoten, Tokyo*, 2003 (in Japanese).
14. J. Hopcroft, R. Motwani, and J. Ullman, Introduction to Automata Theory, Languages and Computation, Second Edition, *Addison-Wesley, Reading*, 2001.
15. J. Savage, Models of Computations, *Addison-Wesley, Reading*, 1998.
16. S. Arora and B. Barak, Computational Complexity—A Modern Approach, *Cambridge University Press, New York*, 2009.
17. T. Kasai, The Theory of Computation, *Kindai Kagaku sha, Tokyo*, 1992 (in Japanese).
18. D. Koren, Automata and Computability, *Springer, Berlin*, 1997.
19. O. Watanabe, An Introduction to Computability and Computational Complexity, *Kindai Kagaku sha, Tokyo*, 1992 (in Japanese).
20. K. Kobayashi, Computational Complexity, *Shokodo, Tokyo*, 1988 (in Japanese).
21. T. Kasai and S. Toda, The Theory of Computation, *Kyoritsu Shuppan, Tokyo*, 1993 (in Japanese).
22. E. Moriya, Formal Languages and Automata, *Saiensu-sha, Tokyo*, 2001 (in Japanese).
23. E. Tomita and T. Yokomori, The Theory of Automaton and Language, *Morikita Publishing, Tokyo*, 1992 (in Japanese).
24. M. Yoneda, S. Hirose, N. Osato, and S. Okawa, The Foundation of the Theory of Automaton and Languages, *Kindai Kagaku sha, Tokyo*, 2003 (in Japanese).
25. G. Takeuti, P and NP, *Nippon-Hyoron-Sha, Tokyo*, 1996.
26. J. Leeuwen, Algorithms and Complexity, Volume A of Handbook of Computer Science, *Elsevier, Amsterdam*, 1990.
27. D. Knuth, The Art of Computer Programming, Vol. I, *Addison-Wesley, Reading*, 1973.
28. D. Knuth, The Art of Computer Programming, Vol. II, *Addison-Wesley, Reading*, 1981.
29. A. Hopcroft and J. Ullman, The Design and Analysis of Computer Algorithms, *Addison-Wesley, Reading*, 1974.

A. Maruoka, *Concise Guide to Computation Theory*,
DOI 10.1007/978-0-85729-535-4, © Springer-Verlag London Limited 2011

30. T. Ibaraki, Algorithms by C and Data Structures, *Shokodo, Tokyo*, 1999 (in Japanese).
31. T. Hirata, Algorithms and Data Structures, *Morikita Publishing, Tokyo*, 1990 (in Japanese).
32. M. Minsky, The Society of Mind, *Simon & Schuster, New York*, 1985.
33. N. Chomsky, Syntactic Structures, *Mouton, The Hague*, 1957.
34. V. Vazirani, Approximation Algorithms, *Springer, Berlin*, 2001.
35. L. Valiant, A Theory of the Learnable, *Communications of the ACM*, Vol. 15, No. 11, pp. 1134–1142, 1984.
36. A. Maruoka and E. Takimoto, Algorithmic Learning Theory, Encyclopedia of Computer Science and Technology, Vol. 45, *Dekker, New York*, 2002.
37. S. Arikawa, M. Sato, T. Sato, A. Maruoka, and Y. Kaneda, Structure and Development of Discovery Science, *Proceedings of the Institute of Artificial Intelligence of Japan*, Vol. 15, No. 4, pp. 595–607, 2000 (in Japanese).

# Index